Practical Power Systems Protection

Other titles in the series

Practical Data Acquisition for Instrumentation and Control Systems (John Park, Steve Mackay)

Practical Data Communications for Instrumentation and Control (Steve Mackay, Edwin Wright, John Park)

Practical Digital Signal Processing for Engineers and Technicians (Edmund Lai)

Practical Electrical Network Automation and Communication Systems (Cobus Strauss)

Practical Embedded Controllers (John Park)

Practical Fiber Optics (David Bailey, Edwin Wright)

Practical Industrial Data Networks: Design, Installation and Troubleshooting (Steve Mackay, Edwin Wright, John Park, Deon Reynders)

Practical Industrial Safety, Risk Assessment and Shutdown Systems for Instrumentation and Control (Dave Macdonald)

Practical Modern SCADA Protocols: DNP3, 60870.5 and Related Systems (Gordon Clarke, Deon Reynders)

Practical Radio Engineering and Telemetry for Industry (David Bailey)

Practical SCADA for Industry (David Bailey, Edwin Wright)

Practical TCP/IP and Ethernet Networking (Deon Reynders, Edwin Wright)

Practical Variable Speed Drives and Power Electronics (Malcolm Barnes)

Practical Centrifugal Pumps (Paresh Girdhar and Octo Moniz)

Practical Electrical Equipment and Installations in Hazardous Areas (Geoffrey Bottrill and G. Vijayaraghavan)

Practical E-Manufacturing and Supply Chain Management (Gerhard Greef and Ranjan Ghoshal)

Practical Grounding, Bonding, Shielding and Surge Protection (G. Vijayaraghavan, Mark Brown and Malcolm Barnes)

Practical Hazops, Trips and Alarms (David Macdonald)

Practical Industrial Data Communications: Best Practice Techniques (Deon Reynders, Steve Mackay and Edwin Wright)

Practical Machinery Safety (David Macdonald)

Practical Machinery Vibration Analysis and Predictive Maintenance (Cornelius Scheffer and Paresh Girdhar)

Practical Power Distribution for Industry (Jan de Kock and Cobus Strauss)

Practical Process Control for Engineers and Technicians (Wolfgang Altmann)

Practical Telecommunications and Wireless Communications (Edwin Wright and Deon Reynders)

Practical Troubleshooting Electrical Equipment (Mark Brown, Jawahar Rawtani and Dinesh Patil)

Practical Power Systems Protection

Les Hewitson

Mark Brown PrEng, DipEE, BSc (ElecEng),
Senior Staff Engineer, IDC Technologies, Perth, Australia

Ben Ramesh Ramesh and Associates, Perth, Australia

Series editor: Steve Mackay FIE(Aust), CPEng, BSc (ElecEng), BSc (Hons), MBA,
Gov. Cert. Comp., Technical Director – IDC Technologies

AMSTERDAM • BOSTON • HEIDELBERG • LONDON
NEW YORK • OXFORD • PARIS • SAN DIEGO
SAN FRANCISCO • SINGAPORE • SYDNEY • TOKYO

Newnes is an imprint of Elsevier

Newnes is an imprint of Elsevier
The Boulevard, Langford Lane, Kidlington, Oxford, OX5 1GB
30 Corporate Drive, Suite 400, Burlington, MA 01803, USA

First edition 2005
Reprinted 2006, 2007, 2009

Copyright © 2005, IDC Technologies. Published by Elsevier Ltd. All rights reserved

No part of this publication may be reproduced, stored in a retrieval system or transmitted in any form or by any means electronic, mechanical, photocopying, recording or otherwise without the prior written permission of the publisher

Permissions may be sought directly from Elsevier's Science & Technology Rights Department in Oxford, UK: phone (+44) (0) 1865 843830; fax (+44) (0) 1865 853333; email: permissions@elsevier.com. Alternatively you can submit your request online by visiting the Elsevier web site at http://elsevier.com/locate/permissions, and selecting *Obtaining permission to use Elsevier material*

Notice
No responsibility is assumed by the publisher for any injury and/or damage to persons or property as a matter of products liability, negligence or otherwise, or from any use or operation of any methods, products, instructions or ideas contained in the material herein. Because of rapid advances in the medical sciences, in particular, independent verification of diagnoses and drug dosages should be made

British Library Cataloguing in Publication Data
A catalogue record for this book is available from the British Library

Library of Congress Cataloging-in-Publication Data
A catalog record for this book is available from the Library of Congress

ISBN: 978-0-7506-6397-7

For information on all Newnes publications
visit our website at www.elsevierdirect.com

Printed and bound in the United Kingdom

Transferred to Digital Print 2010

Working together to grow
libraries in developing countries

www.elsevier.com | www.bookaid.org | www.sabre.org

ELSEVIER BOOK AID International Sabre Foundation

Contents

Preface

This book has been designed to give plant operators, electricians, field technicians and engineers a better appreciation of the role played by power system protection systems. An understanding of power systems along with correct management, will increase your plant efficiency and performance as well as increasing safety for all concerned. The book is designed to provide an excellent understanding on both theoretical and practical level. The book starts at a basic level, to ensure that you have a solid grounding in the fundamental concepts and also to refresh the more experienced readers in the essentials. The book then moves onto more detailed applications. It is most definitely not an advanced treatment of the topic and it is hoped the expert will forgive the simplifications that have been made to the material in order to get the concepts across in a practical useful manner.

The book features an introduction covering the need for protection, fault types and their effects, simple calculations of short circuit currents and system earthing. The book also refers to some practical work such as simple fault calculations, relay settings and the checking of a current transformer magnetisation curve which are performed in the associated training workshop. You should be able to do these exercises and tasks yourself without too much difficulty based on the material covered in the book.

This is an intermediate level book – at the end of the book you will have an excellent knowledge of the principles of protection. You will also have a better understanding of the possible problems likely to arise and know where to look for answers.

In addition you are introduced to the most interesting and 'fun' part of electrical engineering to make your job more rewarding. Even those who claim to be protection experts have admitted to improving their knowledge after attending this book but at worst case perhaps this book will perhaps be an easy refresher on the topic which hopefully you will pass onto your less experienced colleagues.

We would hope that you will gain the following from this book:

- The fundamentals of electrical power protection and applications
- Knowledge of the different fault types
- The ability to perform simple fault and design calculations
- Practical knowledge of protection system components
- Knowledge of how to perform simple relay settings
- Increased job satisfaction through informed decision making
- Know how to improve the safety of your site.

Typical people who will find this book useful include:

- Electrical Engineers
- Project Engineers
- Design Engineers
- Instrumentation Engineers
- Electrical Technicians
- Field Technicians
- Electricians

- Plant Operators
- Plant Operators.

You should have a modicum of electrical knowledge and some exposure to electrical protection systems to derive maximum benefit from this book.

This book was put together by a few authors although initiated by the late Les Hewitson, who must have been one of the finest instructors on the subject and who presented this course in his own right in South Africa and throughout Europe/North America and Australia for IDC Technologies. It is to him that this book is dedicated.

Hambani Kahle (Zulu Farewell)

(*Sources*: *Canciones de Nuestra Cabana* (1980), *Tent and Trail Songs* (American Camping Association), *Songs to Sing & Sing Again by Shelley Gordon*)

Go well and safely.
Go well and safely.
Go well and safely.
The Lord be ever with you.

Stay well and safely.
Stay well and safely.
Stay well and safely.
The Lord be ever with you.

Hambani kahle.
Hambani kahle.
Hambani kahle.
The Lord be ever with you.

Steve Mackay

1

Need for protection

1.1 Need for protective apparatus

A power system is not only capable to meet the present load but also has the flexibility to meet the future demands. A power system is designed to generate electric power in sufficient quantity, to meet the present and estimated future demands of the users in a particular area, to transmit it to the areas where it will be used and then distribute it within that area, on a continuous basis.

To ensure the maximum return on the large investment in the equipment, which goes to make up the power system and to keep the users satisfied with reliable service, the whole system must be kept in operation continuously without major breakdowns.

This can be achieved in two ways:

- The first way is to implement a system adopting components, which should not fail and requires the least or nil maintenance to maintain the continuity of service. By common sense, implementing such a system is neither economical nor feasible, except for small systems.
- The second option is to foresee any possible effects or failures that may cause long-term shutdown of a system, which in turn may take longer time to bring back the system to its normal course. The main idea is to restrict the disturbances during such failures to a limited area and continue power distribution in the balance areas. Special equipment is normally installed to detect such kind of failures (also called 'faults') that can possibly happen in various sections of a system, and to isolate faulty sections so that the interruption is limited to a localized area in the total system covering various areas. The special equipment adopted to detect such possible faults is referred to as 'protective equipment or protective relay' and the system that uses such equipment is termed as 'protection system'.

A protective relay is the device, which gives instruction to disconnect a faulty part of the system. This action ensures that the remaining system is still fed with power, and protects the system from further damage due to the fault. Hence, use of protective apparatus is very necessary in the electrical systems, which are expected to generate, transmit and distribute power with least interruptions and restoration time. It can be well recognized that use of protective equipment are very vital to minimize the effects of faults, which otherwise can kill the whole system.

1.2 Basic requirements of protection

A protection apparatus has three main functions/duties:

1. Safeguard the entire system to maintain continuity of supply
2. Minimize damage and repair costs where it senses fault
3. Ensure safety of personnel.

These requirements are necessary, firstly for early detection and localization of faults, and secondly for prompt removal of faulty equipment from service.

In order to carry out the above duties, protection must have the following qualities:

- *Selectivity*: To detect and isolate the faulty item only.
- *Stability*: To leave all healthy circuits intact to ensure continuity or supply.
- *Sensitivity*: To detect even the smallest fault, current or system abnormalities and operate correctly at its setting before the fault causes irreparable damage.
- *Speed*: To operate speedily when it is called upon to do so, thereby minimizing damage to the surroundings and ensuring safety to personnel.

To meet all of the above requirements, protection must be reliable which means it must be:

- *Dependable*: It *must* trip when called upon to do so.
- *Secure*: It must *not* trip when it is not supposed to.

1.3 Basic components of protection

Protection of any distribution system is a function of many elements and this manual gives a brief outline of various components that go in protecting a system. Following are the main components of protection.

- Fuse is the self-destructing one, which carries the currents in a power circuit continuously and sacrifices itself by blowing under abnormal conditions. These are normally independent *or* stand-alone protective components in an electrical system unlike a circuit breaker, which necessarily requires the support of external components.
- Accurate protection cannot be achieved without properly measuring the normal and abnormal conditions of a system. In electrical systems, voltage and current measurements give feedback on whether a system is healthy or not. Voltage transformers and current transformers measure these basic parameters and are capable of providing accurate measurement during fault conditions without failure.
- The measured values are converted into analog and/or digital signals and are made to operate the relays, which in turn isolate the circuits by opening the faulty circuits. In most of the cases, the relays provide two functions viz., alarm and trip, once the abnormality is noticed. The relays in olden days had very limited functions and were quite bulky. However, with advancement in digital technology and use of microprocessors, relays monitor various parameters, which give complete history of a system during both pre-fault and post-fault conditions.
- The opening of faulty circuits requires some time, which may be in milliseconds, which for a common day life could be insignificant. However, the circuit breakers, which are used to isolate the faulty circuits, are capable of

carrying these fault currents until the fault currents are totally cleared. The circuit breakers are the main isolating devices in a distribution system, which can be said to directly protect the system.

- The operation of relays and breakers require power sources, which shall not be affected by faults in the main distribution. Hence, the other component, which is vital in protective system, is batteries that are used to ensure uninterrupted power to relays and breaker coils.

The above items are extensively used in any protective system and their design requires careful study and selection for proper operation.

1.4 Summary

Power System Protection – Main Functions
1. To safeguard the entire system to maintain continuity of supply. 2. To minimize damage and repair costs. 3. To ensure safety of personnel.

Power System Protection – Basic Requirements
1. *Selectivity*: To detect and isolate the faulty item only. 2. *Stability*: To leave all healthy circuits intact to ensure continuity of supply. 3. *Speed*: To operate as fast as possible when called upon, to minimize damage, production downtime and ensure safety to personnel. 4. *Sensitivity*: To detect even the smallest fault, current or system abnormalities and operate correctly at its setting.

Power System Protection – Speed is Vital!!
The protective system should act fast to isolate faulty sections to prevent: • Increased damage at fault location. Fault energy = $I^2 \times R_f \times t$, where t is time in seconds. • Danger to the operating personnel (flashes due to high fault energy sustaining for a long time). • Danger of igniting combustible gas in hazardous areas, such as methane in coal mines which could cause horrendous disaster. • Increased probability of earth faults spreading to healthy phases. • Higher mechanical and thermal stressing of all items of plant carrying the fault current, particularly transformers whose windings suffer progressive and cumulative deterioration because of the enormous electromechanical forces caused by multi-phase faults proportional to the square of the fault current. Sustained voltage dips resulting in motor (and generator) instability leading to extensive shutdown at the plant concerned and possibly other nearby plants connected to the system.

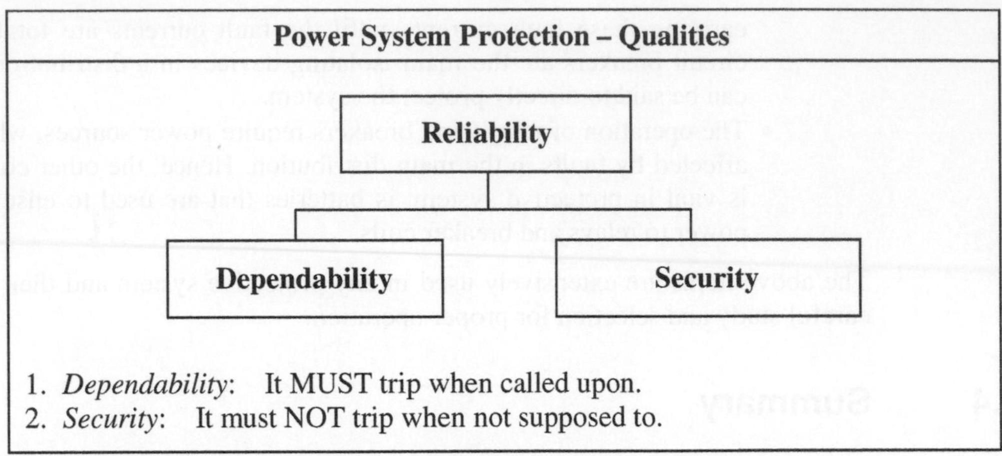

Power System Protection – Basic Components

1. *Voltage transformers and current transformers*: To monitor and give accurate feedback about the healthiness of a system.
2. *Relays*: To convert the signals from the monitoring devices, and give instructions to open a circuit under faulty conditions or to give alarms when the equipment being protected, is approaching towards possible destruction.
3. *Fuses*: Self-destructing to save the downstream equipment being protected.
4. *Circuit breakers*: These are used to make circuits carrying enormous currents, and also to break the circuit carrying the fault currents for a few cycles based on feedback from the relays.
5. *DC batteries*: These give uninterrupted power source to the relays and breakers that is independent of the main power source being protected.

2

Faults, types and effects

2.1 The development of simple distribution systems

When a consumer requests electrical power from a supply authority, ideally all that is required is a cable and a transformer, shown physically as in Figure 2.1.

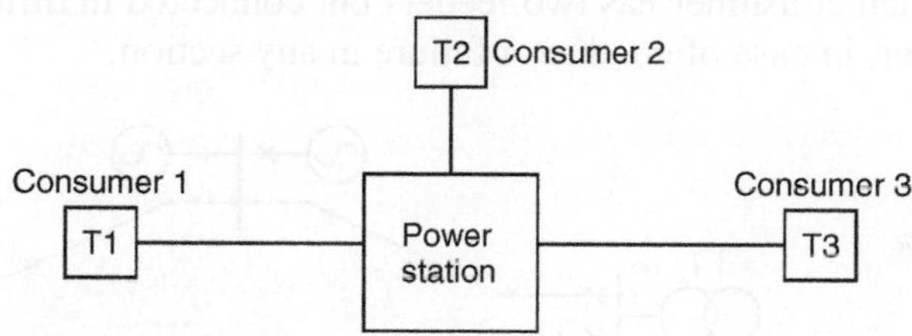

Figure 2.1
A simple distribution system

This is called a radial system and can be shown schematically in the following manner (Figure 2.2).

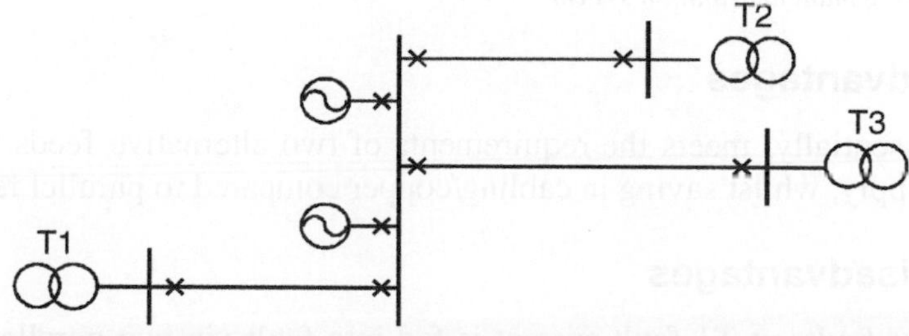

Figure 2.2
A radial distribution system

Advantages

If a fault occurs at T2 then only the protection on one leg connecting T2 is called into operation to isolate this leg. The other consumers are not affected.

Disadvantages

If the conductor to T2 fails, then supply to this particular consumer is lost completely and cannot be restored until the conductor is replaced/repaired.

This disadvantage can be overcome by introducing additional/parallel feeders (Figure 2.3) connecting each of the consumers radially. However, this requires more cabling and is not always economical. The fault current also tends to increase due to use of two cables.

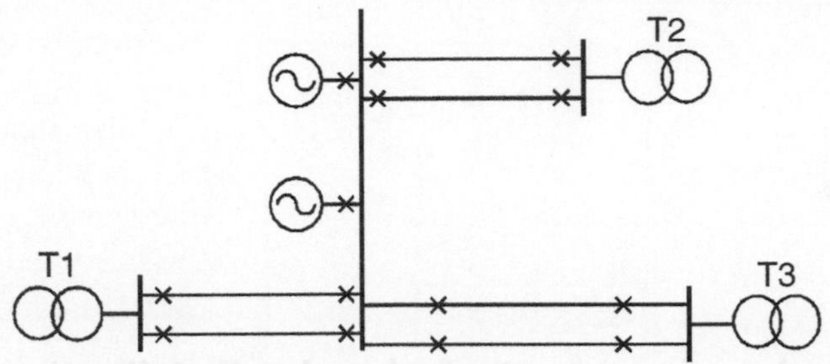

Figure 2.3
Radial distribution system with parallel feeders

The Ring main system, which is the most favored, then came into being (Figure 2.4). Here each consumer has two feeders but connected in different paths to ensure continuity of power, in case of conductor failure in any section.

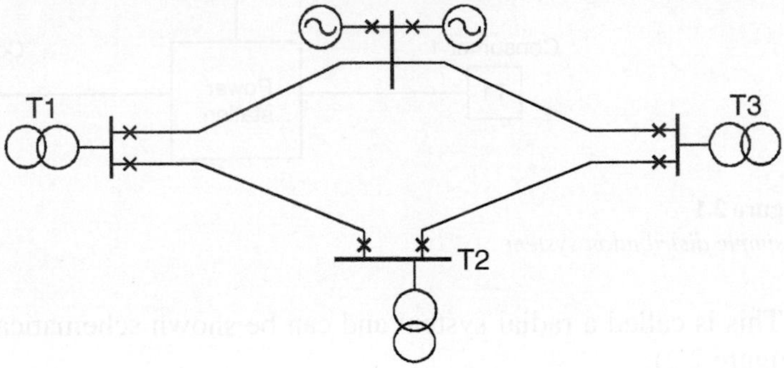

Figure 2.4
A ring main distribution system

Advantages

Essentially, meets the requirements of two alternative feeds to give 100% continuity of supply, whilst saving in cabling/copper compared to parallel feeders.

Disadvantages

For faults at T1 fault current is fed into fault via two parallel paths effectively reducing the impedance from the source to the fault location, and hence the fault current is much higher compared to a radial path. The fault currents in particular could vary depending on the exact location of the fault.

Protection must therefore be fast and discriminate correctly, so that other consumers are not interrupted.

The above case basically covers feeder failure, since cable tend to be the most vulnerable component in the network. Not only are they likely to be hit by a pick or alternatively dug-up, or crushed by heavy machinery, but their joints are notoriously weak, being susceptible to moisture, ingress, etc., amongst other things.

Transformer faults are not so frequent, however they do occur as windings are often strained when carrying through-fault current. Also, their insulation lifespan is very often reduced due to temporary or extended overloading leading to eventual failure. Hence interruption or restriction in the power being distributed cannot be avoided in case of transformer failures. As it takes a few months to manufacture a power transformer, it is a normal practice to install two units at a substation with sufficient spare capacity to provide continuity of supply in case of transformer failure.

Busbars on the other hand, are considered to be the most vital component on a distribution system. They form an electrical 'node' where many circuits come together, feeding in and sending out power.

On E.H.V. systems where mainly outdoor switchgear is used, it is relatively easy and economical to install duplicate busbar system to provide alternate power paths. But on medium-voltage (11 kV and 6.6 kV) and low-voltage (3.3 kV, 1000 V and 500 V) systems, where indoor metal clad switchgear is extensively used, it is not practical or economical to provide standby or parallel switchboards. Further, duplicate busbar switchgear is not immune to the ravages of a busbar fault.

The loss of a busbar in a network can in fact be a catastrophic situation, and it is recommended that this component be given careful consideration from a protection viewpoint when designing network, particularly for continuous process plants such as mineral processing.

2.2 Fault types and their effects

It is not practical to design and build electrical equipment or networks to eliminate the possibility of failure in service. It is therefore an everyday fact that different types of faults occur on electrical systems, however infrequently, and at random locations.

Faults can be broadly classified into two main areas, which have been designated 'active' and 'passive'.

2.2.1 Active faults

The 'active' fault is when actual current flows from one phase conductor to another (phase-to-phase), or alternatively from one phase conductor to earth (phase-to-earth). This type of fault can also be further classified into two areas, namely the 'solid' fault and the 'incipient' fault.

The solid fault occurs as a result of an immediate complete breakdown of insulation as would happen if, say, a pick struck an underground cable, bridging conductors, etc. or the cable was dug up by a bulldozer. In mining, a rockfall could crush a cable, as would a shuttle car. In these circumstances the fault current would be very high resulting in an electrical explosion.

This type of fault must be cleared as quickly as possible, otherwise there will be:

- Increased damage at fault location. Fault energy $= I^2 \times R_f \times t$, where t is time in seconds.
- Danger to operating personnel (flashes due to high fault energy sustaining for a long time).
- Danger of igniting combustible gas in hazardous areas, such as methane in coal mines which could cause horrendous disaster.
- Increased probability of earth faults spreading to healthy phases.

- Higher mechanical and thermal stressing of all items of plant carrying the fault current, particularly transformers whose windings suffer progressive and cumulative deterioration because of the enormous electromechanical forces caused by multi-phase faults proportional to the square of the fault current.
- Sustained voltage dips resulting in motor (and generator) instability leading to extensive shutdown at the plant concerned and possibly other nearby plants connected to the system.

The 'incipient' fault, on the other hand, is a fault that starts as a small thing and gets developed into catastrophic failure. Like for example some partial discharge (excessive discharge activity often referred to as Corona) in a void in the insulation over an extended period can burn away adjacent insulation, eventually spreading further and developing into a 'solid' fault.

Other causes can typically be a high-resistance joint or contact, alternatively pollution of insulators causing tracking across their surface. Once tracking occurs, any surrounding air will ionize which then behaves like a solid conductor consequently creating a 'solid' fault.

2.2.2 Passive faults

Passive faults are not real faults in the true sense of the word, but are rather conditions that are stressing the system beyond its design capacity, so that ultimately active faults will occur. Typical examples are:

- Overloading leading to over heating of insulation (deteriorating quality, reduced life and ultimate failure).
- *Overvoltage*: Stressing the insulation beyond its withstand capacities.
- *Under frequency*: Causing plant to behave incorrectly.
- *Power swings*: Generators going out-of-step or out-of-synchronism with each other.

It is therefore very necessary to monitor these conditions to protect the system against these conditions.

2.2.3 Types of faults on a three-phase system

Largely, the power distribution is globally a three-phase distribution especially from power sources. The types of faults that can occur on a three-phase AC system are shown in Figure 2.5.

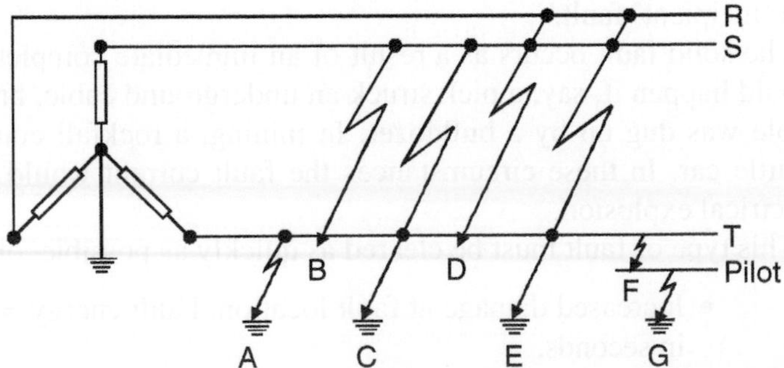

Figure 2.5
Types of faults on a three-phase system: (A) Phase-to-earth fault; (B) Phase-to-phase fault; (C) Phase-to-phase-to-earth fault; (D) Three-phase fault; (E) Three-phase-to-earth fault; (F) Phase-to-pilot fault; (G) Pilot-to-earth fault**
**In underground mining applications only*

It will be noted that for a phase-to-phase fault, the currents will be high, because the fault current is only limited by the inherent (natural) series impedance of the power system up to the point of fault (Ohm's law).

By design, this inherent series impedance in a power system is purposely chosen to be as low as possible in order to get maximum power transfer to the consumer so that unnecessary losses in the network are limited thereby increasing the distribution efficiency. Hence, the fault current cannot be decreased without a compromise on the distribution efficiency, and further reduction cannot be substantial.

On the other hand, the magnitude of earth fault currents will be determined by the manner in which the system neutral is earthed. It is worth noting at this juncture that it is possible to control the level of earth fault current that can flow by the judicious choice of earthing arrangements for the neutral. Solid neutral earthing means high earth fault currents, being limited by the inherent earth fault (zero sequence) impedance of the system, whereas additional impedance introduced between neutral and earth can result in comparatively lower earth fault currents.

In other words, by the use of resistance or impedance in the neutral of the system, earth fault currents can be engineered to be at whatever level desired and are therefore controllable. This cannot be achieved for phase faults.

2.2.4 Transient and permanent faults

Transient faults are faults, which do not damage the insulation permanently and allow the circuit to be safely re-energized after a short period.

A typical example would be an insulator flashover following a lightning strike, which would be successfully cleared on opening of the circuit breaker, which could then be automatically closed. Transient faults occur mainly on outdoor equipment where air is the main insulating medium. Permanent faults, as the name implies, are the result of permanent damage to the insulation. In this case, the equipment has to be repaired and recharging must not be entertained before repair/restoration.

2.2.5 Symmetrical and asymmetrical faults

A symmetrical fault is a balanced fault with the sinusoidal waves being equal about their axes, and represents a steady-state condition.

An asymmetrical fault displays a DC offset, transient in nature and decaying to the steady state of the symmetrical fault after a period of time, as shown in Figure 2.6.

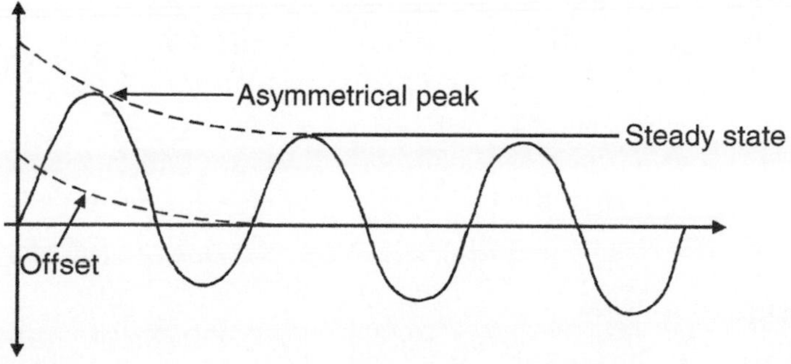

Figure 2.6
An asymmetrical fault

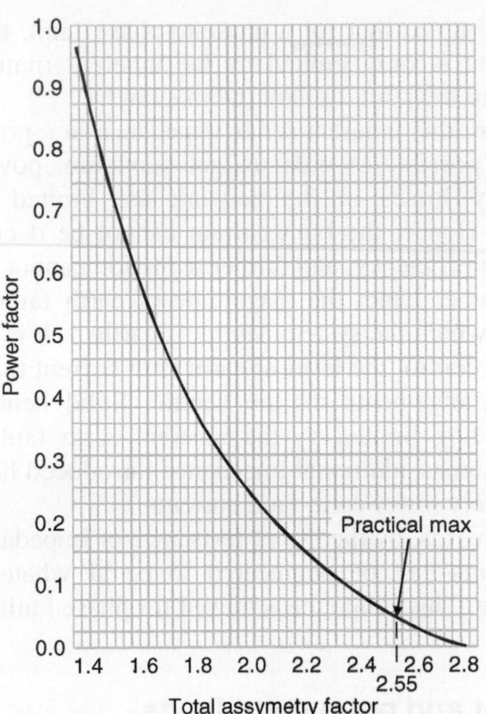

Figure 2.7
Total asymmetry factor chart

The amount of offset depends on the *X/R* (power factor) of the power system and the first peak can be as high as 2.55 times the steady-state level (see Figure 2.7).

3

Simple calculation of short-circuit currents

3.1 Introduction

Before selecting proper protective devices, it is necessary to determine the likely fault currents that may result in a system under various fault conditions. Depending upon the complexity of the system the calculations could also be too much involved. Accurate fault current calculations are normally carried out using an analysis method called symmetrical components. This method is used by design engineers and practicing protection engineers, as it involves the use of higher mathematics. It is based on the principle that any unbalanced set of vectors can be represented by a set of three balanced quantities, namely: positive, negative and zero sequence vectors.

However, for general practical purposes for operators, electricians and men-in-the-field it is possible to achieve a good approximation of three-phase short-circuit currents using some very simple methods, which are discussed below. These simple methods are used to decide the equipment short-circuit ratings and relay setting calculations in standard power distribution systems, which normally have limited power sources and interconnections. Even a complex system can be grouped into convenient parts, and calculations can be made groupwise depending upon the location of the fault.

3.2 Revision of basic formulae

It is interesting to note that nearly all problems in electrical networks can be understood by the application of its most fundamental law viz., Ohm's law, which stipulates,

For DC systems

$$I = \frac{V}{R} \quad \text{i.e. Current} = \frac{\text{Voltage}}{\text{Resistance}}$$

For AC systems

$$I = \frac{V}{Z} \quad \text{i.e. Current} = \frac{\text{Voltage}}{\text{Impedance}}$$

3.2.1 Vectors

Vectors are a most useful tool in electrical engineering and are necessary for analyzing AC system components like voltage, current and power, which tends to vary in line with the variation in the system voltage being generated.

The vectors are instantaneous 'snapshots' of an AC sinusoidal wave, represented by a straight line and a direction. A sine wave starts from zero value at 0°, reaches its peak value at 90°, goes negative after 180° and again reaches back zero at 360°. Straight lines and relative angle positions, which are termed vectors, represent these values and positions. For a typical sine wave, the vector line will be horizontal at 0° of the reference point and will be vertical upwards at 90° and so on and again comes back to the horizontal position at 360° or at the start of the next cycle. Figure 3.1 gives one way of representing the vectors in a typical cycle.

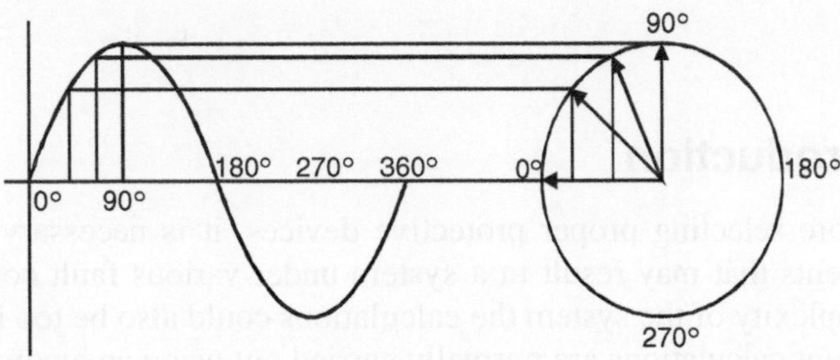

Figure 3.1
Vectors and an AC wave

In an AC system, it is quite common to come across many voltages and currents depending on the number of sources and circuit connections. These are represented in form of vectors in relation to one another taking a common reference base. Then these can be added or subtracted depending on the nature of the circuits to find the resultant and provide a most convenient and simple way to analyze and solve problems, rather than having to draw numerous sinusoidal waves at different phase displacements.

3.2.2 Impedance

This is the AC equivalent of resistance in a DC system, and takes into account the additional effects of reactance. It is represented by the symbol Z and is the vector sum of resistance and reactance (see Figure 3.2).

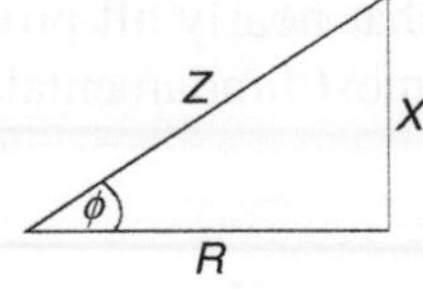

Figure 3.2
Impedance relationship diagram

It is calculated by the formula:

$$Z = R + jX$$

Where R is resistance and X is reactance.

It is to be noted that X is positive for inductive circuits whereas it is negative in capacitive circuits. That means that the Z and X will be the mirror image with R as the base in the above diagram.

3.2.3 Reactance

Reactance is a phenomenon in AC systems brought about by inductance and capacitance effects of a system. Energy is required to overcome these components as they react to the source and effectively reduce the useful power available to a system. The energy, which is spent to overcome these components in a system is thus not available for use by the end user and is termed 'useless' energy though it still has to be generated by the source.

Inductance is represented by the symbol L and is a result of magnetic coupling which induces a back emf opposing that which is causing it. This 'back-pressure' has to be overcome and the energy expended is thus not available for use by the end user and is termed 'useless' energy, as it still has to be generated. L is normally measured in Henries.

The inductive reactance is represented using the formula:

$$\text{Inductive reactance} = 2\pi f L$$

Capacitance is the electrostatic charge required when energizing the system. It is represented by the symbol C and is measured in farads.

To convert this to ohms,

$$\text{Capacitive reactance} = \frac{1}{2\pi f C}$$

Where

f = supply frequency,

L = system inductance and C = system capacitance.

Inductive reactance and capacitive reactance oppose each other vectorally; so to find the net reactance in a system, they must be arithmetically subtracted.

For example, in a system having resistance R, inductance L and capacitance C, its impedance

$$Z = R + (j \times 2\pi f L) - \left(\frac{j}{2\pi f C}\right)$$

When a voltage is applied to a system, which has an impedance of Z, vectorally the voltage is in phase with Z as per the above impedance diagram and the current is in phase with the resistive component. Accordingly, the current is said to be leading the voltage vector in a capacitive circuit and is said to be lagging the voltage vector in an inductive circuit.

3.2.4 Power and power factor

In a DC system, power dissipated in a system is the product of volts × amps and is measured in watts.

$$P = V \times I$$

For AC systems, the power input is measured in volt amperes, due to the effect of reactance and the useful power is measured in watts. For a single-phase AC system, the VA is the direct multiplication of volt and amperes, whereas it is necessary to introduce a $\sqrt{3}$ factor for a three-phase AC system. Hence *VA* power for the standard three-phase system is:

$$VA = \sqrt{3} \times V \times I$$

Alternatively;

$$\text{kVA} = \sqrt{3} \times V \times I$$

Where
V is in kV
I is in amps, or

$$\text{MVA} = \sqrt{3} \times V \times I$$

Where
V is in kV and
I is in kA.

Therefore,

$$I_{\text{amps}} = \frac{\text{kVA}}{\sqrt{3} \times \text{kV}} \quad \text{or} \quad I_{\text{kA}} = \frac{\text{MVA}}{\sqrt{3} \times \text{kV}}$$

From the impedance triangle below, it will be seen that the voltage will be in phase with *Z*, whereas the current will be in phase with resistance *R* (see Figure 3.3).

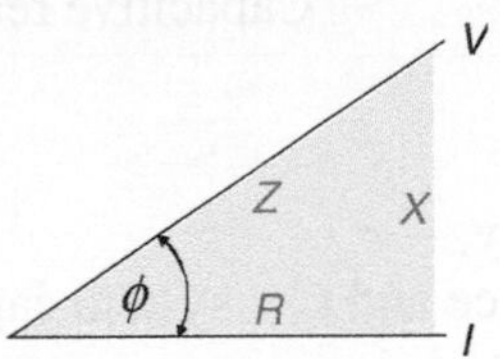

Figure 3.3
Impedance triangle

The cosine of the angle between the two is known as the power factor.

Examples:
When angle = 0°; cosine 0° = 1 (unity)
When angle = 90°; cosine 90° = 0.

The useful kW power in a three-phase system taking into account the system reactive component is obtained by introducing the power factor cos ϕ as below:

$$P = \sqrt{3} \times V \times I \times \cos\phi = \text{kVA} \times \cos\phi$$

It can be noted that kW will be maximum when cos $\phi = 1$ and will be zero when cos $\phi = 0$. It means that the useful power is zero when cos $\phi = 0$ and will tend to increase as the angle increases. Alternatively it can be interpreted, the more the power factor the more would be the useful power.

Put in another way, it is the factor applied to determine how much of the input power is effectively used in the system or simply it is a measure of the efficiency of the system. The 'reactive power' *or* the so called 'useless power' is calculated using the formula

$$p' = \sqrt{3} \times V \times I \times \sin\phi = \text{kVA} \times \sin\phi$$

In a power system, the energy meters normally record the useful power kW, which is directly used in the system and the consumer is charged based on total kW consumed over a period of time (KWH) and the maximum demand required over a period of time. However, the *P' or* the kVAR determines the kVA to be supplied by the source to meet the consumer load after overcoming the reactive components, which will vary depending on the power factor of the system. Hence, it is a usual practice to charge penalties to a consumer whenever the consumer's system has lesser power factor, since it gives the idea of useless power to be generated by the source.

Obviously, if one can reduce the amount of 'useless' power, power that is more 'useful' will be available to the consumer, so it pays to improve the power factor wherever possible. As most loads are inductive in nature, adding shunt capacitance can reduce the inductive reactance as the capacitive reactance opposes the inductive reactance of the load.

3.3 Calculation of short-circuit MVA

We have studied various types and effects of faults that can occur on the system in the earlier chapter. It is important that we know how to calculate the level of fault current that will flow under these conditions, so that we can choose equipment to withstand these faults and isolate the faulty locations without major damages to the system.

In any distribution the power source is a generator and it is a common practice to use transformers to distribute the power at the required voltages. A fault can occur immediately after the generator or after a transformer and depending upon the location of fault, the fault current could vary. In the first case, only the source impedance limits the fault current whereas in the second case the transformer impedance is an important factor that decides the fault current.

Generally, the worst type of fault that can occur is the three-phase fault, where the fault currents are the highest. If we can calculate this current then we can ensure that all equipment can withstand (carry) and in the case of switchgear, interrupt this current. There are simple methods to determine short-circuit MVA taking into account some assumptions.

Consider the following system. Here the source generates a voltage with a phase voltage of E_p and the fault point is fed through a transformer, which has a reactance X_p (see Figure 3.4).

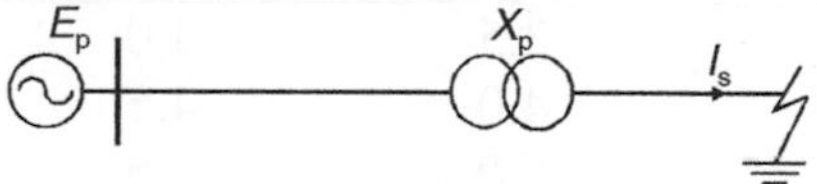

Figure 3.4
Short-circuit MVA calculation

Let I_s = r.m.s. short-circuit current
I = Normal full load current
P = Transformer rated power (rated MVA)
X_p = Reactance per phase
E_p = System voltage per phase.

At the time of fault, the fault current is limited by the reactance of the transformer after neglecting the impedances due to cables up to the fault point. Then from Ohm's law:

$$I_s = \frac{E_p}{X_p}$$

Now,

$$\frac{\text{Short-circuit MVA}}{\text{Rated MVA}} = \frac{\sqrt{3}E_p I_s \times 10^6}{\sqrt{3}E_p I \times 10^6} = \frac{I_s}{I} = \frac{\left(\frac{E_p}{X_p}\right)}{I}$$

Multiplying top and bottom by $X_p/E_p \times 100$

$$\frac{\left(\frac{E_p}{X_p}\right) \times \frac{X_p}{E_p} \times 100}{I \frac{X_p}{E_p} \times 100} = \frac{100}{I \frac{X_p}{E_p} \times 100}$$

But,

$$\frac{IX_p}{E_p} \times 100 = X\% \text{ Reactance per phase}$$

Therefore,

$$\frac{\text{Short-circuit MVA}}{\text{Rated MVA } (P)} = \frac{I_s}{I} = \frac{100}{X\%}$$

Hence,

$$\text{Short-circuit MVA} = \frac{100\,P}{X\%}$$

It can be noted above, that the value of X will decide the short-circuit MVA when the fault is after the transformer. Though it may look that increasing the impedance can lower the fault MVA, it is not economical to choose higher impedance for a transformer. Typical percent reactance values for transformers are shown in the table below.

	Primary Voltage Reactance % at MVA Rating				
MVA Rating	**Up to 11 kV**	**22 kV**	**33 kV**	**66 kV**	**132 kV**
0.25	3.5	4.0	4.5	5.0	6.5
0.5	4.0	4.5	5.0	5.5	6.5
1.0	5.0	5.5	5.5	6.0	7.0
2.0	5.5	6.0	6.0	6.5	7.5
3.0	6.5	6.5	6.5	7.0	8.0
5.0	7.5	7.5	7.5	8.0	8.5
10.0 & above	10.0	10.0	10.0	10.0	10.0

It may be noted that these are only typical values and it is always possible to design transformer with different impedances. However, for design purposes it is customary to consider these standard values to design upstream and downstream protective equipment.

In an electrical circuit, the impedance limits the flow of current and Ohm's law gives the actual current. Alternatively, the voltage divided by current gives the impedance of the system. In a three-phase system which generates a phase voltage of E_p and where the phase current is I_p,

$$\text{Impedance in ohms} = \frac{E_p}{1.732 \times I_p}$$

The above forms the basis to decide the fault current that may flow in a system where the fault current is due to phase-to-phase or phase-to-ground short. In such cases, the internal impedances of the equipment rather than the external load impedances decide the fault currents.

Example:
For the circuit shown below calculate the short-circuit MVA on the LV side of the transformer to determine the breaking capacity of the switchgear to be installed (see Figure 3.5).

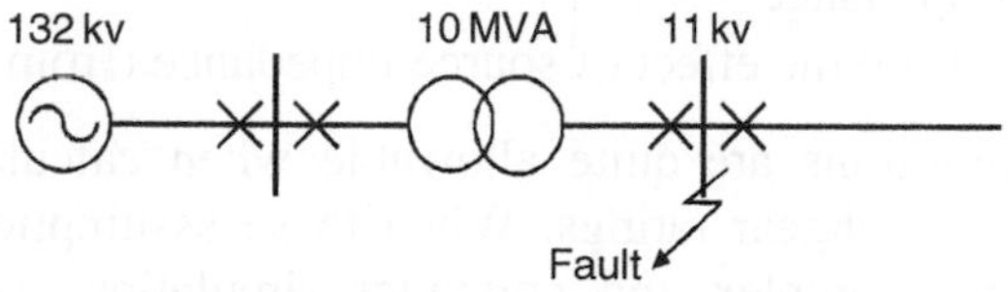

132 kv 10 MVA 11 kv

Fault

Figure 3.5
Short-circuit MVA example

Answer:

$$\text{Short-circuit MVA} = \frac{100P}{X\%} = \frac{100\times 10}{10} = 100\ \text{MVA}$$

Therefore,

$$\text{Fault current} = \frac{100}{\sqrt{3}\times 11} = 5.248\ \text{kA}$$

and

$$\text{Source impedance} = \frac{11}{\sqrt{3}\times 5.248} = 1.21\ \Omega$$

Calculate the fault current downstream after a particular distance from the transformer with the impedance of the line/cable being 1 Ω (see Figure 3.6).

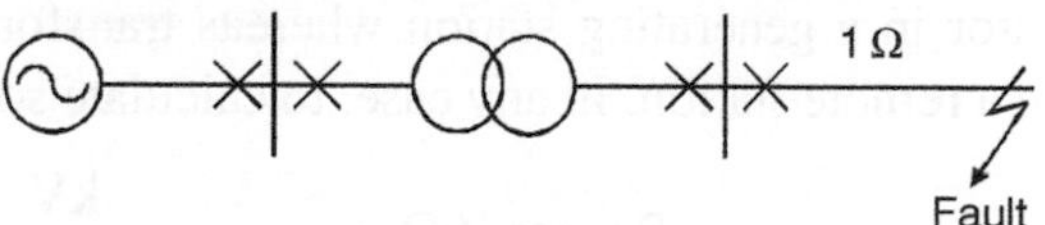

Figure 3.6
Calculation of fault current at end of cable

$$\text{Fault current} = \frac{11}{\sqrt{3}\times(1.21+1)} = 2.874\ \text{kA}$$

It should be noted that, in the above examples, a few assumptions are made to simplify the calculations. These assumptions are the following:

- Assume the fault occurs very close to the switchgear. This means that the cable impedance between the switchgear and the fault may be ignored.
- Ignore any arc resistance.
- Ignore the cable impedance between the transformer secondary and the switchgear, if the transformer is located in the vicinity of the substation. If not,

the cable impedance may reduce the possible fault current quite substantially, and should be included for economic considerations (a lower-rated switchgear panel, at lower-cost, may be installed).

- When adding cable impedance, assume the phase angle between the cable impedance and transformer reactance are zero, hence the values may be added without complex algebra, and values readily available from cable manufacturers' tables may be used.
- Ignore complex algebra when calculating and using transformer internal impedance.
- Ignore the effect of source impedance (from generators or utility).

These assumptions are quite allowable when calculating fault currents for protection settings or switchgear ratings. When these assumptions are not made, the calculations become very complex and computer simulation software should be used for exact answers. However, the answers obtained with making the above assumptions are found to be usually within 5% correct.

3.4 Useful formulae

Following are the methods adopted to calculate fault currents in a power system.

- *Ohmic method*: All the impedances are expressed in Ω.
- *Percentage impedance methods*: The impedances are expressed in percentage with respect to a base MVA.
- *Per unit method*: Is similar to the percentage impedance method except that the percentages are converted to equivalent decimals and again expressed to a common base MVA. For example, 10% impedance on 1 MVA is expressed as 0.1 pu on the same base.

3.4.1 Ohmic reactance method

In this method, all the reactance's components are expressed in actual ohms and then it is the application of the basic formula to decide fault current at any location. It is known that when fault current flows it is limited by the impedance to the point of fault. The source can be a generator in a generating station whereas transformers in a switching station receive power from a remote station. In any case, to calculate source impedance at HV in Ω:

$$\text{Source } Z\,\Omega = \frac{\text{kV}}{\sqrt{3} \times \text{HV fault current}}$$

Transformer impedance is expressed in terms of percent impedance voltage and is defined as the percentage of rated voltage to be applied on the primary of a transformer for driving a full load secondary current with its secondary terminals shorted. Hence, this impedance voltage forms the main factor to decide the phase-to-phase or any other fault currents on the secondary side of a transformer (see Figure 3.7).

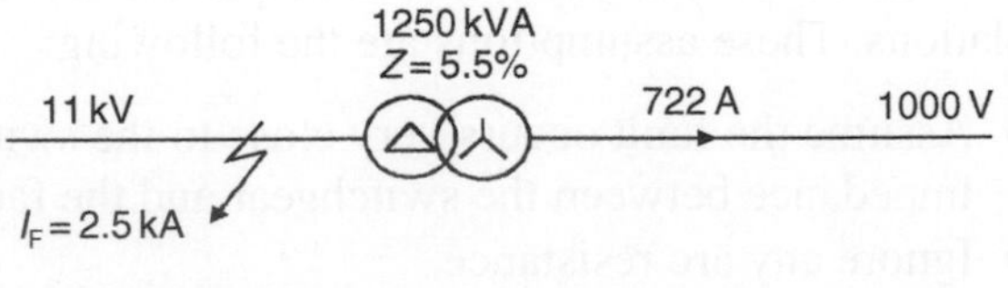

Figure 3.7
Calculation of total impedance and fault currents

To convert transformer impedance in Ω:

$$\text{Transformer } Z\,\Omega = \frac{Z\% \times \text{kV}}{\text{kA}}$$

Where
kV is the rated voltage and
kA is the rated current.

Multiplying by kV on both numerator and denominator, we get:

$$\text{Transformer Z in } \Omega = \frac{\text{Z} \times \text{kV}^2}{100 \times \text{MVA}}$$

Where Z is expressed in percentage impedance value.

In a case consisting of a generator source and a transformer, total impedance at HV including transformer:

$$\text{Total } Z\,\Omega\; HV = \text{Source } Z\,\Omega + \text{Transformer } Z\,\Omega$$

To convert $Z\,\Omega$ from HV to LV:

$$Z\,\Omega\; LV = \frac{\text{Total } Z\,\Omega\; HV \times LV^2}{HV^2}$$

To calculate LV fault current:

$$LV \text{ fault current} = \frac{LV}{\sqrt{3} \times Z\,\Omega\; LV}$$

Note: All voltages to be expressed in kV.

Example:
In the following circuit calculate:

(a) Total impedance in Ω at 1000 V
(b) Fault current at 1000 V.

$$Z\,\Omega \text{ source} = \frac{11}{1.732 \times 2.5} = 2.54\ \Omega$$

$$Z\,\Omega \text{ transformer} = \frac{5.5 \times 11^2}{100 \times 1.25} = 5.324\ \Omega$$

$$Z\,\Omega \text{ total} = 2.54 + 5.324 = 7.864\ \Omega$$

$$Z\,\Omega \text{ total at 1000 V} = \frac{7.864 \times 1^2}{11^2} = 0.065\ \Omega$$

$$\text{Fault current at 1000 V} = \frac{1}{1.732 \times 0.065} = 8.883 \text{ kA}$$

3.4.2 Other formulae in ohmic reactance method

In predominantly inductive circuits, it is usual to neglect the effect of resistant components, and consider only the inductive reactance X and replace the value of Z by X

to calculate the fault currents. Following are the other formulae, which are used in the ohmic reactance method. (These are obtained by multiplying numerators and denominators of the basic formula with same factors.)

1. $\text{Fault value MVA} = \dfrac{E^2}{X}$

2. $X = \dfrac{E^2}{\text{Fault value in MVA}}$

3. $X \text{ at } `B\text{' kV} = \dfrac{(X \text{ at } `A\text{' kV})\, B^2}{A^2}$

3.4.3 Percentage reactance method

In this method, the reactance values are expressed in terms of a common base MVA. Values at other MVA and voltages are also converted to the same base, so that all values can be expressed in a common unit. Then it is the simple circuit analysis to calculate the fault current in a system. It can be noted that these are also extensions of basic formulae.

Formulae for percentage reactance method

4. $\text{Fault value in MVA} = \dfrac{100\,(\text{MVA rating})}{X\,\%}$

5. $X\,\% = \dfrac{100\,(\text{MVA rating})}{\text{Fault value in MVA}}$

6. $X\% \text{ at } `N\text{' MVA} = \dfrac{N\,(X\,\% \text{ at rated MVA})}{\text{rated MVA}}$

For ease of mathematics, the base MVA of N may be taken as 100 MVA, so the formula would now read:

$$X\,\% \text{ at } 100 \text{ MVA} = \frac{100\,(X\,\% \text{ at rated MVA})}{\text{rated MVA}}$$

However, the base MVA can be chosen as any convenient value depending upon the MVA of equipment used in a system.

$$\text{Also, } X\,\% \text{ at } `B\text{' kV} = \frac{(X\,\% \text{ at } `A\text{' kV})\, B^2}{A^2}$$

To calculate the fault current in the earlier example using percentage reactance method:

1250 kVA
$Z = 5.5\%$
11 kV
722 A
1000 V
$I_F = 2.5$ kA

Fault MVA at the source = $1.732 \times 11 \times 2.5 = 47.63$ MVA. Take the transformer MVA (1.25) as the base MVA. Then source impedance at base MVA

$$= \frac{1.25 \times 100}{47.63} = 2.624\% \text{ (using (5))}$$

Transformer impedance = 5.5% at 1.25 MVA. Total percentage impedance to the fault = 2.624 + 5.5 = 8.124%. Hence fault MVA after the transformer = $(1.25 \times 100)/8.124$ = 15.386 MVA. Accordingly fault current at 1 kV = $15.386/(1.732 \times 1)$ = 8.883 kA. It can be noted that the end answers are the same in both the methods.

Formulae correlating percentage and ohmic reactance values

7. $X\% = \dfrac{100 \times \text{MVA rating}}{E^2}$

8. $X = \dfrac{X\%\, E^2}{100\,(\text{MVA rating})}$

3.4.4 Per unit method

This method is almost same as the percentage reactance method except that the impedance values are expressed as a fraction of the reference value.

$$\text{Per unit impedance} = \frac{\text{Actual impedance in ohms}}{\text{Base impedance in ohms}}$$

Initially the base kV (kV_b) and rated kVA or MVA (kVA_b or MVA_b) are chosen in a system. Then,

$$\text{Base current } I_b = \frac{\text{Base kVA}}{1.732 \times \text{Base kV}}$$

$$\text{Base impedance } Z_b = \frac{\text{Base } V}{1.732 \times \text{Base } A} = \frac{kV_b \div 1000}{1.732 \times I_b}$$

Multiplying by kV at top and bottom

$$= \frac{kV_b^2 \times 1000}{kVA_b}$$

Per unit impedance of a source having short-circuit capacity of kVA_{sc} is:

$$Z_{pu} = \frac{kVA_b}{kVA_{sc}}$$

Calculate the fault current for the same example using pu method.

1250 kVA
$Z = 5.5\%$
11 kV
722 A
1000 V
$I_F = 2.5$ kA

Here base kVA is chosen again as 1.25 MVA.

Source short-circuit MVA = 1.732 × 11 × 2.5 = 47.63 MVA

$$\text{Source impedance} = \frac{1.25}{47.63} = 0.02624 \text{ pu}$$

Transformer impedance = 0.055 pu

Impedance to transformer secondary = 0.02624 + 0.055 = 0.08124 pu.

Hence short-circuit current at 1 kV = 1250/(1.732×1×0.08124) = 8.883 kA

Depending upon the complexity of the system, any method can be used to calculate the fault currents.

General formulae

It is quite common that the interconnections in any distribution system can be converted or shown in the combination of series and parallel circuits. Then it would be necessary to calculate the effective impedance at the point of fault by combining the series and parallel circuits using the following well-known formulae. The only care to be taken is that all the values should be in same units and should be referred to the same base.

Series circuits:

9. $X_t = X_1 + X_2 + X_3 + \cdots X_n$ where all values of X are either:

   (a) $X\%$ at the same MVA base; or
   (b) X at the same voltage.

Parallel circuits:

10. $$\frac{1}{X_t} = \frac{1}{X_1} + \frac{1}{X_2} + \frac{1}{X_3} + \cdots \frac{1}{X_n}$$

3.5 Cable information

Though cable impedances have been neglected in the above cases, to arrive at more accurate results, it may be necessary to consider cable impedances in some cases, especially where long distance of transmission lines and cables are involved. Further the cables selected in a distribution system should be capable of withstanding the short-circuit currents expected until the fault is isolated/fault current is arrested.

Cables are selected for their sustained current rating so that they can thermally withstand the heat generated by the current under healthy operating conditions and at the same time, it is necessary that the cables also withstand the thermal heat generated during short-circuit conditions.

The following table will assist in cable selection, which also states the approximate impedance in Ω/km. Current rating and voltage drop of 3 and 4 core PVC insulated cables with stranded copper conductors. Fault current ratings for cables are given in the manufacturers' specifications and tables, and must be modified by taking into account the fault duration.

Sustained Current Rating					
Rated Area (mm^2)	Z Approx. (Ω/km)	Ground	Duct	Air	Voltage Drop per Amp. Meter mV
1.5	13.41	23	19	22	23.2
2.5	8.010	30	25	30	13.86
4.0	5.011	40	34	40	8.67
6.0	3.344	50	42	51	5.79
10.0	2.022	67	55	68	3.46
16.0	1.253	87	72	91	2.17
25.0	0.808	119	95	115	1.40
35.0	0.578	140	114	145	1.00
50.0	0.410	167	135	180	0.71
70.0	0.297	200	166	225	0.51
95.0	0.226	243	205	270	0.39
120.0	0.185	278	231	315	0.32
150.0	0.154	310	257	360	0.27
185.0	0.134	354	294	410	0.23
240.0	0.113	390	347	480	0.20
300.0	0.097	443	392	550	0.17
400.0	0.087	508	448	670	0.15

Example:
Reference to Table 3.1 shows that, a 70 mm^2 copper cable can withstand a short-circuit current of 8.05 kA for 1 s. However, the duration of the fault or the time taken by the protective device to operate has to be considered. This device would usually operate well within 1 s, the actual time being read from the curves showing short-circuit current/tripping time relationships supplied by the protective equipment manufacturer. Suppose the fault is cleared after 0.2 s. We need to determine what short-circuit current the cable can withstand for this time. This can be found from the expression:

$$I_{SC} = \frac{A \times K}{\sqrt{t}}$$

Where
$K = 115$ for PVC/copper cables of 1000 V rating
$K = 143$ for XLPE/copper cables of 1000 V rating
$K = 76$ for PVC/aluminum (solid or stranded) cables of 1000 V rating
$K = 92$ for XLPE/aluminum (solid or stranded) cables of 1000 V rating.
And where
A = the conductor cross-sectional area in mm^2
t = the duration of the fault in seconds.

So in our example:

$$I_{SC} = \frac{70 \times 115}{\sqrt{0.2}} = 18 \text{ kA (for 0.2 s)}$$

Cable bursting is not normally a real threat in the majority of cases where armored cable is used since the armoring gives a measure of reinforcement. However, with larger sizes, in excess of 300 mm^2, particularly when these cables are unarmored, cognizance should be taken of possible bursting effects.

When the short-circuit current rating for a certain time is known, the formula $E = I^2t$ can also be used to obtain the current rating for a different time. In the above example:

$$I_1^2 t_1 = I_2^2 t_2$$

$$\Rightarrow I_2 = \frac{\sqrt{I_1^2 t_1}}{t_2}$$

$$I_2 = \frac{\sqrt{(8.05^2 \times 1)}}{0.2}$$

$$= 18 \text{ kA}$$

Naturally, if the fault is cleared in more than 1 s, the above formula's can also be used to determine what fault current the cable can withstand for this extended period.

Note: In electrical protection, engineers usually cater for failure of the primary protective device by providing back-up protection. It makes for good engineering practice to use the tripping time of the back-up device, which should only be slightly longer than that of the primary device in short-circuit conditions, to determine the short-circuit rating of the cable. This then has a built-in safety margin.

3.6 Copper conductors

	Electrical Properties						Physical Properties							
	Current Ratings						Nominal Diameters						Approx. Mass	
Cable Size (mm^2)	Ground (A)	Ducts (A)	Air (A)	Impedance (Ω /km)	Volt Drop (mV/A/m)	1 s Short-Circuit Rating (kA)	Dl–3c (mm)	Dl–4c (mm)	d–3c (mm)	d–4c (mm)	D2–3c (mm)	D2–4c (mm)	3c (kg/km)	4c (kg/km)
1.5	23	18	18	14.48	25.080	0.17	8.51	9.33	1.25	1.25	14.13	14.95	448	501
2.5	30	24	24	8.87	15.363	0.28	9.61	10.56	1.25	1.25	15.23	16.18	522	597
4.0	38	31	32	5.52	9.561	0.46	11.40	12.57	1.25	1.25	17.02	18.39	667	762
6.0	48	39	40	3.69	6.391	0.69	12.58	13.90	1.25	1.25	18.4	19.72	790	910
10.0	64	52	54	2.19	3.793	1.15	14.59	16.14	1.25	1.25	20.41	21.95	996	1169
16.0	82	67	72	1.38	2.390	1.84	16.55	19.18	1.25	1.60	22.37	25.92	1295	1768
25.0	126	101	113	0.8749	1.515	2.87	19.46	21.34	1.60	1.60	26.46	28.34	1838	2196
35.0	147	120	136	0.6335	1.097	4.02	20.89	23.97	1.60	1.60	27.89	31.17	2215	2732
50.0	176	144	167	0.4718	0.817	5.75	24.26	28.14	1.60	2.00	31.46	36.54	2871	3893
70.0	215	175	207	0.3325	0.576	8.05	27.07	31.29	2.00	2.00	35.47	40.09	3617	4837
95.0	257	210	253	0.2460	0.427	10.92	31.19	35.82	2.00	2.00	39.99	44.62	4901	6115
120.0	292	239	293	0.2012	0.348	13.80	33.38	38.10	2.00	2.00	42.18	47.40	5720	7269
150.0	328	269	336	0.1698	0.294	17.25	36.68	42.05	2.00	2.50	45.98	52.65	6908	9250
185.0	369	303	384	0.1445	0.250	21.27	40.82	46.75	2.50	2.50	51.12	57.45	8690	11039
240.0	422	348	447	0.1220	0.211	27.60	46.43	53.06	2.50	2.50	57.13	64.16	10767	13726
300.0	472	397	509	0.1090	0.189	34.50	51.10	58.53	2.50	2.50	62.20	70.13	12950	16544

Table 3.1
Electrical and physical properties of 3- and 4-core PVC-insulated PVC-bedded SWA PVC-sheathed 600/1000 V cables

4

System earthing

4.1 Introduction

In Chapter 2 we have briefly covered that the phase-to-ground faults in a system can limit the ground fault current depending on adding external impedance between neutral and the earth. This chapter briefly covers the various methods of grounding that are adopted in the electrical systems. In the following clauses, the star-connected transformer is shown which are widely used in power distribution. The grounding methods are also applicable in case of generators, whose windings are also invariably star connected.

The following table highlights the possible problems that can occur in a system due to the common faults and the solutions that can be achieved by adopting system grounding.

Problems
Phase faults: High fault currents Only limited by inherent impedance of power supply. Earth faults: Solid earthing means high earth fault currents Only limited by inherent zero sequence impedance of power system. *Consequence* 1. Heavy currents damage equipment extensively – danger of fire hazard. 2. This leads to long outage times – lost production, lost revenue. 3. Heavy currents in earth bonding gives rise to high touch potentials – dangerous to human life. 4. Large fault currents are more hazardous in igniting gases – explosion hazard.

Solutions
Phase segregation: Eliminates phase-to-phase faults. Resistance earthing: Means low earth fault currents – can be engineered to limit to any chosen value. *Benefits* 1. Fault damage now minimal – reduces fire hazard. 2. Lower outage times – less lost production, less lost revenue. 3. Touch potentials kept within safe limits – protects human life. 4. Low fault currents reduce possibility of igniting gases – minimizes explosion hazard. 5. No magnetic or thermal stresses imposed on plant during fault. 6. Transient overvoltages limited – prevents stressing of insulation, breaker restrikes.

4.2 Earthing devices

4.2.1 Solid earthing

In this case, the neutral of a power transformer is earthed solidly with a copper conductor as shown in Figure 4.1.

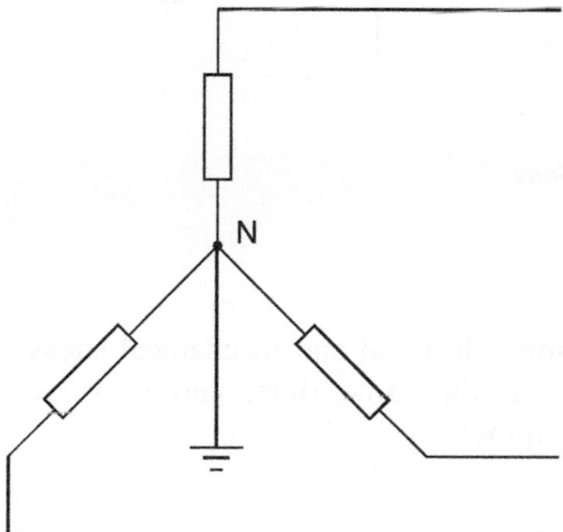

Figure 4.1
Solid earthing of power transformer

Advantages

- Neutral held effectively at earth potential.
- Phase-to-ground faults of same magnitude as phase-to-phase faults; so no need for special sensitive relays.
- Cost of current limiting device is eliminated.
- Grading insulation towards neutral point N reduces size and cost of transformers.

Disadvantages

- As most system faults are phase-to-ground, severe shocks are more considerable than with resistance earthing.
- Third harmonics tend to circulate between neutrals.

4.2.2 Resistance earthing

A resistor is connected between the transformer neutral and earth (see Figure 4.2):

- Mainly used below 33 kV.
- Value is such as to limit an earth fault current to between 1 and 2 times full load rating of the transformer. Alternatively, to twice the normal rating of the largest feeder, whichever is greater.

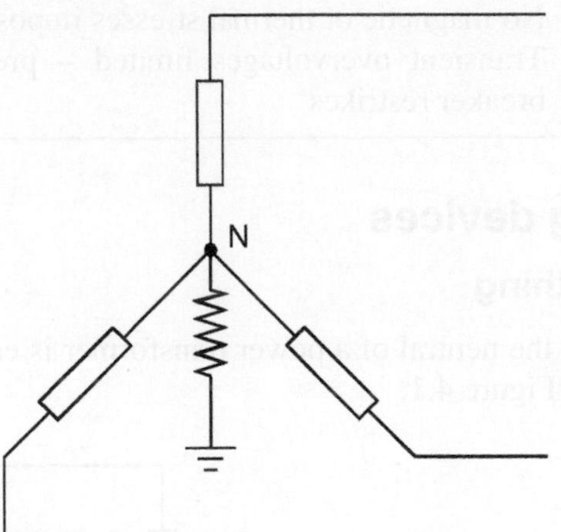

Figure 4.2
Resistance earthing

Advantages

- Limits electrical and mechanical stress on system when an earth fault occurs, but at the same time, current is sufficient to operate normal protection equipment.

Disadvantages

- Full line-to-line insulation required between phase and earth.

4.2.3 Reactance earthing

A reactor is connected between the transformer neutral and earth (see Figure 4.3):

- Values of reactance are approximately the same as used for resistance earthing.
- To achieve the same value as the resistor, the design of the reactor is smaller and thus cheaper.

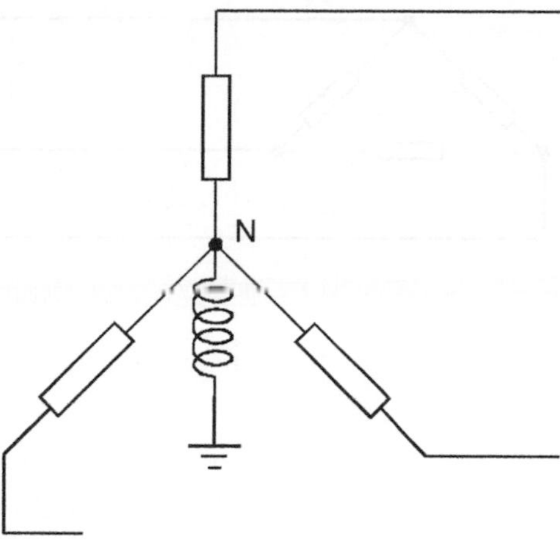

Figure 4.3
Reactance earthing

4.2.4 Arc suppression coil (Petersen coil)

A tunable reactor is connected in the transformer neutral to earth (see Figure 4.4):

- Value of reactance is chosen such that reactance current neutralizes capacitance current. The current at the fault point is therefore theoretically nil and unable to maintain the arc, hence its name.
- Virtually fully insulated system, hence current available to operate protective equipment is so small as to be negligible. To offset this, the faulty section can be left in service indefinitely without damage to the system as most faults are earth faults of a transient nature, the initial arc at the fault point is extinguished and does not restrike.

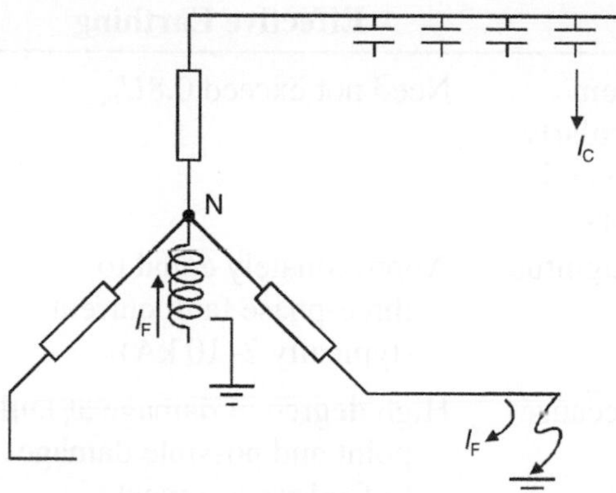

Figure 4.4
Arc suppression coil (Petersen coil)

Sensitive watt-metrical relays are used to detect permanent earth faults.

4.2.5 Earthing via neutral earthing compensator

This provides an earth point for a delta system and combines the virtues of resistance and reactance earthing in limiting earth fault current to safe reliable values (see Figure 4.5).

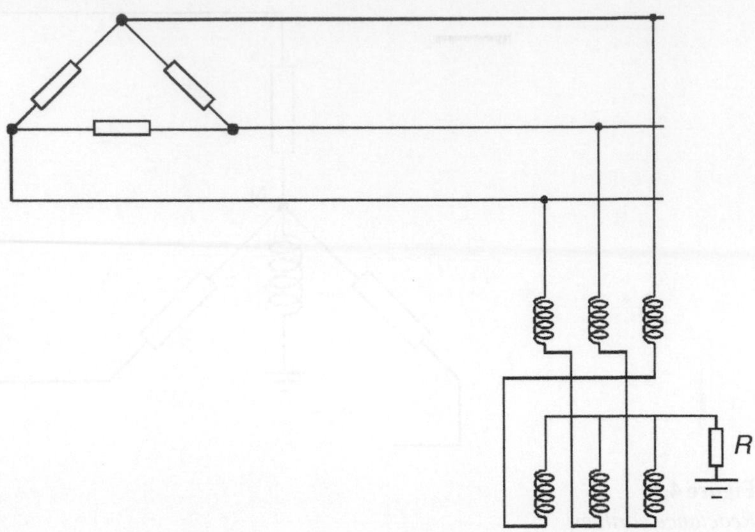

Figure 4.5
Earthing via neutral earthing compensator

4.3 Evaluation of earthing methods

The earthing method is called effectively earthed when it is directly connected to earth (solidly earthed) without any passive component in between. Non-effective earthing refers to, the method of earthing through a resistance, reactance, transformer, etc. Following table compares the earthing methods.

Evaluation of Relative Merits of Effective and Resistive Earthing

1 **Subject**	2 **Effective Earthing**	3 **Resistive Earthing**
Rated voltage of system components, particularly power cables and metal oxide surge arresters	Need not exceed $0.8U_m$	Must be at least $1.0U_m$ for 100 s
Earth fault current magnitude	Approximately equal to three-phase fault current (typically 2–10 kA)	Reduced earth fault current magnitude (typically 300–900 A)
Degree of damage, because of an earth fault	High degree of damage at fault point and possible damage to feeder equipment	Lesser degree of damage at fault point and usually no damage to feeder equipment
Step and touch potentials during earth fault	High step and touch potentials	Reduced step and touch potentials
Inductive interference on and possible damage, to control and other lower-voltage circuits	High probability	Lower probability

(continued)

Evaluation of Relative Merits of Effective and Resistive Earthing		
1	**2**	**3**
Subject	**Effective Earthing**	**Resistive Earthing**
Relaying of fault conditions	Satisfactory	Satisfactory
Cost	Lower initial cost but higher long-term equipment repair cost	Higher initial cost but lower long-term equipment repair cost, usually making resistive earthing more cost-effective

Figure 4.6 gives the touch potential with solid earthing. Figure 4.7 shows the touch potentials with resistance *R* introduced in the neutral. Here the ground fault current is limited by the resistance *R*, so only reduced current flows to the earth. However, it is a normal practice to adopt solid earthing method at low voltages (up to say 600 V) and resistance grounding is adopted for higher voltages (upto 33 kV). The other methods of earthing (reactor, transformer, etc.) are generally adopted in the cases of voltages beyond 33 kV. Cost invariably determines the earthing method.

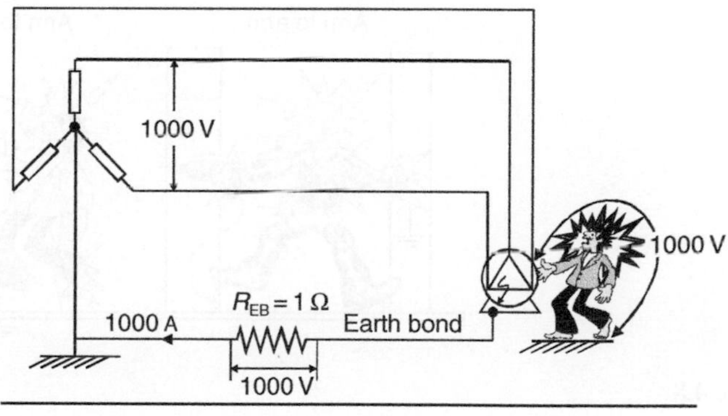

Figure 4.6
Touch potentials – solid earthing

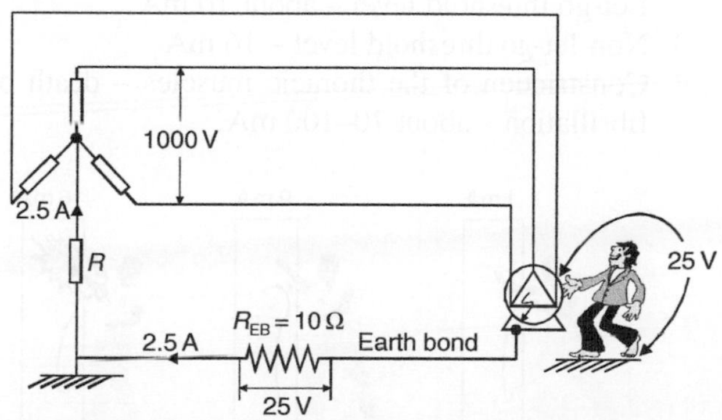

Figure 4.7
Touch potentials – resistive earthing

The main reason for adopting solid earthing is because, the resistance earthing cannot be used for single-phase loads, whereas most of the LV distribution mainly households, etc. comprise of single-phase loads. Nevertheless, resistance grounding is considered at low voltages in industrial environments, where three-phase loads are connected and the process conditions do not accept frequent shutdowns due to ground faults. Though it is true that the power interruptions can be kept low with the use of resistive earthing method, human protection demands that the power to be isolated in case of ground faults. That is one more reason for using solid earthing in utility distribution transformers.

4.4 Effect of electric shock on human beings

4.4.1 Electric shock and sensitive earth leakage protection

There are four major factors, which determine the seriousness of an electric shock:

1. Path taken by the electric current through the body
2. Amount of current
3. Time the current is flowing
4. The body's electrical resistance.

The most dangerous and most common path is through the heart (see Figure 4.8). Persons are not normally electrocuted between phases or phase-to-neutral, almost all accidents are phase-to-earth.

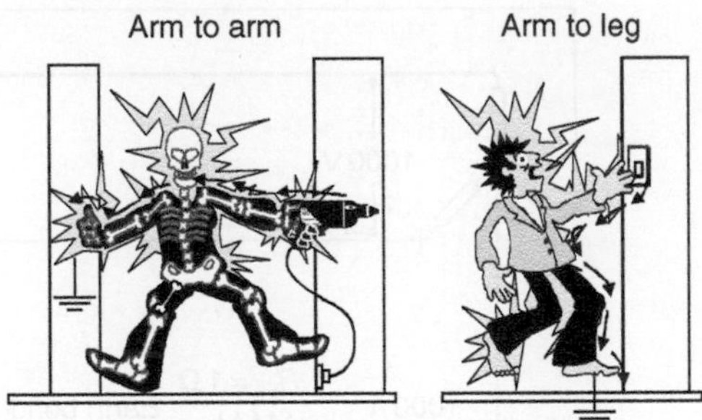

Figure 4.8
Dangerous current flows

Figure 4.9 shows the four stages of the effect of a current flow through the body:

1. Perception – tingling – about 1 mA
2. Let-go threshold level – about 10 mA
3. Non-let-go threshold level – 16 mA
4. Constriction of the thoracic muscles – death by asphyxiation and ventricular fibrillation – about 70–100 mA.

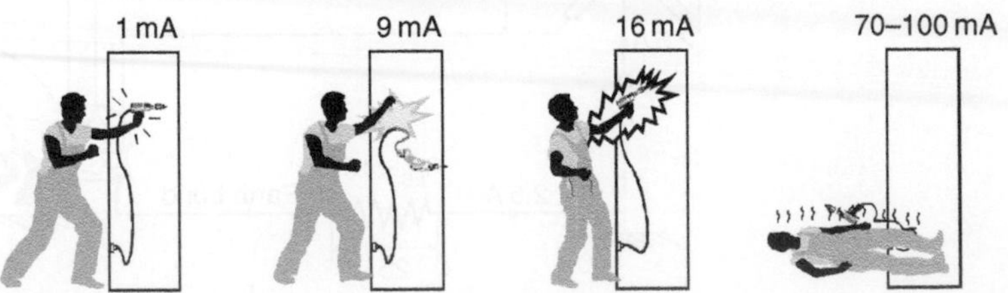

Figure 4.9
Effects of current flow through the body

Figure 4.10 shows the normal electro-cardiogram – one pulse beat – at 80 bpm = 750 ms.

1. *QRS phase*: normal pumping action
2. *T phase*: refractory or rest phase – about 150 ms
3. Death could occur if within this very short period of 150 ms a current flow was at the fibrillation level.

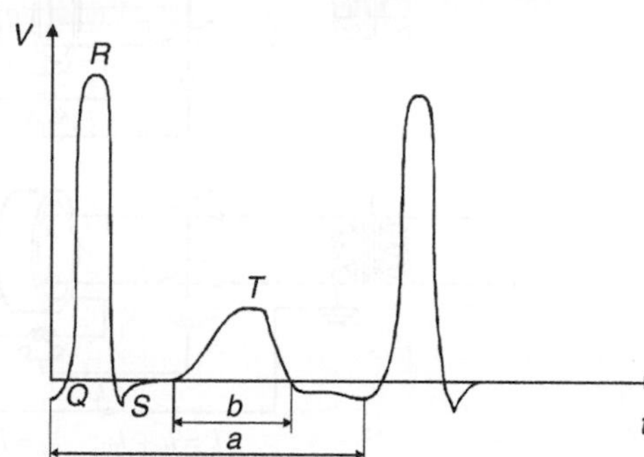

Figure 4.10
Electro-cardiogram

Figure 4.11 shows the resistance of the human body – hand-to-hand or hand to foot. Consider an example of a man working and perspiring, he touches a conductor at 300 V (525 V phase to earth). 300 V divided by 1000 Ω = 300 mA. It is important to remember that, it is the current that kills and not voltage.

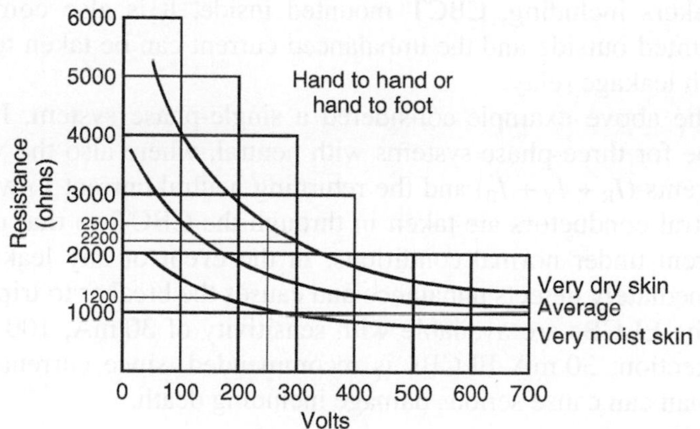

Figure 4.11
Resistance of human body

4.4.2 Sensitive earth leakage protection

Earthing does not ensure that humans will be protected when coming in contact with a live conductor. Though there may be relays, which are set to sense the earth leakages, invariably their settings are high. Hence earth leakage circuit breakers (ELCB) or residual current circuit breakers (RCCB) are adopted where possibility of human interaction to a live conductor is high. These breakers work on the core balance current principle.

Figure 4.12 illustrates the operation of the core balance leakage device. When the system conditions are normal, the phase current and neutral current will be equal and in phase. Hence the CT will not detect any current under normal conditions since $I_L + I_N = 0$ (vector sum).

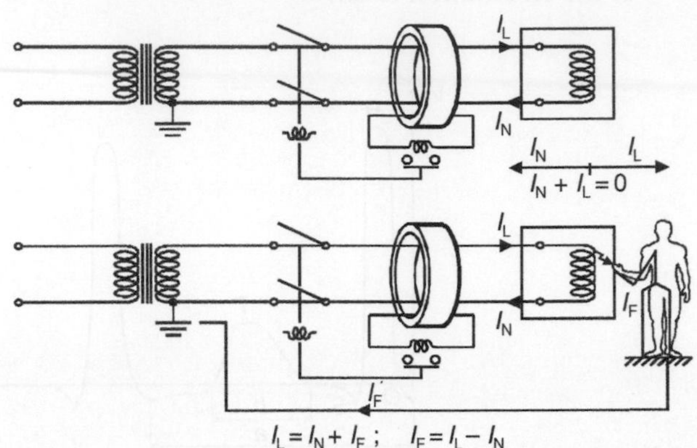

Figure 4.12
Principles of core balance protection

The ELCB comprising of core balance CT is mounted at the source end. When a human comes in contact on any part of the line, a part of the current will start flowing through the body. It will result in unbalance of the currents entering and returning to the CBCT of the ELCB. If the fault current I_F, flows through the human body, I_N is reduced by this amount. Relay is operated by this unbalance quantity, and immediately trips the ELCB.

It is normal that the ELCBs are moulded breakers similar to the miniature circuit breakers including, CBCT mounted inside. It is also common that the CBCT can be mounted outside and the unbalanced current can be taken to trip a separate relay namely earth leakage relay.

The above example considered a single-phase system. However, the principle is the same for three-phase systems with neutral, where also the vector sum of the three-phase currents ($I_R + I_Y + I_B$) and the returning neutral current I_N will be zero. All the phase and neutral conductors are taken in through the CBCT so that the CBCT does not sense any current under normal conditions. In the event of any leakage in any phase, the CBCT immediately detects unbalance and causes the breaker to trip.

The ELCBs are available with sensitivity of 30 mA, 100 mA and 300 mA. For human protection, 30 mA ELCBs is recommended, since currents flowing above 30 mA in a human can cause serious damage including death.

5

Fuses

5.1 Historical

Fuse is the most common and widely used protective device in electrical circuits. Though 'fuseless' concept had been catching on for quite some time, still quite a lot of low-voltage distribution circuits are protected with fuses. Further fuses form a major backup protection in medium-voltage and high-voltage distribution to 11 kV, where switches and contactors with limited short-circuit capacities are used.

In 1881, Edison patented his 'lead safety wire', which was officially recognized as the first fuse. However, it was also said that Swan actually used this device in late 1880 in the lighting circuits of Lord Armstrong's house. He used strips of tin-foil jammed between brass blocks by plugs of woods. The application of the fuse in those days was not to protect the wires and system against short circuit, but to protect the lights which cost 25 shillings a time (a fortune in those days).

Later, as electrical distribution systems grew, it was found that after short circuits, certain conductors failed. This was due to the copper conductors, not being accurately drawn out (extruded) to a constant diameter throughout the cable length; faults always occurring at the smallest cross-sectional area.

Fuses were often considered as casual devices until not so long ago. The open tin-foil (rewireable) sometimes came in for a lot of abuse. If it blew constantly, then the new fuse was just increased until it stayed in permanently. Sometimes hairpins were used. Greater precision only became possible with the introduction of the Cartridge fuse.

5.2 Rewireable type

As the name indicates the fuse can be replaced or 'rewired' once it fails. Fusible wire used to be contained in an asbestos tube to prevent splashing of volatile metal.

Disadvantages

1. Open to abuse due to incorrect rating of replacement elements hence affording incorrect protection.
2. Deterioration of element as it is open to the atmosphere.

5.3 Cartridge type

Silver element, specially shaped, enclosed in a barrel of insulating material, filled with quartz. Silver and quartz combine to give a very good insulator and prevent arc from re-striking (see Figure 5.1).

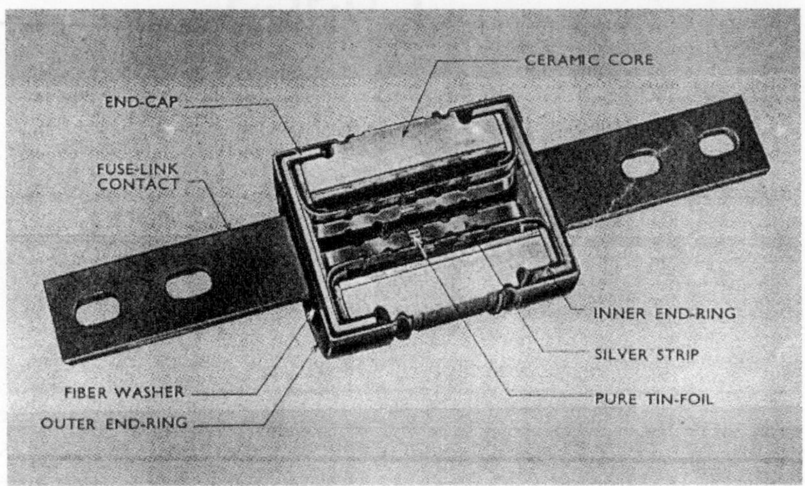

Figure 5.1
Sectional view of a typical class – GP type 5 Cartridge fuse – link

Advantages

1. Correct rating and characteristic fuse always fitted to a circuit-not open to abuse as rewireable type.
2. Arc and fault energy contained within insulating tube-prevents damage.
3. Normally sealed therefore not affected by atmosphere hence gives more stable characteristic-reliable grading.
4. Can operate considerably faster, suitable for higher short-circuit duty:
   - Cartridge type can handle 100 000 A
   - Semi-open type can handle 4000 A.

Normal currents carried continuously are much closer to fusing current due to special design of element. These fuses are most widely used in electrical systems and are named as HRC (high rupturing capacity) fuses, with the name synonymous with their short-circuit current breaking capacity.

5.4 Operating characteristics

All fuses irrespective of the type have inverse characteristic as shown in graph that follows. Inverse means that they can withstand their nominal current rating almost indefinitely but as the currents increases their withstanding time starts decreasing making them 'blow'. The blowing time decreases as the flowing currents increase. The thermal characteristic or withstand capacity of a fuse is indicated in terms of '$I^2 t$' where I is the current and t is the withstand time (see Figure 5.2). The prospective current is the I_{rms} that would flow on the making of a circuit when the circuit is equipped for insertion of a fuse but that the fuse is replaced with a solid link. The curves are very important when determining the application of fuses as they allow the correct ratings to be chosen to give grading.

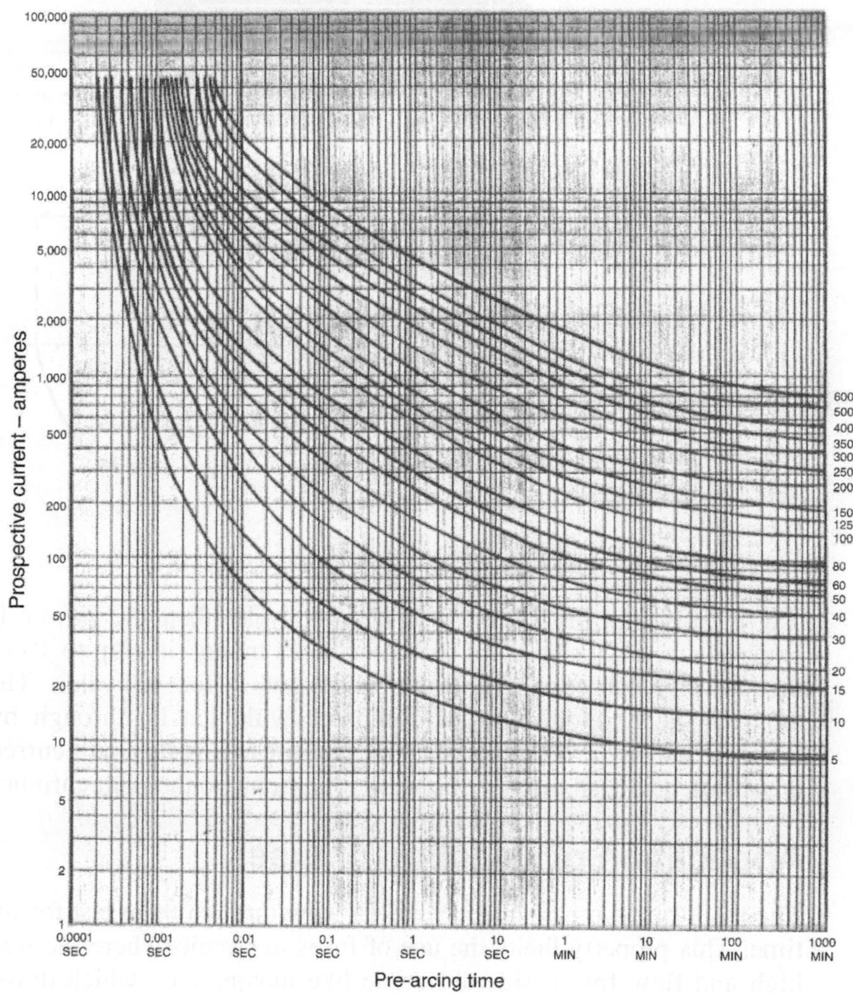

Figure 5.2
Inverse characteristic of fuse

5.5 British standard 88:1952

This standard lays down definite limits of:

(a) Temperature rise
(b) Fusing factor = minimum fusing current/current rating = 1.4
(c) Breaking capacity.

These are all dependent on one another and by careful balancing of factors a really good fuse can be produced. For example, a cool working fuse may be obtained at the expense of breaking capacity. Alternatively, too low a fusing factor may result in too high a temperature, therefore too close protection and possibilities of blowing are more.

5.6 Energy 'let through'

Fuses operate very quickly and can cut-off fault current long before it reaches its first peak (see Figure 5.3).

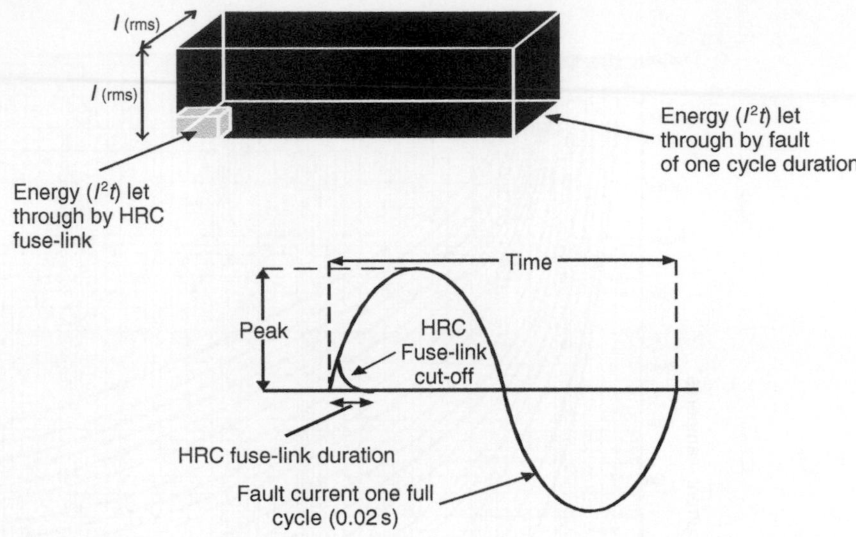

Figure 5.3
Energy 'let through'

If a fuse cuts off in the first quarter cycle, then the power let-through is I^2t. By comparison, circuit breakers can clear faults in any time up to 10 cycles and in this case the power let-through is the summation of I^2 for 10 cycles. The energy released at the fault is therefore colossal compared with that let through by a fuse. Damage is therefore extensive. In addition, all apparatus carrying this fault current (transformers, etc.) is subjected to high magnetic forces proportional to the fault current squared ($I_F{}^2$)!

5.7 Application of selection of fuses

The fuses blow in case the currents flowing through them last for more than its withstand time. This property limits the use of fuses in circuits where the inrush currents are quite high and flow for considerable time like motors, etc., which draw more than six times their full load current for a short time ranging from milliseconds to few seconds depending on the capacity. Hence, it is not possible to use fuses as overload protection in such circuits, since it may be necessary to select higher-rated fuse to withstand inrush currents. Accordingly, the fuses are mostly used as short-circuit protection rather than as overload protection in such circuits.

The fuses can be used as either for overload and short-circuit protection or for short-circuit protection as noted below:

- Circuits where the load does not vary much above normal value during switching on and operating conditions. Resistive circuits like lamps show such characteristics. Hence, it is possible to use fuses as overload protection in such circuits. They also protect against short circuits.
- Circuits where loads vary considerably compared to the normal rating e.g.
  - Direct-on-line motors
  - Cranes

- Rolling mills
- Welding set, etc. In these cases, fuses are used to provide short-circuit protection only as it is not possible to select a size meeting both overload and inrush conditions.

Fuse selection depends on a number of factors:

- Maximum fault kVA of circuit to be protected
- Voltage of circuit.

The above factors help to calculate the prospective current of circuit to be protected. The full prospective current is usually never reached due to rapid operation of the fuse and hence the following factors need to be considered.

1. *Full load current of circuit*: Short-circuit tests show that the cut-off current increases as the rating increases. Hence if a higher-rated fuse is used it may take longer time to blow under short circuits which may affect the system depending upon the value and duration. Hence, greater benefit is derived from use of correct or nearest rating cartridge fuses compared to the circuit rating.
2. *Degree of overcurrent protection required*: It is necessary to consider slightly higher rating for the fuses compared to maximum normal current expected in a system. This factor is called the fusing factor (refer clause 5.5) and can be anywhere between 1.25 and 1.6 times the normal rating.
3. Level of overcurrent required to be carried for a short time without blowing or deteriorating e.g. motor starting currents. This point is important for motor circuits. Fuses must be able to carry starting surge without blowing or deteriorating.
4. Whether fuses are required to operate or grade in conjunction with other protective apparatus. This factor is necessary to ensure that only faulty circuits are isolated during fault conditions without disturbing the healthy circuits.

Depending on the configuration used, discrimination must be achieved between:

- Fuses and fuses
- Fuses and relays, etc.

There is no general rule laid down for the application of fuses and each problem must be considered on its own merits.

5.8 General 'rules of thumb'

5.8.1 Short-circuit protection

Transformers, fluorescent lighting circuits

Transient switching surges – take next highest rating above full-load current.

Capacitor circuits

Select fuse rating of 25% or greater than the full-load rating of the circuit to allow for the extra heating by capacitance effect.

Motor circuits

Starting current surge normally lasts for 20 s. Squirrel cage induction motors:

- Direct-on-line takes about 7 times full-load current
- 75% tap auto-transformer takes about 4 times full-load current
- 60% tap auto-transformer takes about 2.5 times full-load current
- Star/delta starting takes about 2.5 times full-load current.

5.8.2 Overload protection

Recommend 2:1 ratio to give satisfactory discrimination.

5.9 Special types

5.9.1 Striker pin

This type is most commonly used on medium- and low-voltage circuits. When the fuse blows, a striker pin protrudes out of one end of the cartridge. This is used to hit a tripping mechanism on a three-phase switch-fuse unit, so tripping all three phases. This prevents single phasing on three-phase motors.

Note: On switch-fuse LV distribution gear, it is a good policy to have an isolator on the incoming side of the fuse to facilitate changing.

5.9.2 Drop-out type

Used mainly on rural distribution systems. Drops out when fuse blows, isolating the circuit and giving line patrolman easy indication of fault location.

5.10 General

The fuse acts as both fault detector and interrupter. It is satisfactory and adequate for both of these functions in many applications. Its main virtue is speed.

However, as a protective device it does have a number of limitations, the more important of which are as follows:

- It can only detect faults that are associated with excess current.
- Its operating characteristic (i.e. current/time relationship) cannot be adjusted or set.
- It requires replacement after each operation.
- It can be used only at low and medium voltages.

Because of these limitations, fuses are normally used only on relatively unimportant, small power, low- and/or medium-voltage circuits (see Figure 5.4).

5.10.1 Series overcurrent AC trip coils

These are based on the principle of working of fuses where the coils are connected to carry the normal current and operated as noted below:

- A coil (instead of a fuse) is connected into the primary circuit and magnetism is used to lift a plunger to trip a circuit breaker.
- Refinement on this arrangement is the dashpot, which gives a time/current characteristic like a fuse.

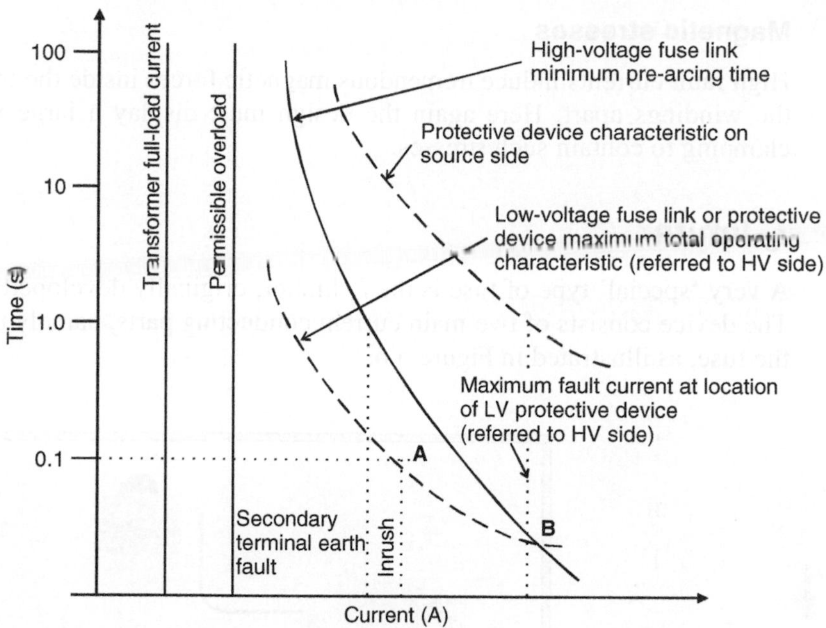

Figure 5.4
Characteristic of transformer HV/LV

- Amps-turns are a measurement of magnetism; therefore for the same flux (i.e. lines of magnetism necessary to lift the tripping plunger) say 50 Amp-turns, a 50 A coil would have 1 turn, whereas a 10 A coil would have 5 turns (see Figure 5.5).

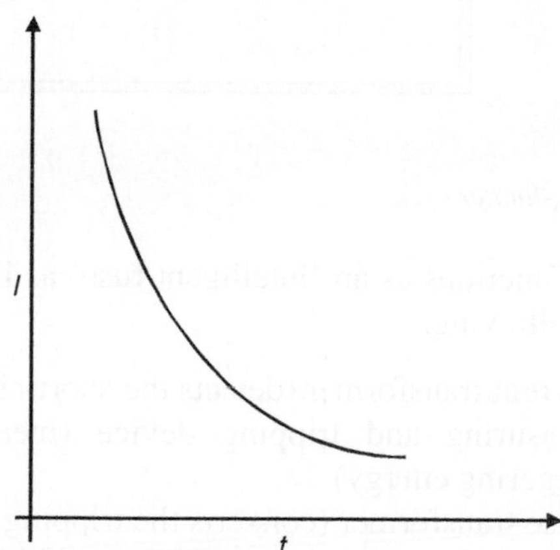

Figure 5.5
Series overcurrent AC trip coil characteristic

Limitation of this type of arrangement is:

Thermal rating

This coil must carry the full fault current and if this is high then the heating effect (I^2) may be so great as to burn out the insulation. The design must therefore be very conservative.

Magnetic stresses

High fault currents induce tremendous magnetic forces inside the trip coil tending to force the windings apart. Here again the design must display a large margin of support and clamping to contain such stresses.

5.11 I_S-limiter

A very 'special' type of fuse is the I_S-limiter, originally developed by the company ABB. The device consists of two main current conducting parts, namely the main conductor and the fuse, as illustrated in Figure 5.6.

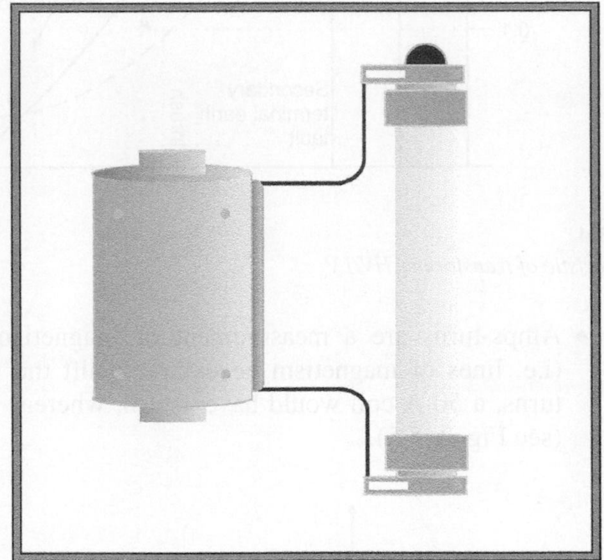

Figure 5.6
Construction of I_S*-limiter*

The device functions as an 'intelligent fuse' as illustrated in Figure 5.7. The functional parts are the following:

1. Current transformer (detects the short-circuit current)
2. Measuring and tripping device (measures the current and provides the triggering energy)
3. Pulse transformer (converts the tripping pulse to busbar potential)
4. Insert holder with insert (conducts the operating current and limits the short-circuit current).

The I_S-limiter is intended to interrupt very high short-circuit currents very quickly, in order to protect the system against these high currents. Currents of values up to 210 kA (11 kV) can be interrupted in 1 ms. This means that the fault current is interrupted very early in the first cycle, as illustrated in Figure 5.8.

When a fault current is detected, the main conductor is opened very swiftly. The current then flows through the fuse, which interrupts the fault current. The overvoltage occurring due to the interruption of current is relatively low due to the fact that the magnitude of current on the instant of interruption is still relatively low. The main conductor and parallel fuse have to be replaced after each operation.

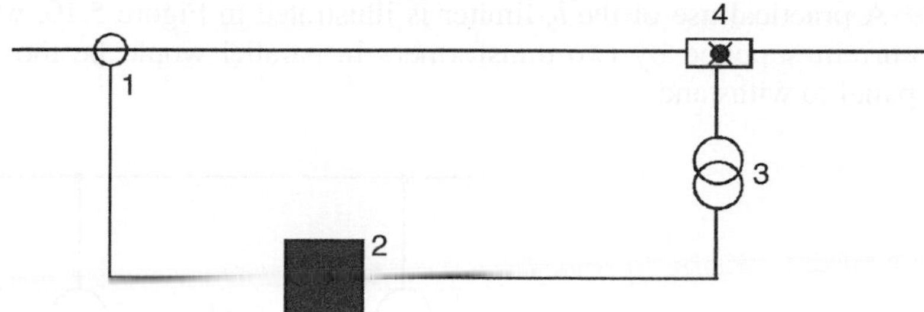

Figure 5.7
Functional diagram of I_S*-limiter*

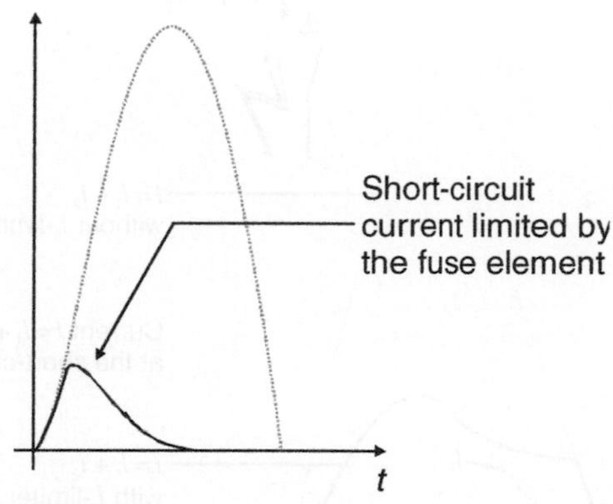

Figure 5.8
Fault current cycle

The I_S-limiter is only intended to interrupt high fault currents, leaving the lower fault currents to be interrupted by the circuit breakers in the system. This is achieved by the measuring device detecting the instantaneous current level and the rate of current rise. The rate of current rise (di/dt), is high with high fault currents, and lower with lower fault currents as illustrated in Figure 5.9. The I_S-limiter only trips when both set response values are reached.

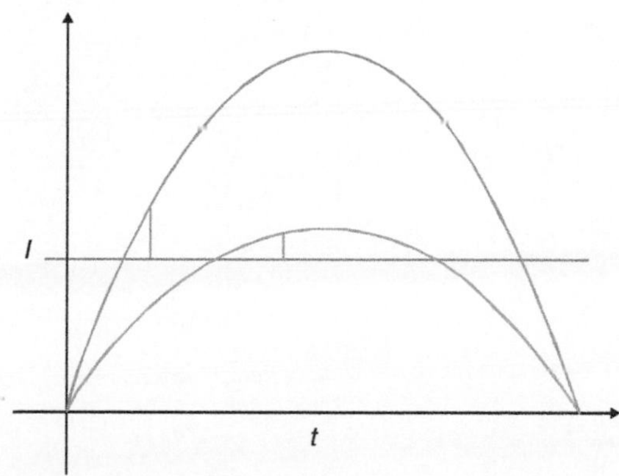

Figure 5.9
Rate of current rise

A practical use of the I_S-limiter is illustrated in Figure 5.10, where the combined fault current supplied by two transformers in parallel would be too high for the switchgear panel to withstand.

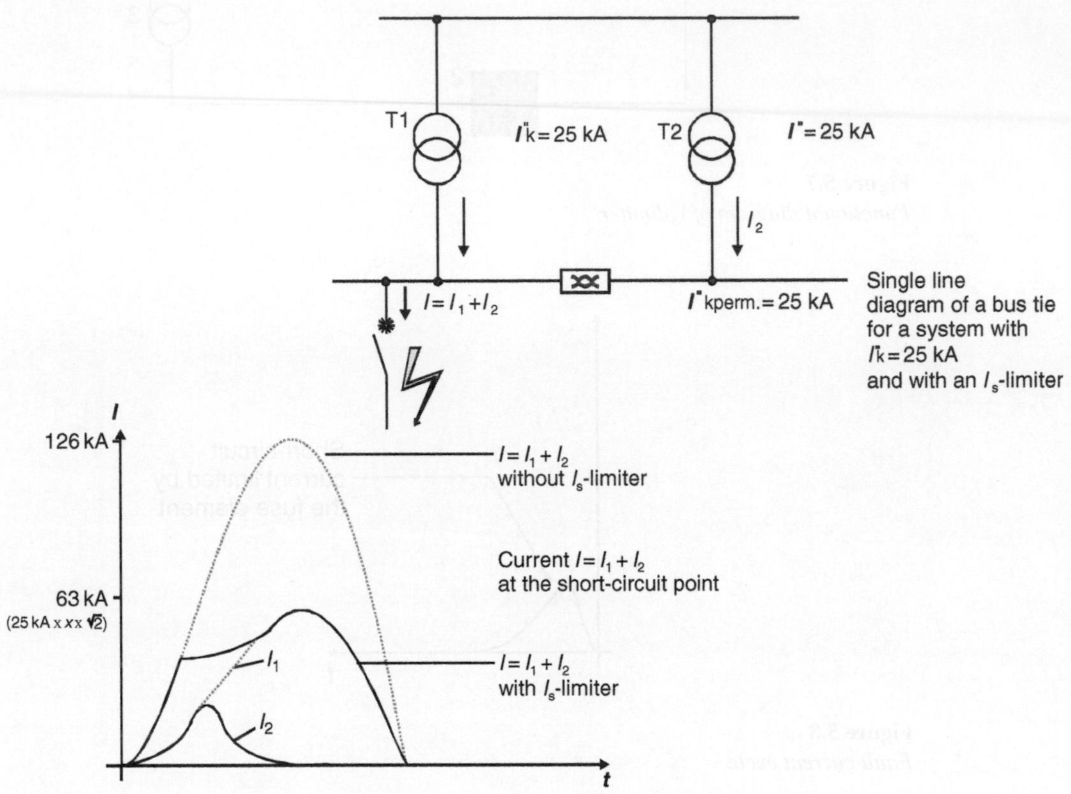

Figure 5.10
Practical use of I_S*-limiter*

Here I_2 is interrupted first thereby decreasing the fault current to the value of I_1, and I_1 is interrupted subsequently. The net resultant fault current follows the path of I_1 once the limiter functions thereby limiting the overall fault current.

6

Instrument transformers

6.1 Purpose

The voltage transformers and current transformers continuously measure the voltage and current of an electrical system and are responsible to give feedback signals to the relays to enable them to detect abnormal conditions. The values of actual currents in modern distribution systems varies from a few amperes in households, small industrial/commercial houses, etc. to thousands of amperes in power-intensive plants, national grids, etc., which also depend on the operating voltages. Similarly, the voltages in electrical systems vary from few hundreds of volts to many kilo volts. However, it is impossible to have monitoring relays designed and manufactured for each and every distribution system and to match the innumerable voltages and currents being present. Hence the voltage transformers and current transformers are used which enable same types of relays to be used in all types of distribution systems ensuring the selection and cost of relays to be within manageable ranges.

The main tasks of instrument transformers are:

- To transform currents or voltages from usually a high value to a value easy to handle for relays and instruments.
- To insulate the relays, metering and instruments from the primary high-voltage system.
- To provide possibilities of standardizing the relays and instruments, etc. to a few rated currents and voltages.

Instrument transformers are special versions of transformers in respect of measurement of current and voltages. The theories for instrument transformers are the same as those for transformers in general.

6.2 Basic theory of operation

The transformer is one of the high efficient devices in electrical distribution systems, which are used to convert the generated voltages to convenient voltages for the purpose of transmission and consumption. A transformer comprises of two windings viz., primary and secondary coupled through a common magnetic core.

When the primary winding is connected to a source and the secondary circuit is left open, the transformer acts as an inductor with minimum current being drawn from the source. At the same time, a voltage will be produced in the secondary open-circuit winding due to the magnetic coupling. When a load is connected across the secondary terminals, the current will start flowing in the secondary, which will be decided by the load impedance and the open-circuit secondary voltage. A proportionate current is drawn in the primary winding depending upon the turns ratio between primary and secondary. This principle of transformer operation is used in transfer of voltage and current in a circuit to the required values for the purpose of standardization.

A voltage transformer is an open-circuited transformer whose primary winding is connected across the main electrical system voltage being monitored. A convenient proportionate voltage is generated in the secondary for monitoring. The most common voltage produced by voltage transformers is 100–120 V (as per local country standards) for primary voltages from 380 V to 800 kV or more.

However, the current transformer is having its primary winding directly connected in series with the main circuit carrying the full operating current of the system. An equivalent current is produced in its secondary, which is made to flow through the relay coil to get the equivalent measure of the main system current. The standard currents are invariably 1 A and 5 A universally.

6.3 Voltage transformers

There are basically, two types of voltage transformers used for protection equipment.

1. Electromagnetic type (commonly referred to as a VT)
2. Capacitor type (referred to as a CVT).

The electromagnetic type is a step down transformer whose primary (HV) and secondary (LV) windings are connected as below (see Figure 6.1).

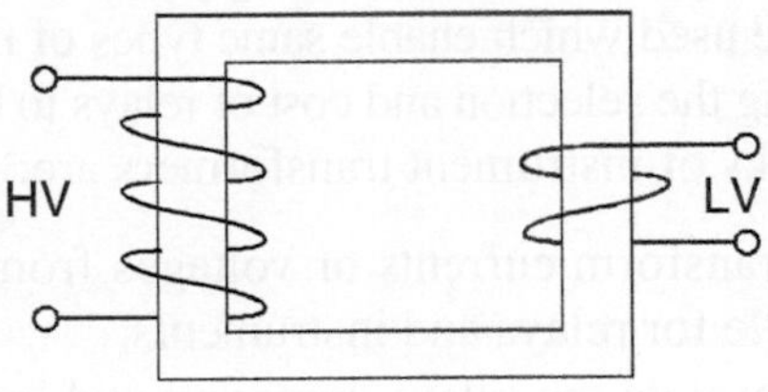

Figure 6.1
Electromagnetic type

The number of turns in a winding is directly proportional to the open-circuit voltage being measured or produced across it. The above diagram is a single-phase VT. In the three-phase system it is necessary to use three VTs at one per phase and they being connected in star or delta depending on the method of connection of the main power source being monitored. This type of electromagnetic transformers are used in voltage circuits upto 110/132 kV.

For still higher voltages, it is common to adopt the second type namely the capacitor voltage transformer (CVT). Figure 6.2 below gives the basic connection adopted in this type. Here the primary portion consists of capacitors connected in series to split the primary voltage to convenient values.

The magnetic voltage transformer is similar to a power transformer and differs only so far as a different emphasis is placed on cooling, insulating and mechanical aspects. The primary winding has larger number of turns and is connected across the line voltage; either phase-to-phase or phase-to-neutral. The secondary has lesser turns however, the volts per turn on both primary and secondary remains same.

The capacitor VT is more commonly used on extra high-voltage (EHV) networks. The capacitors also allow the injection of a high-frequency signals onto the power line conductors to provide end-to-end communications between substations for distance relays, telemetry/supervisory and voice communications. Hence, in EHV national grid networks of utilities, the CVTs are commonly used for both protection and communication purposes.

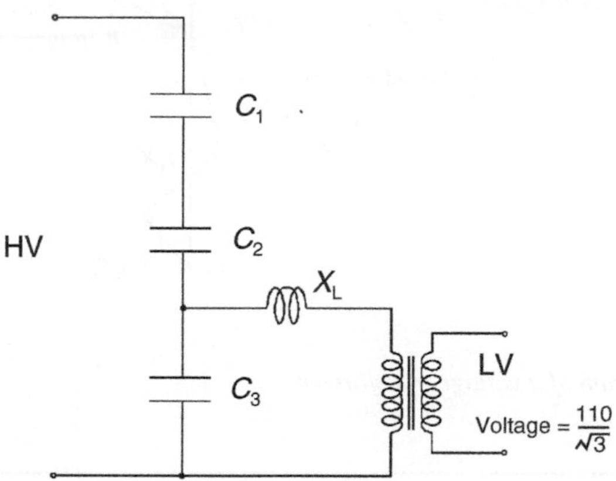

Figure 6.2
Capacitor-type VT

It should be remembered that these voltage transformers are also used for measuring purposes. It is possible to have one common primary winding and two or more secondary windings in one unit. The voltage transformers having this kind of arrangement are referred to as two core or three core VT depending on the number of secondary windings.

6.3.1 Vector diagram

The vector diagram for a single-phase voltage transformer is as follows. The primary parameters are suffixed with p while the secondary parameters have suffix s. It is to be noted that the vector diagram for a three-phase connection will be identical, except for the phase shift introduced in each phase in relation to the other phases (Figure 6.3). The capacity of a voltage transformer is normally represented in VA rating, which indicates the maximum load that can be connected across its secondary. The other common name for this VA rating is 'burden'. Output burdens of 500 VA per phase are common.

6.3.2 Accuracy of voltage transformers

The voltage transformers shall be capable to produce secondary voltages, which are proportionate to the primary voltages over the full range of input voltage expected in a system. Voltage transformers for protection are required to maintain reasonably good accuracy over a large range of voltage from 0 to 173% of normal.

However, the close accuracy is more relevant for metering purposes, while for protection purposes the margin of accuracy can be comparatively less. Permissible errors vary depending on the burden and purpose of use and typical values as per IEC are as follows (see Table 6.1).

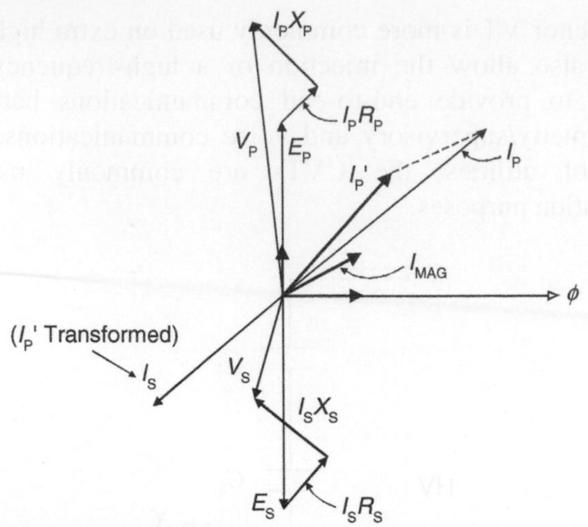

Figure 6.3
Vector diagram of a voltage transformer

		Range		Limits of Errors		
Standard	Class	Burden %	Voltage %	Ratio %	Phase Displacement Min	Application
IEC 186	0.1	25–100	80–120	0.1	5	Laboratory
	0.2	25–100	80–120	0.2	10	Precision metering, revenue metering
	0.5	25–100	80–120	0.5	20	Standard revenue metering
	1.0	25–100	80–120	1.0	40	Industrial grade meters
	3.0	25–100	80–120	3.0	–	Instruments
	3P	25–100	5-V_f*	3.0	120	Protection
	6P	25–100	5-V_f*	6.0	240	Protection
* V_f = Voltage factor						

Table 6.1
Accuracy class, voltage transformers

The accuracy is not a major cost-deciding factor for a voltage transformer due to the high efficiency of the transformers, which normally ensures that there is no major voltage drop in the secondary leads. Thus, it is common to select voltage transformers based on the loads (choosing appropriate rated burden). The question of accuracy of VT's used in protection circuits can be ignored and is generally neglected in practice.

6.3.3 Connection of voltage transformers

Electromagnetic voltage transformers may be connected interphase or between phase and earth. However, capacitor voltage transformers can only be connected phase-to-earth. Voltage transformers are commonly used in three-phase groups, generally in star–star configuration. Typical connection is as per Figure 6.4. With this arrangement, the secondary voltages provide a complete replica of the primary voltages as shown below and any voltage (phase-to-phase or phase-to-earth) may be selected for monitoring at the secondary (see Figure 6.5).

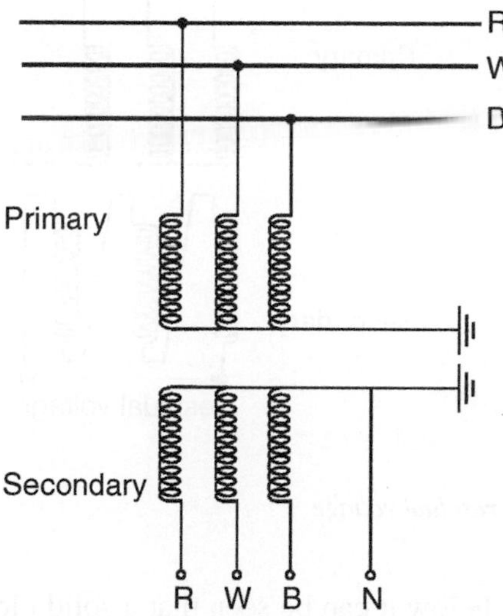

Figure 6.4
Voltage transformers connected in star–star configuration

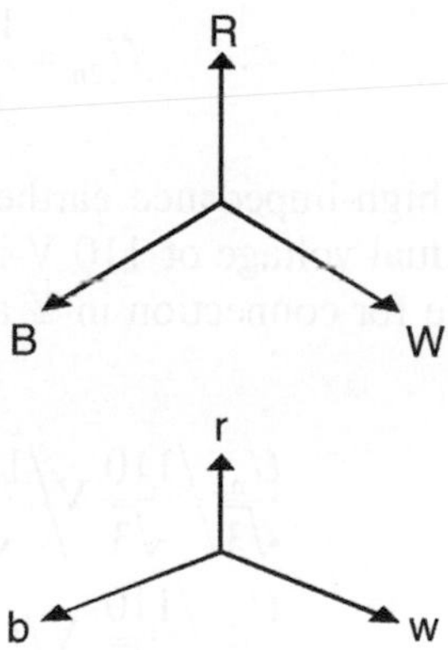

Figure 6.5
Vector diagram for VTs in star–star configuration

6.3.4 Connection to obtain the residual voltage

It is common to detect earth faults in a three-phase system using the displacement that occurs in the neutral voltage when earth faults take place. The residual voltage (neutral displacement voltage, polarizing voltage) for earth fault relays can be obtained from a VT between neutral and earth, for instance at a power transformer neutral. It can also be obtained from a three-phase set of VTs, which have their primary winding connected phase to earth and one of the secondary windings connected in a broken delta. Figure 6.6 below illustrates the measuring principle for the broken delta connecting during an earth fault in a high-impedance earthed (or unearthed) and an effectively earthed power system respectively.

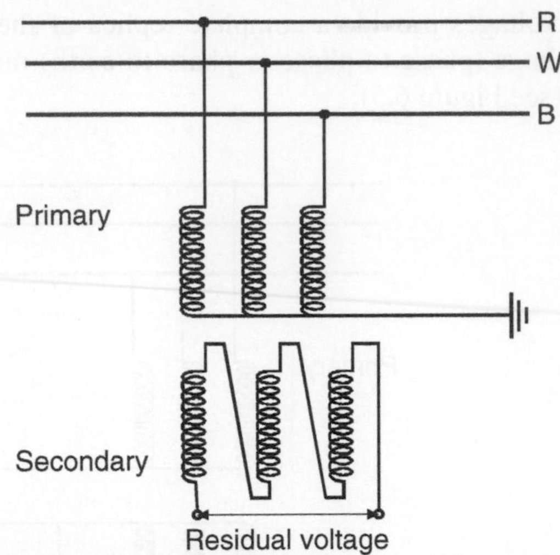

Figure 6.6
Connection to source residual voltage

From the figure below it can be seen that a solid close-up earth fault produces an output voltage of $U_{rsd} = 3 \times U_{sn}$ in a high-impedance earthed system and $U_{rsd} = U_{2n}$ in an effectively earthed system (see Figure 6.7). Therefore a VT secondary normal voltage of:

$$U_{2n} = \frac{110}{\sqrt{3}}\ \text{V}$$

which is often used in high-impedance earthed systems and $U_{2n} = 110$ V in effectively earthed systems. A residual voltage of 110 V is obtained in both the cases. VTs with two secondary windings, one for connection in Y and the other in broken delta can then have the ratio:

$$\frac{U_n}{\sqrt{3}} \Big/ \frac{110}{\sqrt{3}}\ \text{V} \Big/ \frac{110}{\sqrt{3}}\ \text{V} \quad \text{and}$$

$$\frac{U_n}{\sqrt{3}} \Big/ \frac{110}{\sqrt{3}}\ \text{V}/110\ \text{V}$$

for high-impedance and effective earthed systems respectively. Other nominal voltages than 110 V e.g. 100 V or 120 V are also used depending on national standards and practice.

Ferro-resonance in magnetic voltage transformer

When the ferro-resonance in a CVT is an internal oscillation between the capacitor and the magnetic IVT, the ferro-resonance in a magnetic voltage transformer is an oscillation between the magnetic voltage transformer and the network. The oscillation can only occur in a network having an insulated neutral. An oscillation can occur between the network's capacitance to ground and the non-linear inductance in the magnetic voltage transformer. The oscillation can be triggered by a sudden change in the network voltage.

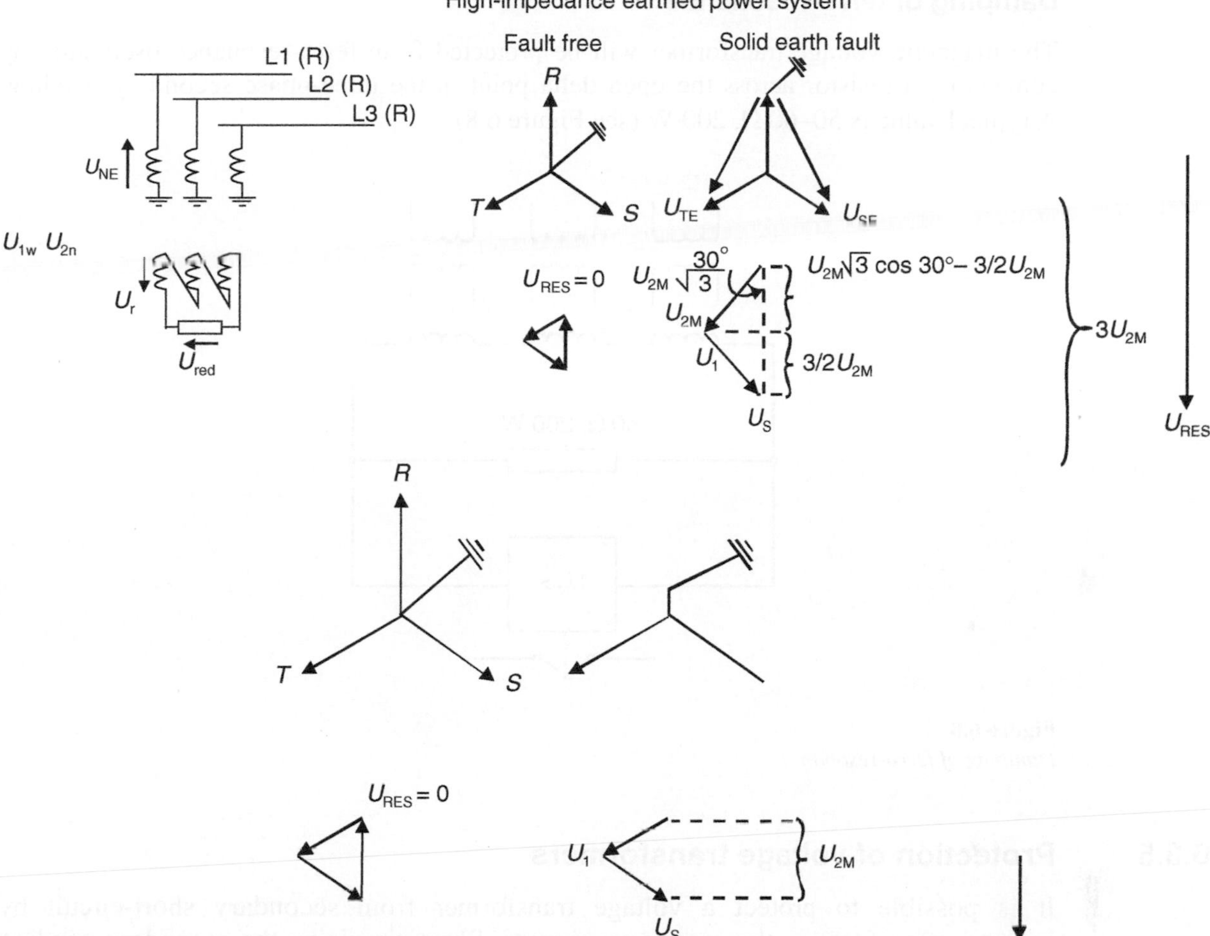

Figure 6.7
Residual voltage (neutral displacement voltage) from an open delta circuit

It is difficult to give a general figure of a possible risk of ferro-resonance. It depends on the design of the transformer. We can roughly calculate that there will be a risk of resonance when the zero sequence capacitance expressed in S km transmission line

$$S < \frac{42\,000}{U_n^2}\ \text{km}$$

U_n = System voltage in kV.

The corresponding value for cable is:

$$S < \frac{1400}{U_n^2}\ \text{km}$$

Damping of ferro-resonance

The magnetic voltage transformer will be protected from ferro-resonance oscillation by connecting a resistor across the open delta point in the three-phase secondary winding. A typical value is 50–60 Ω, 200 W (see Figure 6.8).

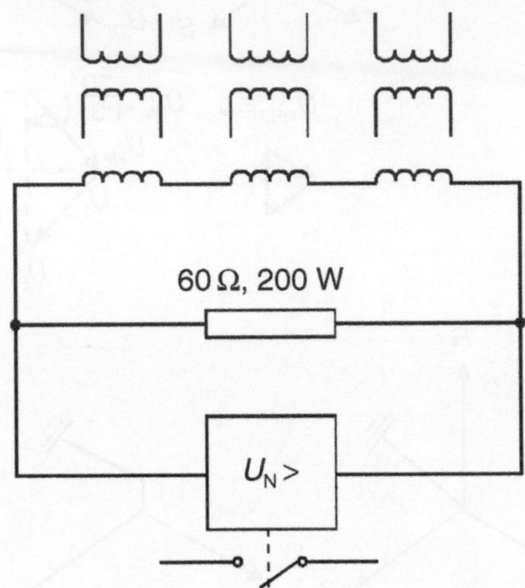

Figure 6.8
Damping of ferro-resonance

6.3.5 Protection of voltage transformers

It is possible to protect a voltage transformer from secondary short-circuit by incorporating fuses in the secondary circuits. Short-circuit on the secondary winding gives only a few amperes in the primary winding and is not sufficient to rupture a high-voltage fuse. Hence high-voltage fuses on the primary side do not protect the transformers, they protect only the network in case of any short-circuit on the primary side.

6.3.6 Voltage drop in voltage transformers

The voltage drop in the secondary circuit is of importance. The voltage drop in the secondary fuses and long connection wires can change the accuracy of the measurement. It is especially important for revenue metering windings of high accuracy (class 0.2, 0.3). The total voltage drops in this circuit must not be more than 0.1%.

Typical values of resistance in fuses:

6 A	0.048
10 A	0.024
16 A	0.0076
25 A	0.0042

A 6–10 A is a typical value for safe rupture of the fuses.

The voltage drop in the leads from the VT to the associated equipment must be considered as this, in practice, can be alarming mainly in case of measuring circuits. This is the one that separates the metering circuits (with low burden) from protective circuits (with higher burdens). Refer Figure 6.9.

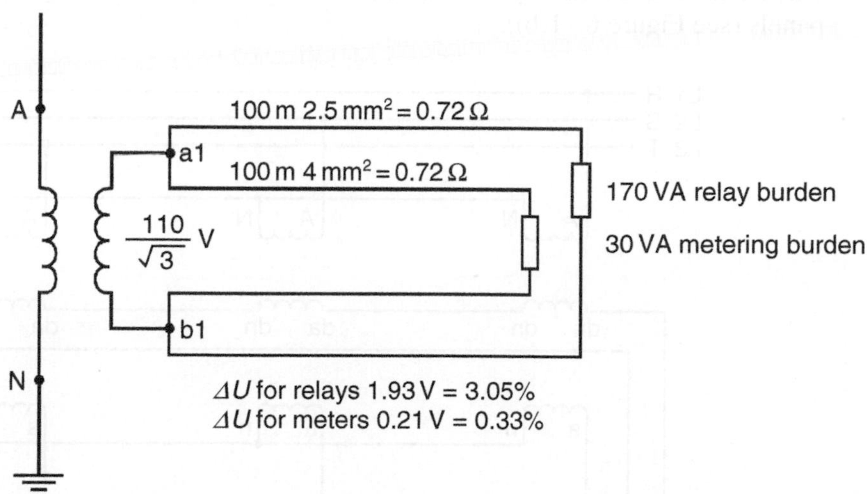

Figure 6.9
The accuracy of a voltage transformer is guaranteed at the secondary terminals

6.3.7 Secondary earthing of voltage transformers

To prevent secondary circuits from reaching dangerous potential, the circuits should be earthed. Earthing should be made at only one point of a VT secondary circuit or galvanically interconnected circuits. A VT with the primary connected phase-to-earth shall have the secondary earthed at terminal n. A VT with the primary winding connected across two-phases, shall have that secondary terminal earthed which has a voltage lagging the other terminal by 120°. Windings not under use shall also be earthed (see Figure 6.10).

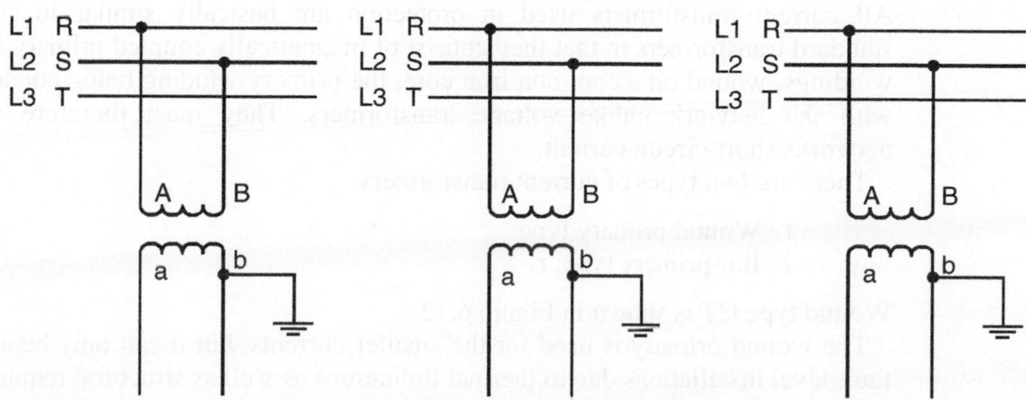

Figure 6.10
VTs connected between phases

Figure 6.11(a) shows the methods of connection in a three-phase system with primary connected in star and secondary connected in two different ways viz., star and broken delta. Alternatively, it is often a common practice to earth the white phase as shown. This practice stems from metering where the two wattmeter method requires two CTs and two line voltages. With this arrangement the red and blue phases now at line potential to the white and it saves the expense and bother of running a neutral conductor throughout the panels (see Figure 6.11(b)).

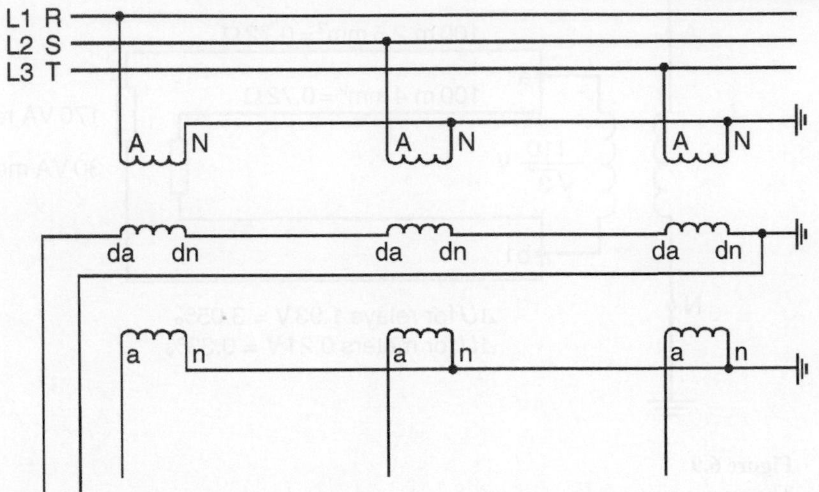

Figure 6.11(a)
A set of VTs with one Y-connected and one broken delta secondary circuit

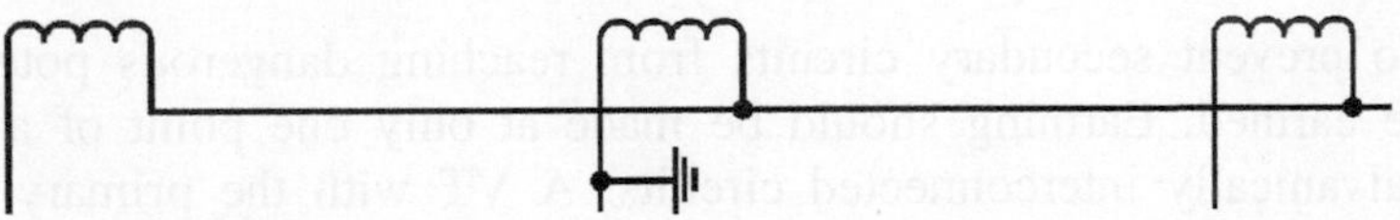

Figure 6.11(b)
VT secondary earthed on white phase

6.4 Current transformers

All current transformers used in protection are basically similar in construction to standard transformers in that they consist of magnetically coupled primary and secondary windings, wound on a common iron core, the primary winding being connected in series with the network unlike voltage transformers. They must therefore withstand the networks short-circuit current.

There are two types of current transformers:

1. Wound primary type
2. Bar primary type.

Wound type CT is shown in Figure 6.12.

The wound primary is used for the smaller currents, but it can only be applied on low fault level installations due to thermal limitations as well as structural requirements due to high magnetic forces. For currents greater than 100 A, the bar primary type is used as shown in Figure 6.13. If the secondary winding is evenly distributed around the complete iron core, its leakage reactance is eliminated (see Figure 6.14).

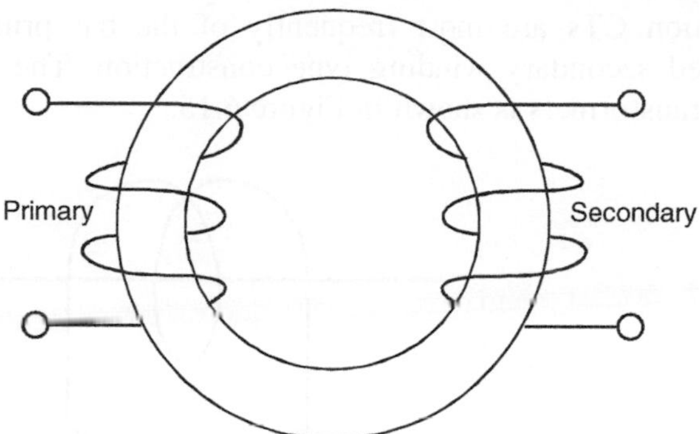

Figure 6.12
Wound primary

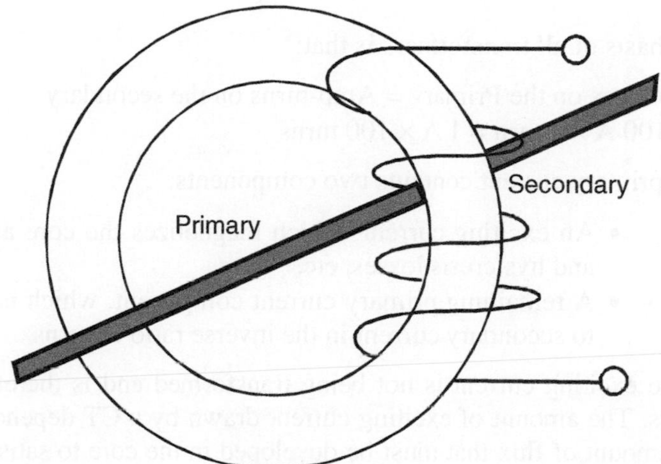

Figure 6.13
Bar primary

Figure 6.14
Secondary winding is evenly distributed around iron core

Protection CTs are most frequently of the bar primary, toroidal core with evenly distributed secondary winding type construction. The standard symbol used to depict current transformers is shown in Figure 6.15.

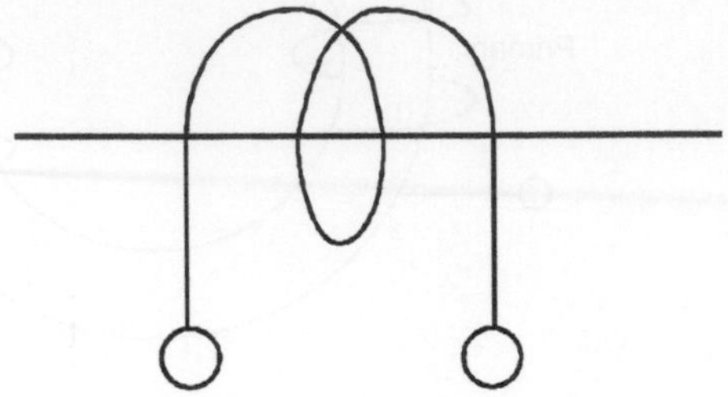

Figure 6.15
Standard symbol for current transformers

The basis of all transformers is that:

Amp-turns on the Primary = Amp-turns on the secondary
e.g. 100 A × 1 turn = 1 A × 100 turns

The primary current contains two components:

- An exciting current, which magnetizes the core and supplies the eddy current and hysteresis losses, etc.
- A remaining primary current component, which is available for transformation to secondary current in the inverse ratio of turns.

The exciting current is not being transformed and is therefore the cause of transformer errors. The amount of exciting current drawn by a CT depends upon the core material and the amount of flux that must be developed in the core to satisfy the output requirements of the CT. that is, to develop sufficient driving voltage required, pushing the secondary current through its connected load or burden. This can be explained vectorally in Figure 6.16.

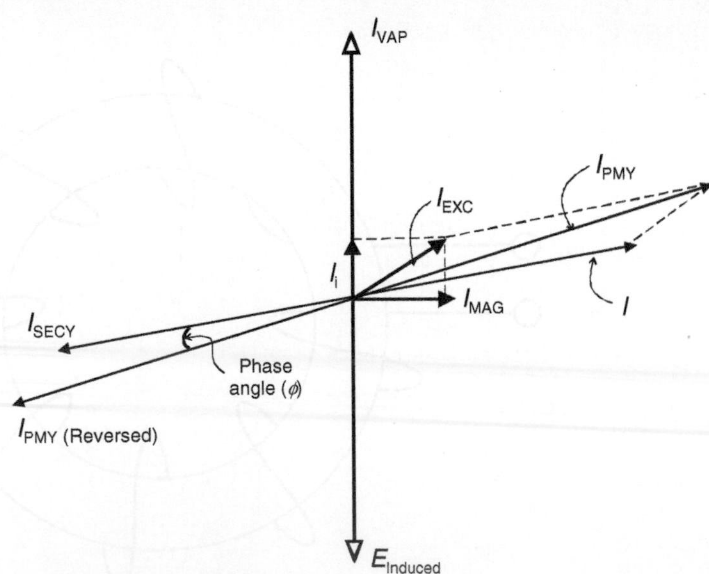

Figure 6.16
Vector diagram for a current transformer

6.4.1 Magnetization curve

This curve is the best method of determining a CTs performance. It is a graph of the amount of magnetizing current required to generate an open-circuit voltage at the terminals of the unit. Due to the non-linearity of the core iron, it follows the B-H loop characteristic and comprises three regions, namely the initial region, unsaturated region and saturated region (see Figure 6.17).

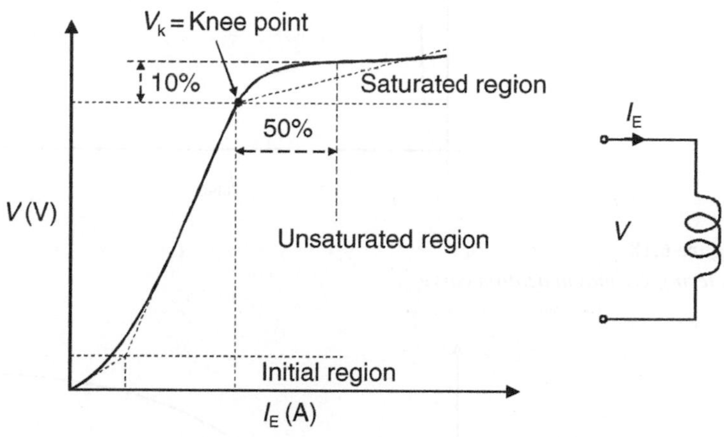

Figure 6.17
Typical CT magnetization curve

6.4.2 Knee-point voltage

The transition from the unsaturated to the saturated region of the open-circuit excitation characteristic is a rather gradual process in most core materials. This transition characteristic makes a CT not to produce equivalent primary current beyond certain point. This transition is defined by 'knee-point' voltage in a CT, which decides its accurate working range.

It is generally defined as the voltage at which a further 10% increase in volts at the secondary side of the CT requires more than 50% increase in excitation current. For most applications, it means that current transformers can be considered as approximately linear up to this point.

6.4.3 Metering CTs

Instruments and meters are required to work accurately up to full-load current, but above this, it is advantageous to saturate and protect the instruments under fault conditions. Hence, it is common to have metering CTs with a very sharp knee-point voltage. A special nickel-alloy metal having a very low magnetizing current is used in order to achieve the accuracy.

Following curve shows the magnetization curve of metering CT (see Figure 6.18).

6.4.4 Protection CTs

Protective relays are not normally expected to give tripping instructions under normal conditions. On the other hand these are concerned with a wide range of currents from acceptable fault settings to maximum fault currents many times normal rating. Larger errors may be permitted and it is important that saturation is avoided wherever possible to ensure positive operation of the relays mainly when the currents are many times the normal current (see Figure 6.19).

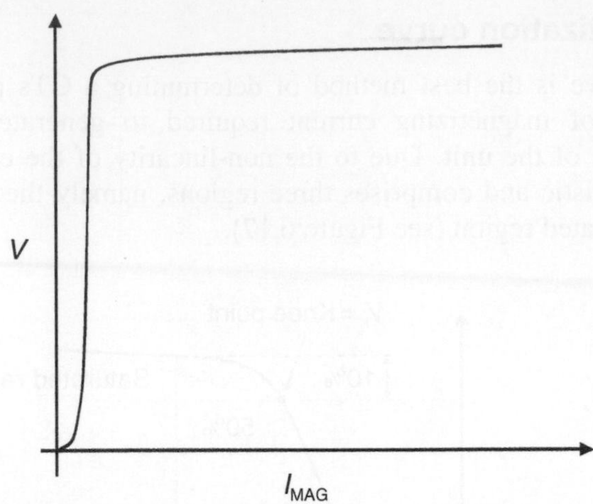

Figure 6.18
Metering CT magnetization curve

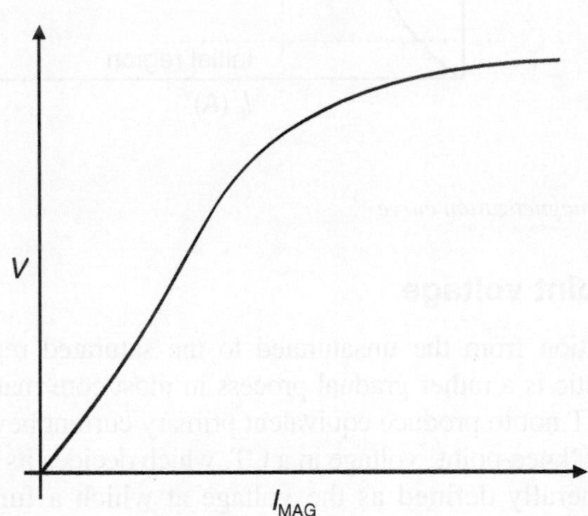

Figure 6.19
Protection CT magnetization curve

Test set-up for the CT magnetic curve

It is necessary to test the characteristics of a CT before it is put into operation, since the results produced by the relays and meters depend on how well the CT behaves under normal and fault conditions.

Figure 6.20 shows a simple test connection diagram that is adopted to find the magnetic curve of a CT.

6.4.5 Polarity

Polarity in a CT is similar to the identification of +ve and –ve terminals of a battery. Polarity is very important when connecting relays, as this will determine correct operation or not depending on the types of relays. The terminals of CT are marked by P1 and P2 on the primary, and S1 and S2 on the secondary as per Figure 6.21(a). BS 3938 states that at the instant when current is flowing from P1 to P2 in primary, then current, in secondary must flow from S1 to S2 through the external circuit.

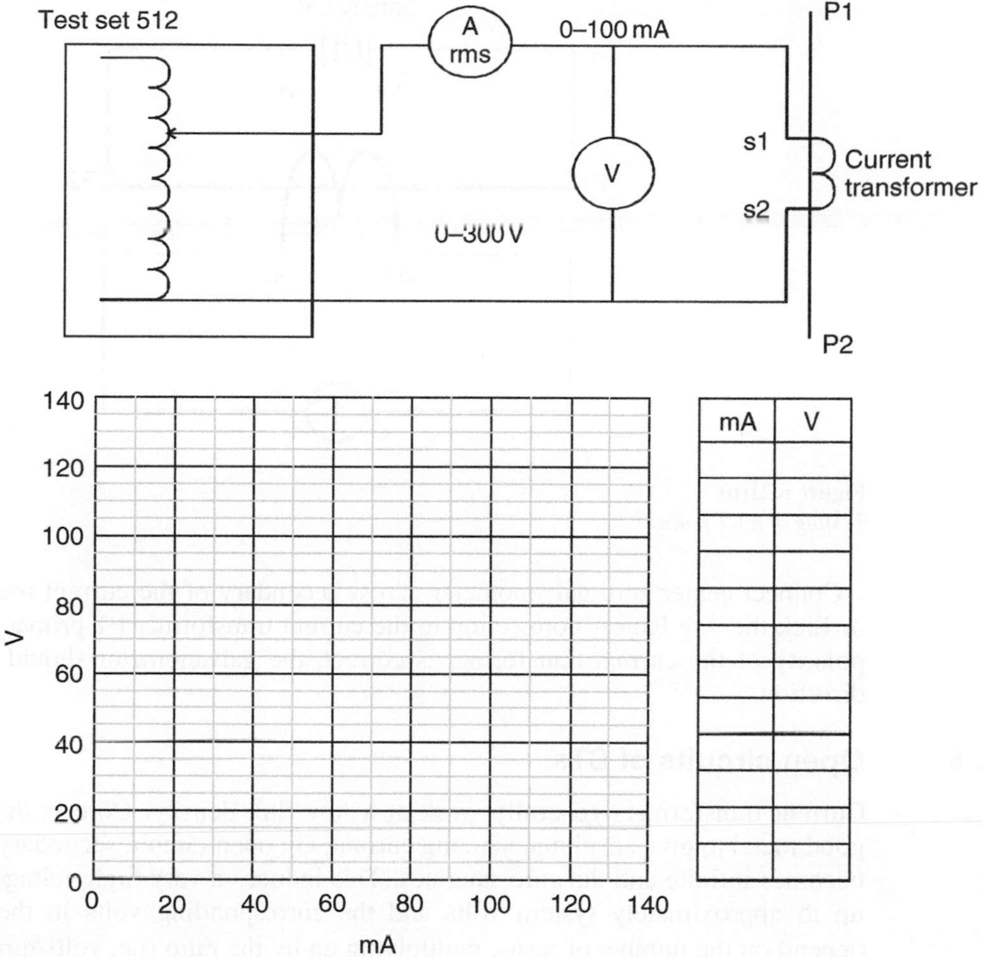

Figure 6.20
Circuit to test magnetization curve

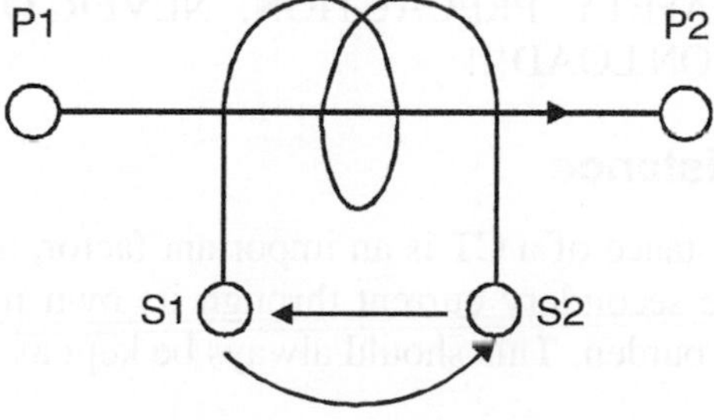

Figure 6.21(a)
Polarity markings of a CT

Figure 6.21(b) shows the simple testing arrangement for cross-checking the CT polarity markings at the time of commissioning electrical systems.

Connect battery –ve terminal to the current transformer P2 primary terminal. This arrangement will cause current to flow from P1 to P2 when +ve terminal is connected to P1 until the primary is saturated. If the polarities are correct, a momentary current will flow from S1 to S2.

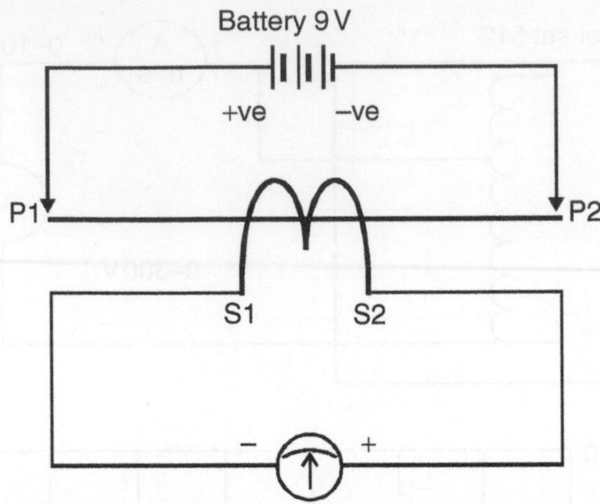

Figure 6.21(b)
Testing of a CT polarity

Connect center zero galvanometer across secondary of the current transformer. Touch or flick the +ve battery connection to the current transformer P1 primary terminal. If the polarity of the current transformer is correct, the galvanometer should flick in the +ve direction.

6.4.6 Open circuits of CTs

Current transformers generally work at a low flux density. Core is then made of very good metal to give small magnetizing current. On open-circuit, secondary impedance now becomes infinite and the core saturates. This induces a very high voltage in the primary-up to approximately system volts and the corresponding volts in the secondary will depend on the number of turns, multiplying up by the ratio (i.e. volts/turn × no. of turns). Since CT normally has much more turns in secondary compared to the primary, the voltage generated on the open-circuited CT will be much more than the system volts, leading to flashovers.

HENCE AS A SAFETY PRECAUTION, NEVER OPEN-CIRCUIT A CURRENT TRANSFORMER ON LOAD!!!

6.4.7 Secondary resistance

The secondary resistance of a CT is an important factor, as the CT has to develop enough voltage to push the secondary current through its own internal resistance as well as the connected external burden. This should always be kept as low as possible.

6.4.8 CT specification

A current transformer is normally specified in terms of:

- A rated burden at rated current
- An accuracy class
- An upper limit beyond which accuracy is not guaranteed (known as the accuracy limit factor, ALF), which is more vital in case of protection CTs.

In the relevant BSS 3938 the various accuracy classes are in accordance with the following tables (see Tables 6.2 and 6.3):

Class	± Percentage Current (Ratio) Error at Percentage of Rated Current Shown Below			± Phase Displacement at Percentage of Rated Current Shown Below					
				Minutes			Centiradians		
	10 up to but not incl. 20	20 up to but not incl. 100	100 up to 120	10 up to but not incl. 20	20 up to but not incl. 100	100 up to 120	10 up to but not incl. 20	20 up to but not incl. 100	100 up to 120
0.1	0.25	0.2	0.1	10	8	5	0.3	0.24	0.15
0.2	0.5	0.35	0.2	20	15	10	0.6	0.45	0.3
0.5	1.0	0.75	0.5	60	45	30	1.8	1.35	0.9
1.0	2.0	1.5	1.0	120	90	60	3.6	2.7	1.8

Table 6.2
Limits of error for accuracy classes 0.1–1 (metering CT)

Accuracy Class	Current Error at Rated Primary Current %	Phase Displacement at Rated Primary Current		Composite Error at Rated Accuracy Limited Primary Current %
		Minutes	Centiradians	
5P	±1	±60	±1.8	5
10P	±3			10

Table 6.3
Limits of error for accuracy class 5P and class 10P (protection CT)

In terms of the specification a current transformer would, for example, be briefly referred to as 15 VA 5P20 if it were a protection CT or 15 VA Class 0.5 if it is a metering CT. The meanings of these figures are as below:

	Protection	Metering
Rated burden	15 VA	15 VA
Accuracy class	5P	0.5
Accuracy limit factor	20	Class 1,0

(ALF is 20 times normal or rated current)

6.4.9 Class X current transformers

These are normally specified for special purpose applications such as busbar protection, where it is important that CTs have matching characteristics.

For this type of CT an exact point on the magnetization curve is specified, e.g.

1. Rated primary current
2. Turns ratio
3. Rated knee-point emf at maximum secondary turns
4. Maximum exciting current at rated knee-point emf
5. Maximum resistance of secondary winding.

In addition, the error in the turns ratio shall not exceed ±0.25%.

6.4.10 Connection of current transformers

Current transformers for protection are normally provided in groups of three, one for each phase. They are most frequently connected in 'star' as illustrated in Figure 6.22. The secondary currents obtainable with this connection are the three individual phase currents and the residual or neutral current. The residual current is the vector sum of the three-phase currents, which under healthy conditions would be zero. Under earth fault conditions, this would be the secondary equivalent of the earth fault current in the primary circuit.

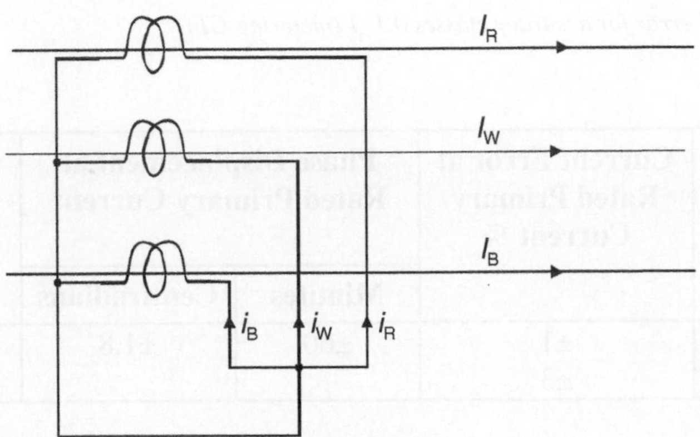

Figure 6.22
Star connection of current transformers

Sometimes, current transformers are connected in 'delta'. The reasons for adopting this connection are one or more of the following:

- To obtain the currents $I_r–I_w$, $I_w–I_b$, $I_b–I_r$
- To eliminate the residual current from the relays
- To introduce a phase-shift of 30° under balanced conditions, between primary and relay currents (see Figures 6.23 and 6.24).

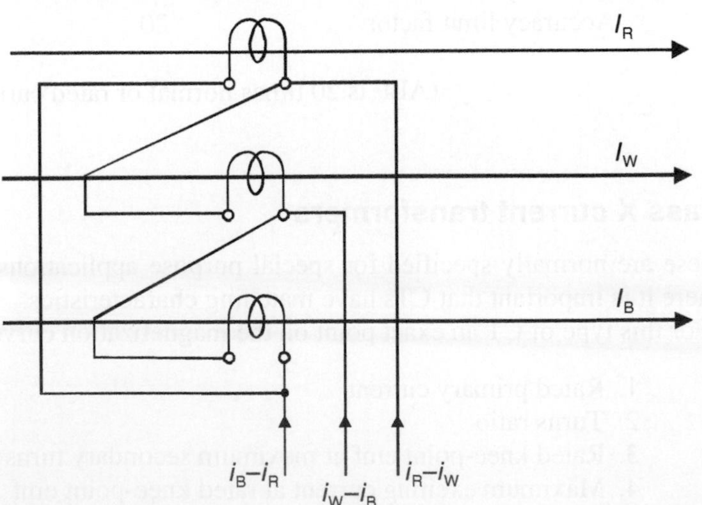

Figure 6.23
Delta connection of current transformers

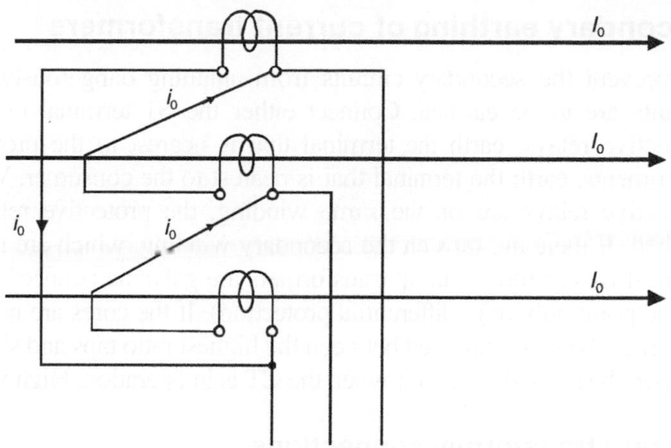

Figure 6.24
Current distribution under earth fault conditions (I_0 circulating inside the delta)

6.4.11 Terminal designations for current transformers

According to IEC publication 185, the terminals are to be designated as shown in Figure 6.25. All terminals that are marked P1, S1 and C1 should have the same polarity.

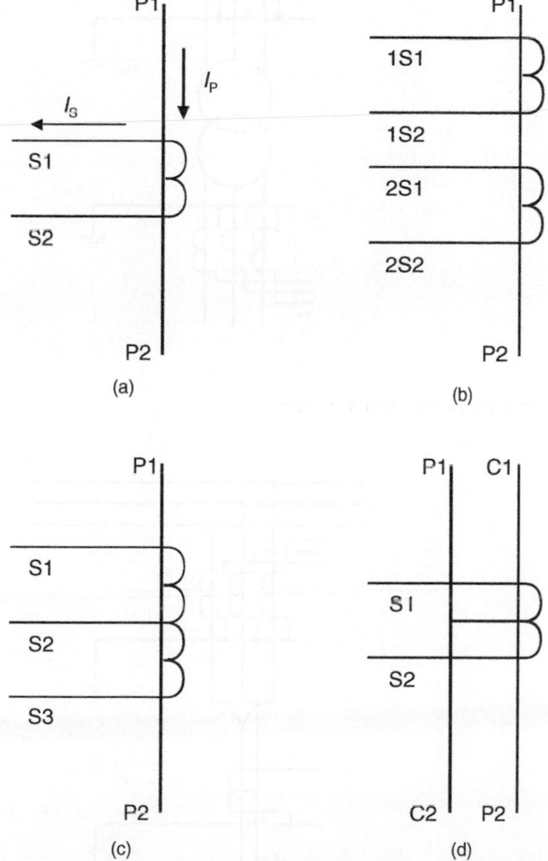

Figure 6.25
Marking of current transformers: (a) one secondary winding; (b) two secondary windings; (c) one secondary winding which has an extra tapping; (d) two primary windings and one secondary winding

6.4.12 Secondary earthing of current transformers

To prevent the secondary circuits from attaining dangerously high potential to earth, these circuits are to be earthed. Connect either the S1 terminal or the S2 terminal to earth. For protective relays, earth the terminal that is nearest to the protected objects. For meters and instruments, earth the terminal that is nearest to the consumer. When metering instruments and protective relays are on the same winding, the protective relay determines the point to be earthed. If there are taps on the secondary winding, which are not used, then they must be left open. If two or more current transformers are galvanic connected together they shall be earthed at one point only (e.g. differential protection). If the cores are not used in a current transformer, they must be short-circuited between the highest ratio taps and should be earthed. It is dangerous to open the secondary circuit when the CT is in operation. High voltage will be induced.

Current transformer connections

Figures 6.26(a)–(c) shows the various ways of connecting current transformers at various parts of an electrical system.

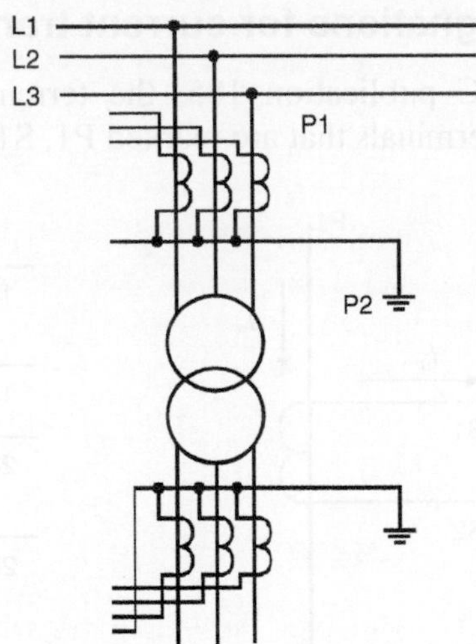

Figure 6.26(a)
Current transformers for a power transformer

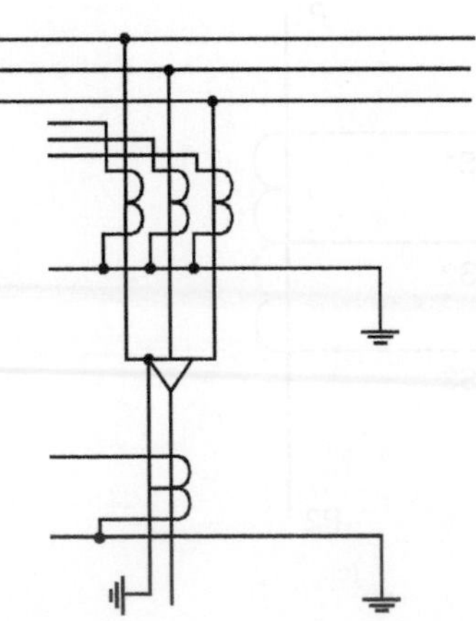

Figure 6.26(b)
Current transformers monitoring cable currents

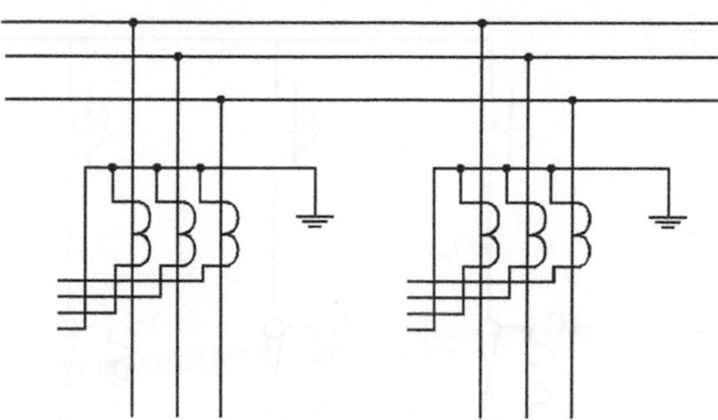

Figure 6.26(c)
CT connections in busbars

6.4.13 Test windings

It is often necessary to carry out on-site testing of current transformers and the associated equipment but it is not always possible to do primary injection because of access of test sets not being large enough to deliver the high value of current required.

Additional test windings can be provided to make such tests easier. These windings are normally rated at 10 A and when injected with this value of current produce the same output as the rated primary current passed through the primary winding.

It should be noted that when energizing the test winding, the normal primary winding should be open circuited, otherwise the CT will summate the effects of the primary and test currents. Conversely, in normal operation the test winding should be left open-circuited. Test windings do, however, occupy an appreciable amount of additional space and therefore increase the cost. Alternatively, for given dimensions they will restrict the size and hence the performance of the main current transformer.

6.5 Application of current transformers

In earlier chapters, we have come across AC trip coils, which are to be designed to carry the normal and fault currents. However, it is difficult to use the same for higher current circuits. In order to overcome the limitations as experienced by series trip coils, current transformers are used so that the high primary currents are transformed down to manageable levels that can be handled comfortably by protection equipment.

A typical example would be fused AC trip coils. These use current transformers, which must be employed above certain limits i.e., when current rating and breaking capacity becomes excessively high. Some basic schemes are:

6.5.1 Overcurrent

In this application, the fuses bypass the AC trip coils as shown in Figure 6.27. Under normal conditions, the fuses carry the maximum secondary current of the CT due to the low-impedance path. Under fault conditions, I_{sec} having reached the value at which the fuse blows and operates, trip coil TC to trip the circuit breaker. Characteristic of the fuse is inverse to the current, so a limited degree of grading is achieved.

6.5.2 Overcurrent and earth fault

There are two methods of connection and the second one shown in Figures 6.28 and 6.29 are the most economical arrangement for this protection.

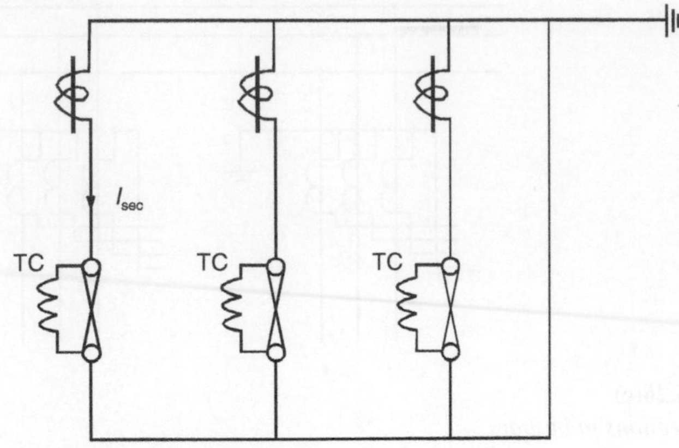

Figure 6.27
CTs for overcurrent use in series trip coils

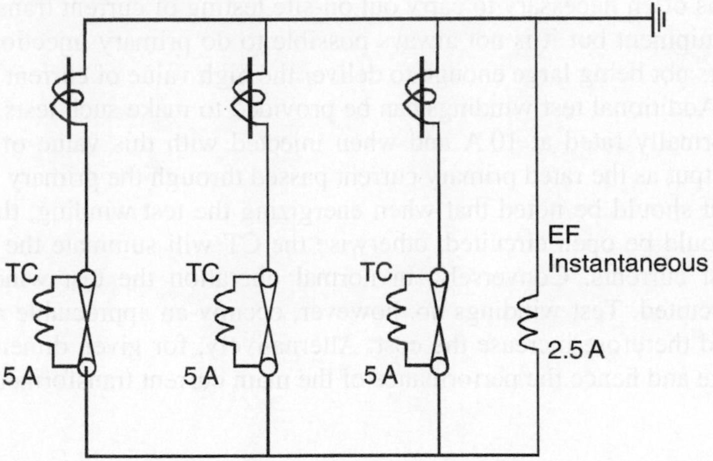

Figure 6.28
CTs for overcurrent and earth fault protection using series trip coils

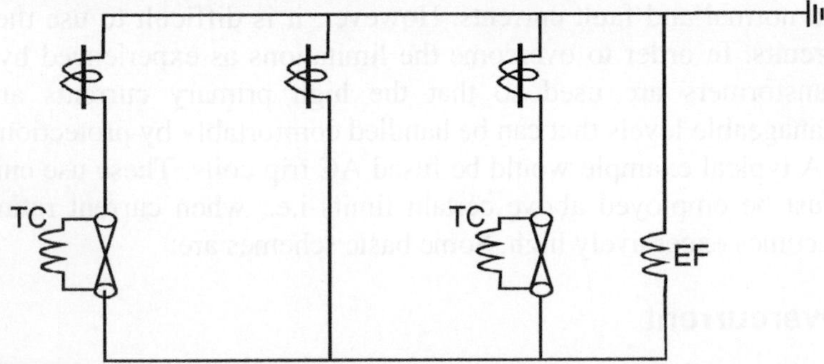

Figure 6.29
Economical use of overcurrent and earth fault configuration

6.6 Introducing relays

The electromechanical relays basically comprise of armature coils with mechanical contacts to energize a tripping coil of a breaker. The CTs are connected in the same way as seen in the earlier pictures. Figures 6.30–6.32 gives an overview of the various connections commonly adopted.

6.6.1 Relays in conjunction with fuses

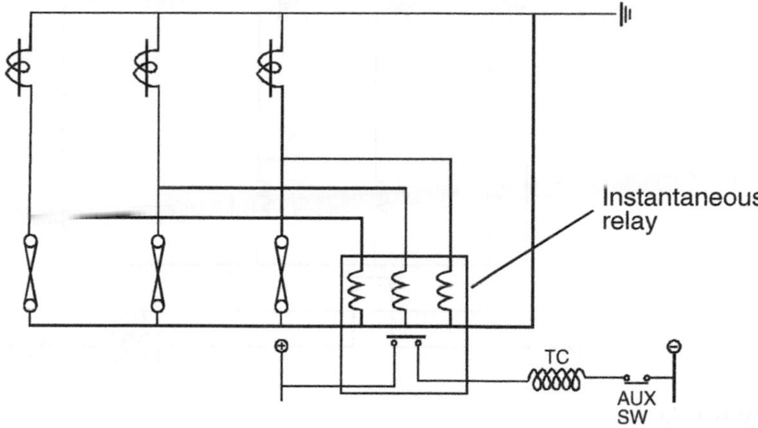

Figure 6.30
Inverse overcurrent tripping characteristic

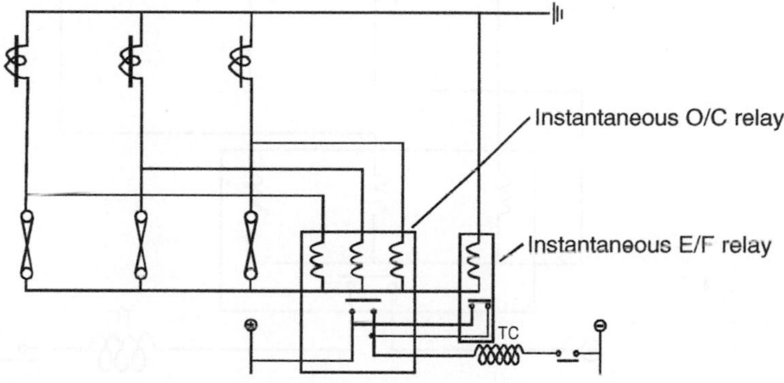

Figure 6.31
Inverse overcurrent + instantaneous earth fault

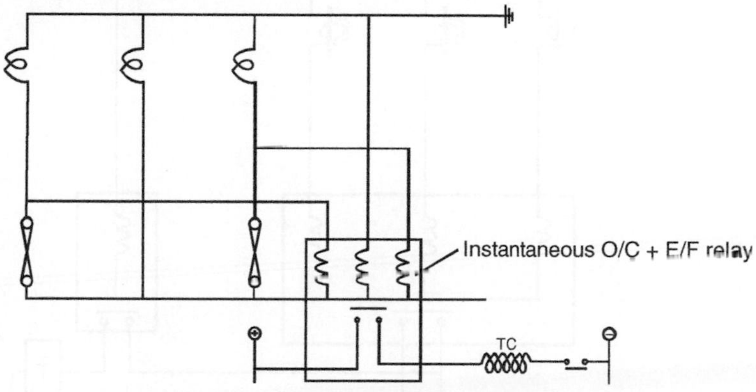

Figure 6.32
More economic method

6.7 Inverse definite minimum time lag (IDMTL) relay

Here the relays are designed to have inverse characteristics similar to that of fuses and hence the fuses are eliminated in the CT connections as shown in Figures 6.33–6.36.

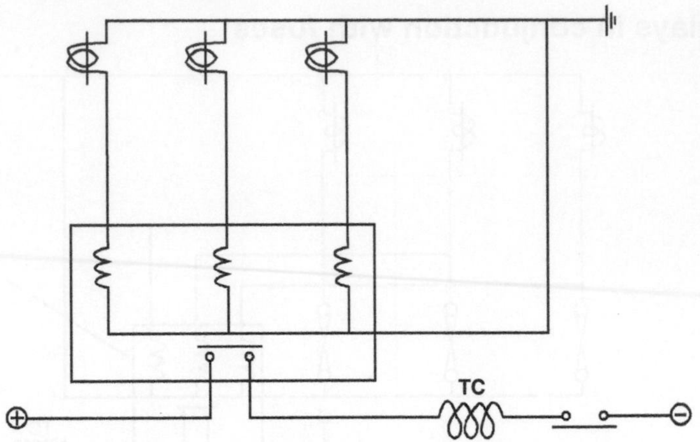

Figure 6.33
Overcurrent

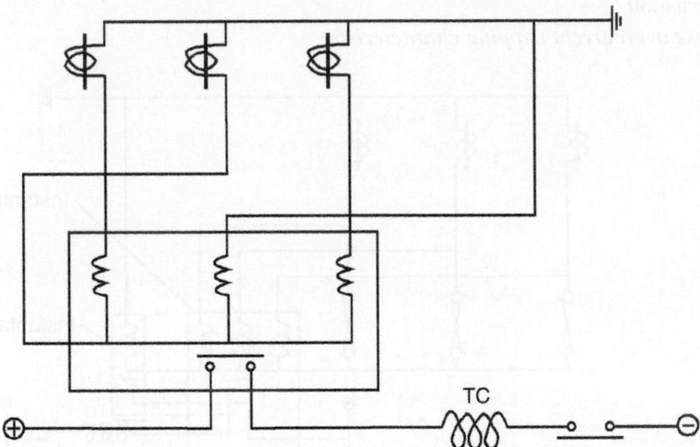

Figure 6.34
Overcurrent + earth fault

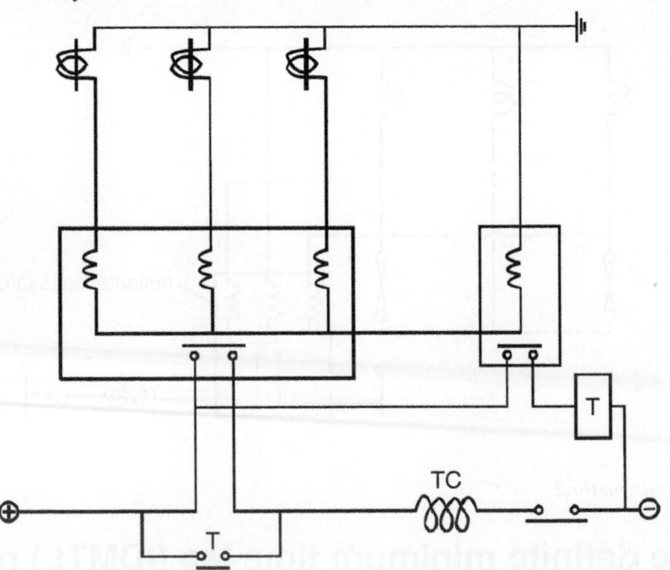

Figure 6.35
IDMTL overcurrent + time lag earth fault

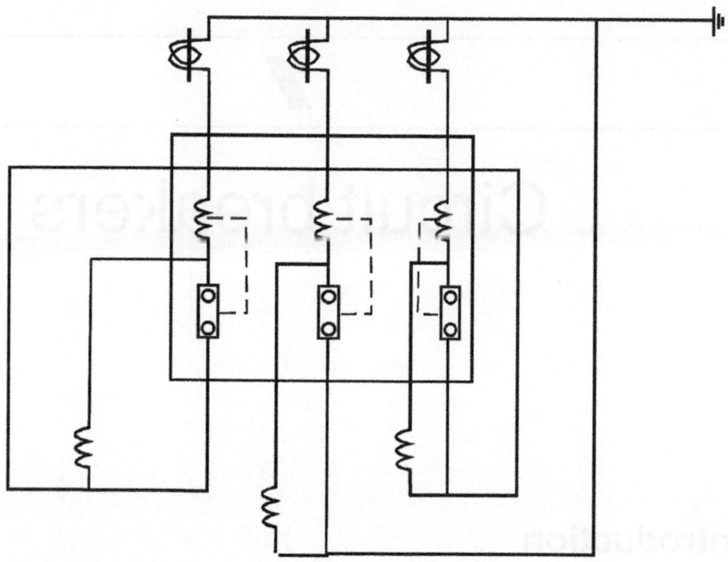

Figure 6.36
IDMTL OC + EF relay with AC trip coils

7

Circuit breakers

7.1 Introduction

Where fuses are unsuitable or inadequate, protective relays and circuit breakers are used in combination to detect and isolate faults. Circuit breakers are the main making and breaking devices in an electrical circuit to allow or disallow flow of power from source to the load. These carry the load currents continuously and are expected to be switched ON with loads (making capacity). These should also be capable of breaking a live circuit under normal switching OFF conditions as well as under fault conditions carrying the expected fault current until completely isolating the fault side (rupturing/breaking capacity).

Under fault conditions, the breakers should be able to open by instructions from monitoring devices like relays. The relay contacts are used in the making and breaking control circuits of a circuit breaker, to prevent breakers getting closed or to trip breaker under fault conditions as well as for some other interlocks.

7.2 Protective relay–circuit breaker combination

The protective relay detects and evaluates the fault and determines when the circuit should be opened. The circuit breaker functions under control of the relay, to open the circuit when required. A closed circuit breaker has sufficient energy to open its contacts stored in one form or another (generally a charged spring). When a protective relay signals to open the circuit, the store energy is released causing the circuit breaker to open. Except in special cases where the protective relays are mounted on the breaker, the connection between the relay and circuit breaker is by hard wiring.

Figure 7.1 indicates schematically this association between relay and circuit breaker. From the protection point of view, the important parts of the circuit breaker are the trip coil, latching mechanism, main contacts and auxiliary contacts.

The roles played by these components in the tripping process is clear from Figure 7.1 and the following step by step procedure takes place while isolating a fault (the time intervals between each event will be in the order of a few electrical cycles i.e. milliseconds):

- The relay receives information, which it analyzes, and determines that the circuit should be opened.
- Relay closes its contacts energizing the trip coil of the circuit breaker.

- The circuit breaker is unlatched and opens its main contacts under the control of the tripping spring.
- The trip coil is deenergized by opening of the circuit breaker auxiliary contacts.

Circuit breakers are normally fitted with a number of auxiliary contacts, which are used in a variety of ways in control and protection circuits (e.g. to energize lamps on a remote panel to indicate whether the breaker is open or closed).

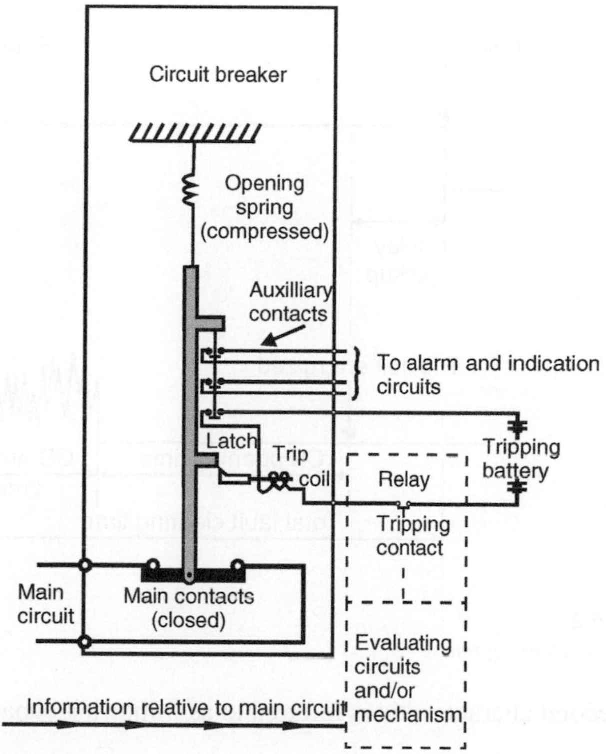

Figure 7.1
Relay–circuit breaker combination

7.3 Purpose of circuit breakers (switchgear)

The main purpose of a circuit breaker is to:

- Switch load currents
- Make onto a fault
- Break normal and fault currents
- Carry fault current without blowing itself open (or up!) i.e. no distortion due to magnetic forces under fault conditions.

The important characteristics from a protection point of view are:

- The speed with which the main current is opened after a tripping impulse is received
- The capacity of the circuit that the main contacts are capable of interrupting.

The first characteristic is referred to as the 'tripping time' and is expressed in cycles. Modern high-speed circuit breakers have tripping times between three and eight cycles. The tripping or total clearing or break time is made up as follows:

- *Opening time*: The time between instant of application of tripping power to the instant of separation of the main contacts.
- *Arcing time*: The time between the instant of separation of the main circuit breaker contacts to the instant of arc extinction of short-circuit current.
- *Total break or clearing time*: The sum of the above (see Figure 7.2).

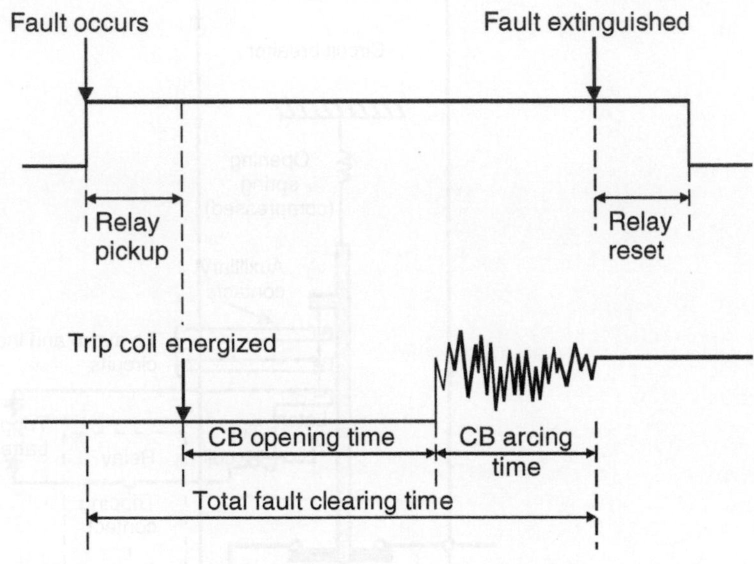

Figure 7.2
Total fault clearing time

The second characteristic is referred to as 'rupturing capacity' and is expressed in MVA.

$$\text{MVA rating (breaking capacity)} = \frac{\sqrt{3} \times \text{System volts} \times \text{SC current}}{10^6}$$

$$= \frac{\sqrt{3} \times V_{\text{L}} \times I_{\text{F}}}{10^6}$$

Typical rupturing capacities of modern circuit breakers are as follows:

kV	MVA	kA
3.3	50	8.8
	75	13.1
	150	26.3
6.6	150	13.1
	250	21.9
	350	31.5
11.0	150	7.9
	250	13.1
	500	26.3
	750	40.0

(continued)

kV	MVA	kA
33.0	500	8.8
	750	13.1
	1500	26.3
66.0	1500	13.1
	2500	21.9

The selection of the breaking capacity depends on the actual fault conditions expected in the system and the possible future increase in the fault level of the main source of supply. In the earlier chapters we have studied simple examples of calculating the fault currents expected in a system. These simple calculations are applied with standard ratings of transformers, etc., to select the approximate rupturing capacity duty for the circuit breakers.

7.4 Behavior under fault conditions

Before the instant of short-circuit, load current will be flowing through the switch and this can be regarded as zero when compared to the level of fault current that would flow (see Figure 7.3).

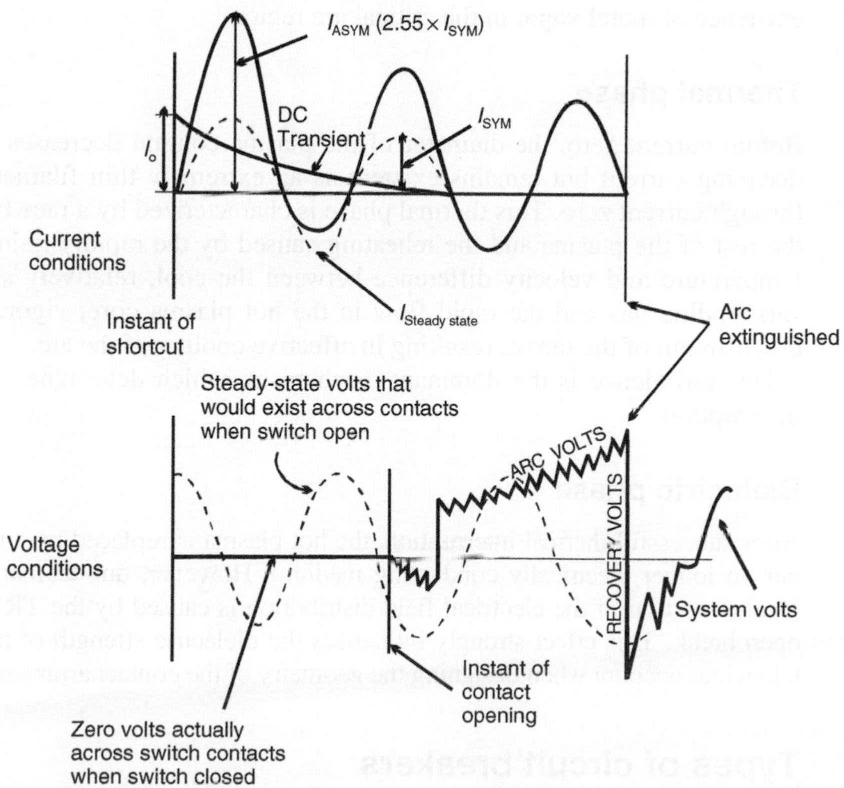

Figure 7.3
Behavior under fault conditions

7.5 Arc

The arc has three parts:

1. *Cathode end (–ve)*: There is approximately 30–50 V drop due to emission of electrons.
2. *Arc column*: Ionized gas, which has a diameter proportional to current. Temperature can be in the range of 6000–25 000 °C.
3. *Anode end (+ve)*: Volt drop 10–20 V.

When short-circuit occurs, fault current flows, corresponding to the network parameters. The breaker trips and the current is interrupted at the next natural current zero. The network reacts by transient oscillations, which gives rise to the transient recovery voltage (TRV) across the circuit breaker main contacts.

All breaking principles involve the separation of contacts, which initially are bridged by a hot, highly conductive arcing column. After interruption at current zero, the arcing zone has to be cooled to such an extent that the TRV is overcome and it cannot cause a voltage breakdown across the open gap.

Three critical phases are distinguished during arc interruption, each characterized by its own physical processes and interaction between system and breaker.

7.5.1 High current phase

This consists of highly conductive plasma at a very high temperature corresponding to a low mass density and an extremely high flow velocity. Proper contact design prevents the existence of metal vapor in the critical arc region.

7.5.2 Thermal phase

Before current zero, the diameter of the plasma column decreases very rapidly with the decaying current but remains existent as an extremely thin filament during the passage through current zero. This thermal phase is characterized by a race between the cooling of the rest of the plasma and the reheating caused by the rapidly rising voltage. Due to the temperature and velocity difference between the cool, relatively slow axial flow of the surrounding gas and the rapid flow in the hot plasma core, vigorous turbulence occurs downstream of the throat, resulting in effective cooling of the arc.

This turbulence is the dominant mechanism, which determines thermal re-ignition or interruption.

7.5.3 Dielectric phase

After successful thermal interruption, the hot plasma is replaced by a residual column of hot, but no longer electrically conducting medium. However, due to marginal ion-conductivity, local distortion of the electrical field distribution is caused by the TRV appearing across the open break. This effect strongly influences the dielectric strength of the break and has to be taken into account when designing the geometry of the contact arrangement.

7.6 Types of circuit breakers

The types of breakers basically refer to the medium in which the breaker opens and closes. The medium could be oil, air, vacuum or SF6. The further classification is single break and double break. In a single break type only the busbar end is isolated but in a double break type, both busbar (source) and cable (load) ends are broken. However, the double break is the most common and accepted type in modern installations.

7.6.1 Arc control device

A breaker consists of moving and fixed contact, and during the breaker operation, the contacts are broken and the arc created during such separation needs to be controlled. The arc control devices, otherwise known as turbulator or explosion pot achieves this:

1. Turbulence caused by arc bubble.
2. Magnetic forces tend to force main contacts apart and movement causes oil to be sucked in through ports and squirted past gap.
3. When arc extinguished (at current zero), ionized gases get swept away and prevents restriking of the arc (see Figure 7.4).

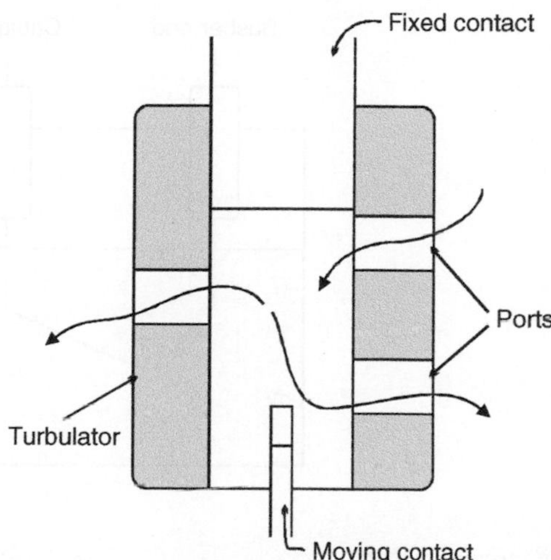

Figure 7.4
Arc control device

7.6.2 Oil circuit breakers

In modern installations, oil circuit breakers, which are becoming obsolete, are being replaced by vacuum and SF6 breakers. However there are many installations, which still employ these breakers where replacements are found to be a costly proposition. In this design, the main contacts are immersed in oil and the oil acts as the ionizing medium between the contacts. The oil is mineral type, with high dielectric strength to withstand the voltage across the contacts under normal conditions.

(a) Double break (used since 1890), see Figure 7.5.
(b) Single break (more popular in earlier days as more economical to produce – less copper, arc control devices, etc., see Figure 7.6).

Arc energy decomposes oil into 70% hydrogen, 22% acetylene, 5% methane and 3% ethylene. Arc is in a bubble of gas surrounded by oil.

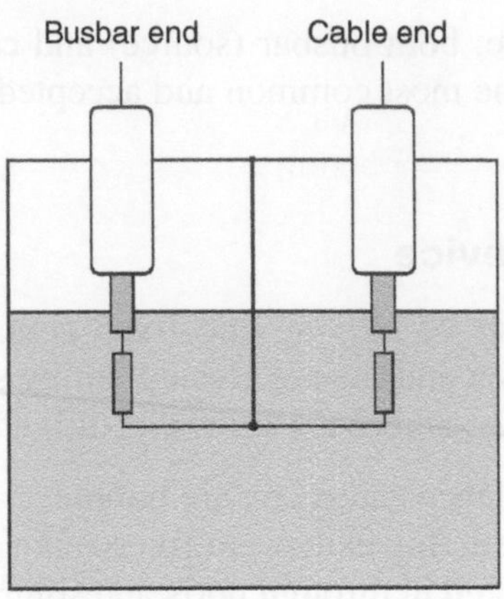

Figure 7.5
Double break oil circuit breaker

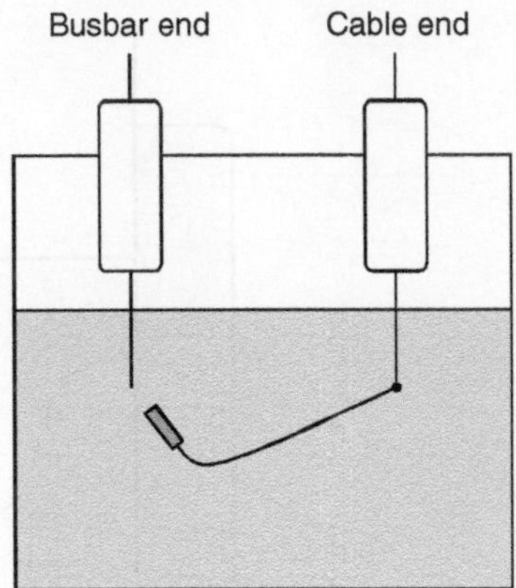

Figure 7.6
Single break oil circuit breaker

Oil has the following advantages:

- Ability of cool oil to flow into the space after current zero and arc goes out
- Cooling surface presented by oil
- Absorption of energy by decomposition of oil
- Action of oil as an insulator lending to more compact design of switchgear.

Disadvantages:

- Inflammability (especially if there is any air near hydrogen)
- Maintenance (changing and purifying).

In the initial stages, the use of high-volume (bulk) oil circuit breakers was more common. In this type, the whole breaker unit is immersed in the oil. This type had the disadvantage of production of higher hydrogen quantities during arcing and higher maintenance requirements. Subsequently these were replaced with low oil (minimum oil) types, where the arc and the bubble are confined into a smaller chamber, minimizing the size of the unit.

7.6.3 Air break switchgear

Interrupting contacts situated in air instead of any other artificial medium (see Figure 7.7). Arc is chopped into a number of small arcs by the Arc-shute as it rises due to heat and magnetic forces. The air circuit breakers are normally employed for 380~480 V distribution.

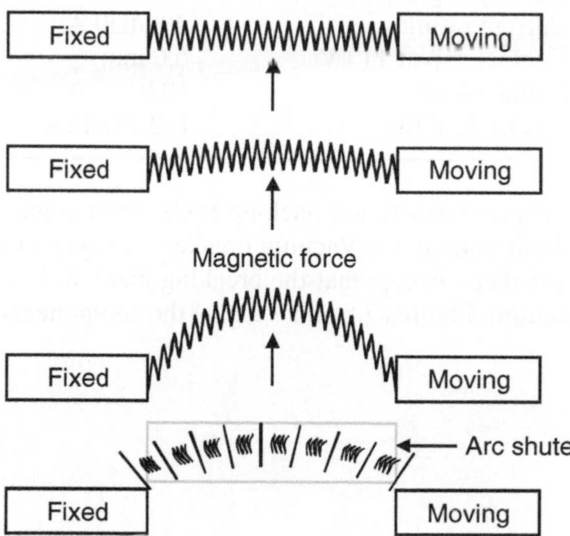

Figure 7.7
Air break switchgear

7.6.4 SF6 circuit breakers

Sulphur-hexaflouride (SF6) is an inert insulating gas, which is becoming increasingly popular in modern switchgear designs both as an insulating as well as an arc-quenching medium.

Gas insulated switchgear (GIS) is a combination of breaker, isolator, CT, PT, etc., and are used to replace outdoor substations operating at the higher voltage levels, namely 66 kV and above.

For medium- and low-voltage installations, the SF6 circuit breaker remains constructionally the same as that for oil and air circuit breakers mentioned above, except for the arc interrupting chamber which is of a special design, filled with SF6.

To interrupt an arc drawn when contacts of the circuit breaker separate, a gas flow is required to cool the arcing zone at current interruption (i.e. current zero). This can be achieved by a gas flow generated with a piston (known as the 'puffer' principle), or by heating the gas of constant volume with the arc's energy The resulting gas expansion is directed through nozzles to provide the required gas flow.

The pressure of the SF6 gas is generally maintained above atmospheric; so good sealing of the gas chambers is vitally important. Leaks will cause loss of insulating medium and clearances are not designed for use in air.

7.6.5 Vacuum circuit breakers and contactors

Vacuum circuit breakers and contactors were introduced in the late 1960s. A circuit breaker is designed for high through-fault and interrupting capacity and as a result has a low mechanical life. On the other hand, a contactor is designed to provide large number of operations at typical rated loads of 200/400/600 A at voltages of 1500/3300/6600/11 000 V.

The following table illustrates the main differences between a contactor and a circuit breaker.

	Contactor	Circuit Breaker
Interrupting capacity	4.0 kA	40 kA
Current rating	400/630 A	630/3000 A
Contact gap at 11 kV	6.0 mm	16.0 mm
Contact force	10 kg	80 kg
Mechanical life	1–2.5 million	10 000

Hence, it is necessary to use back-up fuses when contactors are employed to take care of the high fault conditions. Vacuum breakers are also similar in construction like the other types of breakers, except that the breaking medium is vacuum and the medium sealed to ensure vacuum. Figures 7.8 and 7.9 give the components of a vacuum circuit breaker.

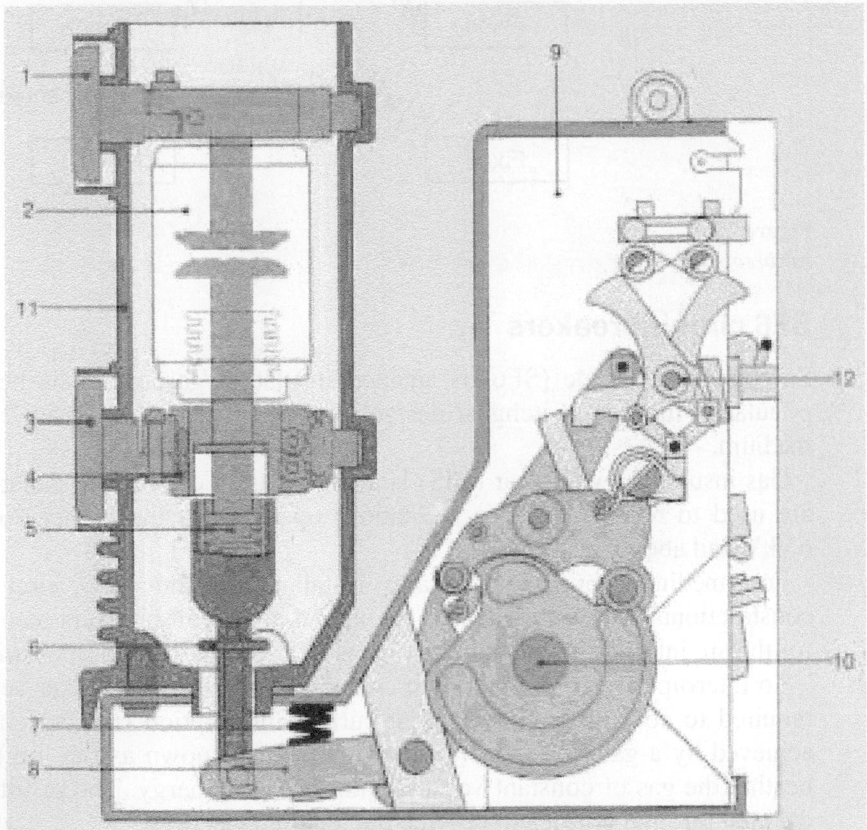

1 Upper connection	7 Opening spring
2 Vacuum interrupter	8 Shift lever
3 Lower connection	9 Mechanism housing with spring operating mechanism
4 Roller contact (swivel contact for 630 A)	10 Drive shaft
5 Contact pressure spring	11 Pole tube
6 Insulated coupling rod	12 Release mechanism

Figure 7.8
General construction of a vacuum circuit breaker

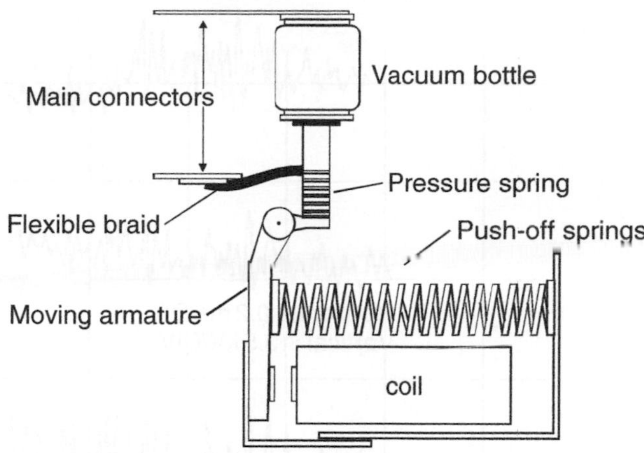

Figure 7.9
Diagrammatic representation

The modern vacuum bottle, which is used in both breakers and contactors, is normally made from ceramic material. It has pure oxygen-free copper main connections, stainless steel bellows and has composite weld-resistant main contact materials. A typical contact material comprises a tungsten matrix impregnated with a copper and antimony alloy to provide a low melting point material to ensure continuation of the arc until nearly current zero.

Because it is virtually impossible for electricity to flow in a vacuum, the early designs displayed the ability of current chopping i.e. switching off the current at a point on the cycle other than current zero. This sudden instantaneous collapse of the current generated extremely high-voltage spikes and surges into the system, causing failure of equipment.

Another phenomenon was pre-strike at switch on. Due to their superior rate of dielectric recovery, a characteristic of all vacuum switches was the production of a train of pulses during the closing operation. Although of modest magnitude, the high rate of rise of voltage in pre-strike transients can, under certain conditions produce high-insulation stresses in motor line end coils.

Subsequent developments attempted to alleviate these shortcomings by the use of 'softer' contact materials, in order to maintain metal vapor in the arc plasma so that it did not go out during switching. Unfortunately, this led to many instances of contacts welding on closing.

Restrike transients produced under conditions of stalled motor switch off was also a problem. When switching off a stalled induction motor, or one rotating at only a fraction of synchronous speed, there is little or no machine back emf, and a high voltage appears across the gap of the contactor immediately after extinction. If at this point of time the gap is very small, there is the change that the gap will break down and initiate a restrike transient, puncturing the motor's insulation (see Figures 7.10 and 7.11).

Modern designs have all but overcome these problems. In vacuum contactors, higher operating speeds coupled with switch contact material are chosen to ensure high gap breakdown strength, produce significantly shorter trains of pulses.

In vacuum circuit breakers, operating speeds are also much higher which, together with contact materials that ensure high dielectric strength at a small gap, have ensured that pre-strike transients have ceased to become a significant phenomenon. These have led to the use of vacuum breakers more common in modern installations.

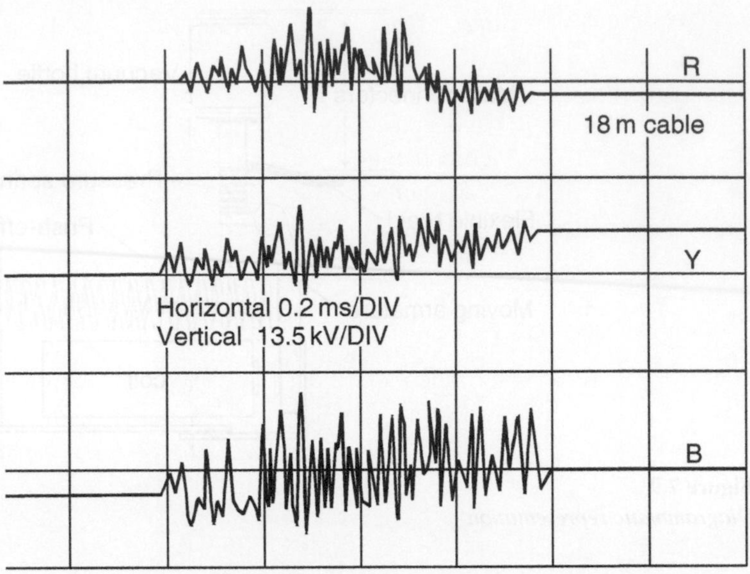

Figure 7.10
Typical pre-strike transient at switch on of 6.6 kV 200 kW motor

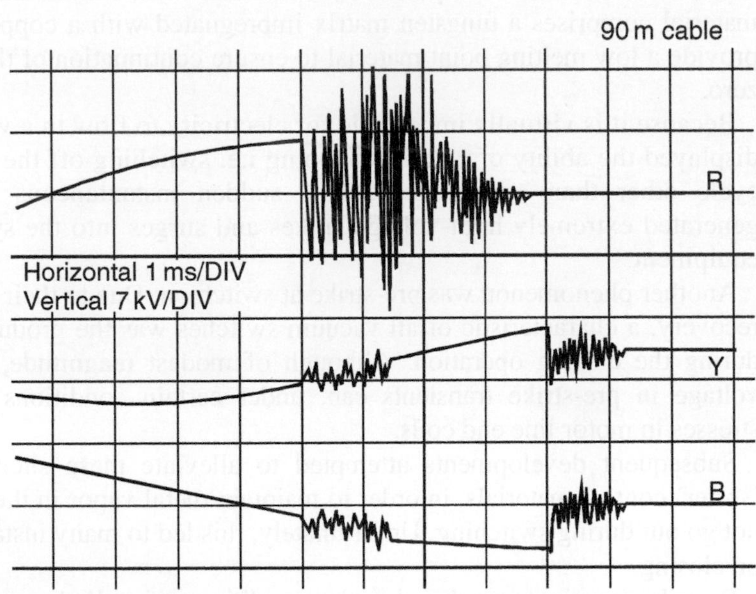

Figure 7.11
Switch off of stalled 6.6 kV 200 kW motor-escalating restrike on R phase

7.6.6 Types of mechanisms

The mechanisms are required to close and break the contacts with high speed. Following are the types of mechanisms employed.

1. *Hand operated*: Cheap but losing popularity. Speed depends entirely on operator. Very limited use in modern installations that too for low-voltage applications only.
2. *Hand operated spring assisted*: Hand movement compresses spring over top deadcentre. Spring takes over and closes the breaker.

3. *Quick make*: Spring charged-up by hand, then released to operate mechanism.
4. *Motor wound spring*: Motor charges spring, instead of manual. Mainly useful when remote operations are employed, which are common in modern installations because of computer applications.
5. *Solenoid*: As name implies.
6. *Pneumatic*: Used at 66 kV and above. Convenient when drying air is required.

7.6.7 Dashpots

In oil circuit breakers, when the breaker is closed, if the operation is not damped then contact bounce may occur and the breaker may kick open. Dashpots prevent this. They also prevent unnecessary physical damage to the contacts on impact. Their use of course depends on the design.

7.6.8 Contacts

Fixed contacts normally have an extended finger for arc control purposes. Moving contacts normally have a special tip (Elkonite) to prevent burning from arcing.

Comparison of insulating methods for CBs

Property	Air	Oil	SF6	Vacuum
Number of operations	Medium	Low	Medium	High
'Soft' break ability	Good	Good	Good	Fair
Monitoring of medium	N/A	Manual test	Automatic	Not possible
Fire hazard risk	None	High	None	None
Health hazard risk	None	Low	Low	None
Economical voltage range	Up to 1 kV	3.3–22 kV	3.3–800 kV	3.3–36 kV

7.7 Comparison of breaker types

Following curve gives the requirement of electrode gaps for circuit breakers with different insulating mediums (see Figure 7.12).

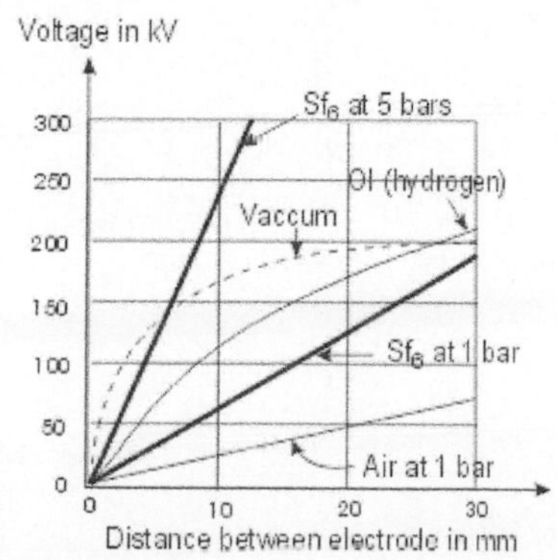

Figure 7.12
Influence of electrode gap for different mediums

The following table highlights the features for different types of circuit breakers.

Factor	Oil Breakers	Air Breakers	Vacuum/SF6
Safety	Risk of explosion and fire due to increase in pressure during multiple operations	Emission of hot air and ionized gas to the surroundings	No risk of explosion
Size	Quite large	Medium	Smaller
Maintenance	Regular oil replacement	Replacement of arcing contacts	Minimum lubrication for control devices
Environmental factors	Humidity and dust in the atmosphere can change the internal properties and affect the dielectric		Since sealed, no effect due to environment
Endurance	Below average	Average	Excellent

8

Tripping batteries

8.1 Tripping batteries

8.1.1 Importance

The operation of monitoring devices like relays and the tripping mechanisms of breakers require independent power source, which does not vary with the main source being monitored. Batteries provide this power and hence they form an important role in protection circuits.

The relay/circuit breaker combination depends entirely on the tripping battery for successful operation. Without this, relays and breakers will not operate, becoming 'solid', making their capital investment very useless and the performance of the whole network unacceptable.

It is therefore necessary to ensure that batteries and chargers are regularly inspected and maintained at the highest possible level of efficiency at all times to enable correct operation of relays at the correct time.

8.1.2 How a battery works?

A battery is an assembly of cells. Whether it is used to make a call using mobile phone or to trip a circuit breaker, every cell has three things in common – positive and negative electrodes and an electrolyte. Whereas some of the dry cell batteries drain out their energy and are to be discarded, a stationary or storage battery used in the switchgear protection has the capability to be recharged.

There are two types of batteries used in an electrical control system:

1. Lead acid type
2. Nickel cadmium type.

Both the above types can be classified further into flooded type and sealed maintenance free type. The flooded cell construction basically refer to the electrodes of the cell in the electrolyte medium, which can be topped up with distilled water as the electrolyte gets diluted due to charging and discharging cycles.

The batteries also discharge hydrogen during these cycles and it is very necessary to restrict this discharge to less than 4% by volume to air, to avoid the surroundings becoming hazardous. The higher discharge of H_2 in lead acid cells have resulted in the manufacture of sealed maintenance free or valve-regulated lead acid (VRLA) batteries. Here the H_2 discharge is restricted to be below the hazardous limit.

Nickel cadmium batteries are comparatively costlier though they are considered more reliable with lesser maintenance and lesser environmental issues that go with lead acid types. In addition, the hydrogen discharge in a nickel cadmium cell is comparatively less. Hence, for conventional switchgear protection applications, sealed nickel cadmium batteries are not required. As such, the sealed nickel cadmium cells are only used for small battery cells used in modern electronic gadgets.

The rechargeable lead acid cells as used in switchgear/relay applications are generally of the Plante type and has an electrical voltage of 2 V. The cell contains a pure lead (Pb) positive plate, a lead oxide (PbO_2) negative plate, and an electrolyte of dilute sulphuric acid. The nickel cadmium cell has an electrical voltage of 1.2 V containing nickel compound (+) and cadmium compound (–) plates with potassium hydroxide solution as the electrolyte.

The following table briefly gives the advantages and disadvantages of nickel cadmium batteries over lead acid type, the most common types being used for protection application (see Table 8.1).

Advantages	Disadvantages
Better mechanical strength	Lower cell voltage (1.2 vs 2.0)
Easy maintenance	More expensive
Long life	Higher current consumption for charging
Space and weight low	Not recommended at higher ambient temperatures
Low H2 discharge and no spill over issues	Higher distilled water consumption

Table 8.1
Nickel cadmium vs lead acid cells

Discharging and recharging

When a load is connected, across plate terminals of a charged cell an electrical current flows and the lead and lead oxide start to change into lead sulphate. A similar phenomenon occurs with nickel cadmium cell. The result is the dilution and weakening of the electrolyte. It is thus possible to measure the state of the battery's charge by measuring the electrolyte's specific gravity with a hydrometer.

The cell is recharged by injecting a direct current in the opposite direction using another source to restore its plates and electrolyte to their original state.

Application guide (see Table 8.2)

	Plante	Flat Plate	Tubular
Substations	*		
Telephone exchanges	*		
Mobile telephone exchanges			*
Emergency lighting	*	*	
Alarms	*	*	
Computer emergency	*	*	*
Engine starting	*	*	
Oil rings		*	*

Table 8.2
Application for different electrode types

8.1.3 Life expectancy

Plante25 – 30 years
Flat Plate5 – 6 years
Tubular10 – 12 years

8.1.4 Construction

For long life with very high reliability needed in places like power stations and substations, batteries are made up of cells of the kind named after Plante. Figure 8.1 gives typical construction of a lead acid battery.

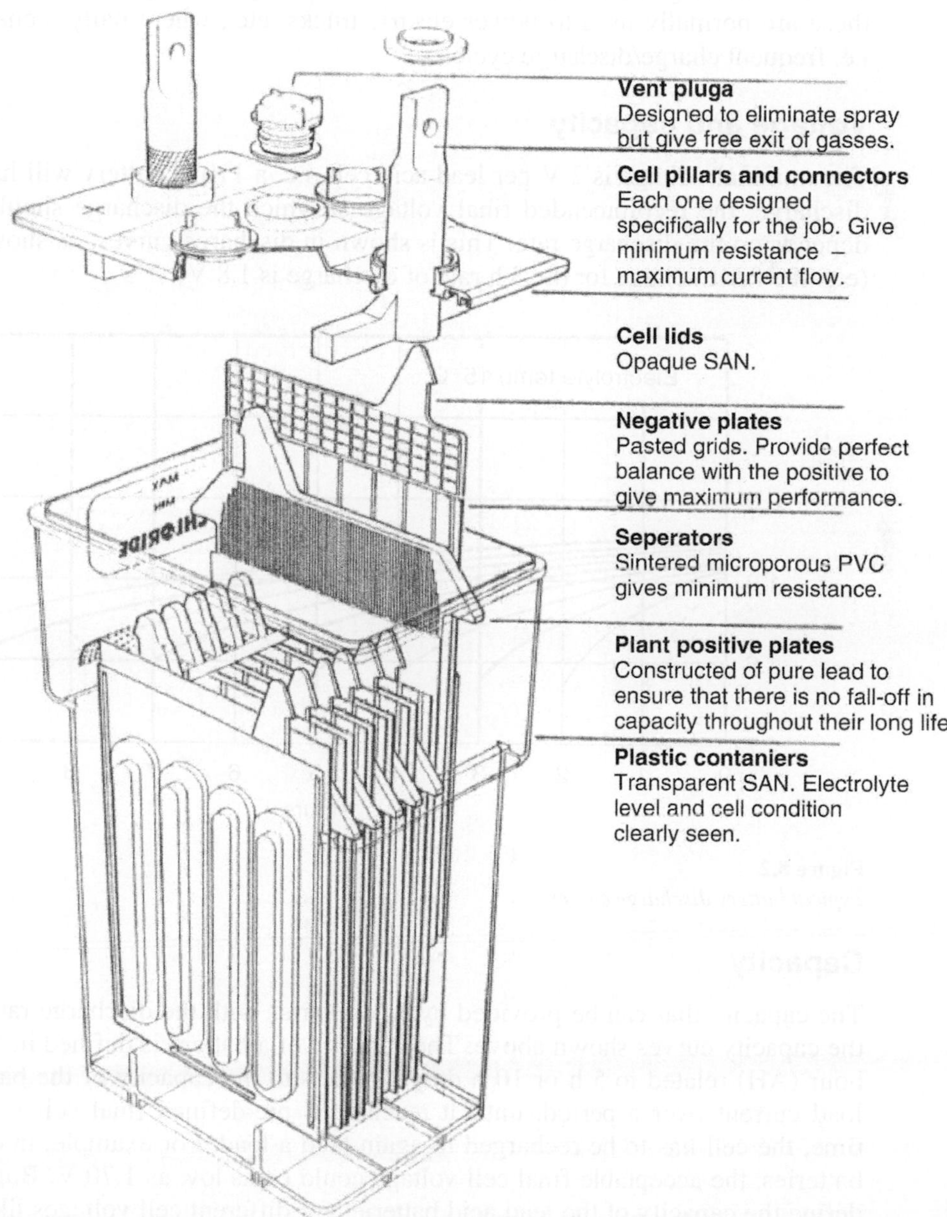

Figure 8.1
Typical construction of a lead acid cell

The positive plate is cast from pure lead in a form which gives it a surface twelve times its apparent area. The negative plate is of the pasted grid type made by forcing lead oxide paste into a cast lead alloy grid.

The positive and negative plates are interleaved and insulated from each other to prevent short circuits, and are mounted in transparent plastic containers to allow visual checking of the acid level and general condition.

Because of the high initial cost of Plante cells, specially designed flat plate cells have been developed to provide a cheaper but shorter-lived alternative source of standby power. Although this is the basis of the modern car battery, it is totally unsuitable for switch-tripping duty because it has been designed to give a high current for a short time as when starting a car engine. Cells with tubular positive plates are also available but these are normally used to power electric trucks, etc., where daily recharging is needed i.e. frequent charge/discharge cycles.

8.1.5 Voltage and capacity

The nominal voltage is 2 V per lead acid cell, i.e. a 110 V battery will have 55 cells. On discharge, the recommended final voltage at which the discharge should be terminated depends on the discharge rate. This is shown in discharge curves, as shown in Figure 8.2 (e.g. the final voltage for the 3 h rate of discharge is 1.8 V).

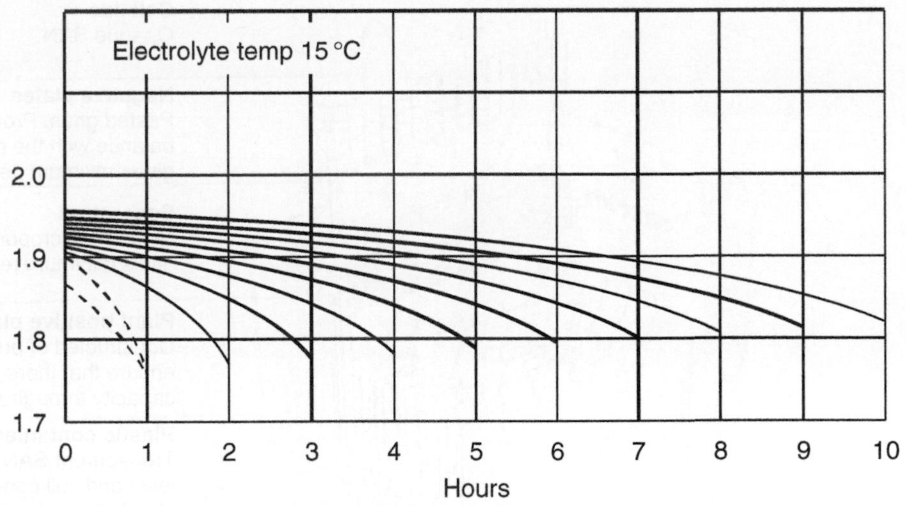

Figure 8.2
Typical battery discharge curves

8.1.6 Capacity

The capacity that can be provided by a cell varies with the discharge rate as indicated in the capacity curves shown above. The capacity of a battery is defined in terms of ampere-hour (AH) related to 5 h or 10 h duty. It refers to the capacity of the battery to supply a load current over a period, until it reaches its pre-defined final cell voltage. After this time, the cell has to be recharged to again feed a load. For example, in case of lead acid batteries, the acceptable final cell voltage could be as low as 1.70 V. But it is common to define the capacity of the lead acid batteries for different cell voltages like 1.75, 1.80 and 1.85 V. Accordingly, the discharge curves of a battery vary showing comparatively higher time to reach the lowest acceptable cell voltage. Table 8.3 gives the current that can be drawn from a battery depending upon the 10 h rating.

Time in hours	1	2	3	4	5	6	7	8	9	10
Capacity in %	60	73	80	84	88	91	93	96	98	100
Final cell voltage	1.75	1.78	1.80	1.81	1.82	1.83	1.835	1.84	1.845	1.85
Current in % of 10 h rating	600	365	267	210	176	151	133	120	109	100

Table 8.3
Capacity variation of a lead acid cell with load current

The above table typically refers to a cell, which can supply 100% of its rated amperes for 10 h at the end of which it reaches an end voltage of 1.85 V. The cell will reach 1.85 V if 100% rated current is continuously drawn for 10 h. Alternatively, if the current drawn is 600% of its rating, the cell will reach 1.75 V at the end of 1 h itself. Hence, while designing the capacity of the cell, proper margins should be taken into account based on the nature of loads and the likely currents to be drawn over a cycle.

Capacity is also affected by ambient temperature. The lower the ambient temperature, the capacity will be comparatively higher.

8.1.7 Battery charger

In a protection system, it is necessary that the control DC voltage shall remain constant for as much time as possible, so that the system works without interruption. Hence, the batteries are normally kept on charge continuously by a battery charger. The charger is a rectifier, which produces a slightly higher voltage compared to the nominal cell voltage of a battery. The main power source is derived from the normally available AC source, which is rectified by the charger. Typical connection is as seen in Figure 8.3.

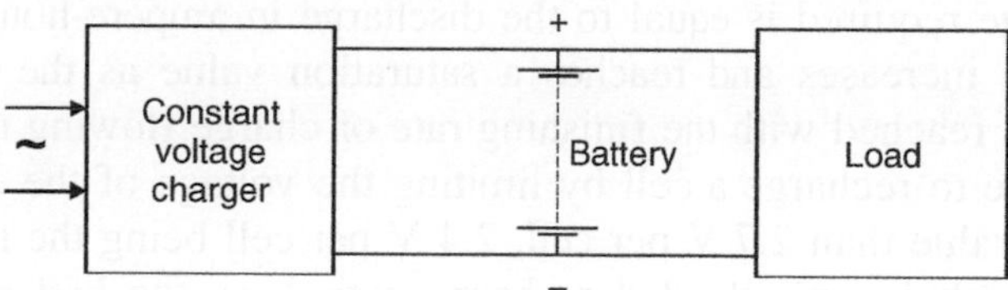

Figure 8.3
Charger/battery/load connection

Here the battery is a combination of multiple cells connected in series to get the nominal DC tripping/control voltage required for the operation of relays and breakers and could be from 24 to 220 V, depending on the loads and the capacity requirement.

8.1.8 Trickle charge

Trickle charging is a method of keeping the cells in a fully charged condition by passing a small current through them. The correct trickle charge current is that which does not allow the cell to discharge gas and does not allow the specific gravity to fall over a period. The cell voltage will be approximately 2.25 V for lead acid cell and 1.35 V for nickel cadmium cell.

8.1.9 Float charge

Float charging is keeping the voltage applied to the battery at 2.25 V per lead acid cell or 1.35 V for nickel cadmium cell, i.e. maintaining a constant voltage across each cell. This method is usually adopted in conjunction with supplying continuous and variable DC loads from the charging equipment, as would typically happen for a substation battery.

The loads in a substation normally comprise of small continuous load consisting of pilot lamps, relays, etc., and momentary short time loads of comparatively high values such as those for circuit breaker tripping and closing operations, motor wound springs and so on. Since the charger, battery and load are all connected in parallel as per Figure 8.3, the continuous load is carried by the charger at normal floating voltage and the battery draws its own maintenance current at the same time. Any load that exceeds the charger capacity will lower its voltage slightly, to the point where the battery discharges to supply the remainder. If there should be a complete power failure the battery will supply the entire load for a period depending on the AH capacity and the load, until AC power is restored and then automatically starts being recharged. Typical float currents will be in the range of 30–50 mA per 100 AH of rated capacity, increasing to about 10 times towards the end of the battery's life.

8.1.10 Specific gravity

A simple hydrometer reading indicates the state of charge in a cell. A fully charged cell will have a specific gravity reading of between 1.205 and 1.215. As a rule of thumb, in a lead acid battery,

$$\text{Open-circuit volts} = \text{specific gravity} + 0.84$$

Thus, the open-circuit voltage of a cell with a SG of 1.21 will be 2.05 V; one with an SG of 1.28 will be 2.12 V.

8.1.11 Recharge

The ampere-hour efficiency of the cells is 90%; therefore, on recharge, the amount of recharge required is equal to the discharge in ampere-hours plus 11%. On recharge, the voltage increases and reaches a saturation value as the charge proceeds. The highest voltage reached with the finishing rate of charge flowing is 2.7 V per lead acid cell. It is possible to recharge a cell by limiting the voltage of the charging equipment to a much lower value than 2.7 V per cell, 2.4 V per cell being the minimum desirable value. This will result in an extended recharge period, as the battery will automatically limit the charge current irrespective of the charger output.

8.2 Construction of battery chargers

Battery charging is accomplished with sophisticated electronically controlled rectifiers that permit a battery to be continuously maintained on a floating charge and to recharge a discharged battery as fast as possible. It also has to handle the electrical load.

Chargers presently used in stationary applications are normally of the constant voltage type. Voltage adjustments can be made with precision to 0.01 of a volt per cell. This is necessary because floating voltage and equalizing voltage levels critically affect battery performance and life expectancy. Voltage level specifications are normally expressed to two decimal places, i.e. 2.16–2.33 V for a lead acid cell.

Proper specifications and correct adjustments of the battery charger are the most important factors affecting the satisfactory performance and life of the battery cells. Voltage levels from the charger also usually serve the electrical load, so changes in charger voltage output affect the load.

Chargers are normally equipped to accommodate normal float voltages and the higher voltages for equalizing charges when required. As the charger DC output is rectified from AC there will be a ripple on the output unless smoothing techniques are employed.

Care must be taken to ensure that the maximum rms current value of the AC component does not exceed 7% of the battery capacity expressed in amperes. Failure to do this will result in a phenomenon called AC corrosion, where the negative peak of the AC component reverses the direction of the charging current, leading to corrosion and ultimate destruction of the plates in the cells.

8.3 Maintenance guide

Following are brief guidelines for maintaining the lead acid battery system, which are also applicable for nickel cadmium batteries.

- Check batteries and chargers regularly, at least once per month.
- Take specific gravity reading – should be 1.215 minimum.
- Check electrolyte level – top up if necessary to keep level between its normal and upper limits. Use distilled water wherever possible. Impurities in normal tap water such as chlorine and iron tend to increase internal losses. Frequent topping up of electrolyte means excessive gassing brought about by overcharging.
- Check for gassing: This normally starts when lead acid cell voltage reaches 2.30–2.35 V per cell (1.30–1.35 for Ni-Cd cell) and increases as charge progresses. At full charge, most energy goes into gas, oxygen being liberated at the positive plate and hydrogen at the negative. Four percent hydrogen in the air may be hazardous. The room employing lead acid batteries must be well ventilated. However, the nickel cadmium batteries discharge very low hydrogen.
- Look between the plates for any signs of mossing or treeing. This is a build-up of a sponge like layer of lead on the negative plates, which can accumulate to such an extent as to bridge over or around the separators and cause a short circuit to the adjacent positive plate. This condition is usually an indication of overcharging.
- Also look for sediment build-up on the floor of the container. If this increases to the point where it reaches the bottom of the plates, it will short them out to cause failure. Overcharging can also accelerate the accumulation of sediment and shorten the useful life of the battery.
- Battery cells must be kept clean and dry to the extent that no corrosion, dust or moisture offers a conducting path to partially short-circuit the cell or contact ground.
- Finally, proper charging is the most important factor in battery service and life. A short boost charge is beneficial on a regular basis to prevent stratification, to freshen the electrolyte and to equalize the cells. Ensure that the charger is working properly and that it is operating in accordance with the manufacturers recommended settings.
- The VRLA batteries cannot be inspected visually like the flooded cells but their healthiness is ascertained by measuring the cell voltage with/without the charger input. Cells reaching end voltage after disconnection of charger comparatively faster than the other cells need to be replaced completely, as there is no question of changing the electrolyte.

- The VRLA batteries last longer if their ambient temperature is controlled to around 25 °C by use of air conditioning, as otherwise they may lose their charge more quickly.
- The flooded type lead acid batteries are normally prone for spill over of electrolyte and it is highly recommended to keep these batteries in separate room with acid proof tiles to take care of the spill over conditions. The nickel cadmium batteries on the other hand do not pose such major hazards though they are quite costlier compared to lead acid type.

8.3.1 Arrangement of DC supplies

For strategic switch boards it is sometimes worthwhile to fit two trip coils to each circuit breaker to ensure positive tripping. Two batteries and chargers should then be installed to ensure the integrity of each tripping system, DC fail relays being installed on each panel to monitor the continuity of each supply (see Figure 8.4). For breakers fitted with only one trip coil, a single battery and charger should be installed and trip coil supervision relays fitted to monitor each circuit (see Figure 8.5).

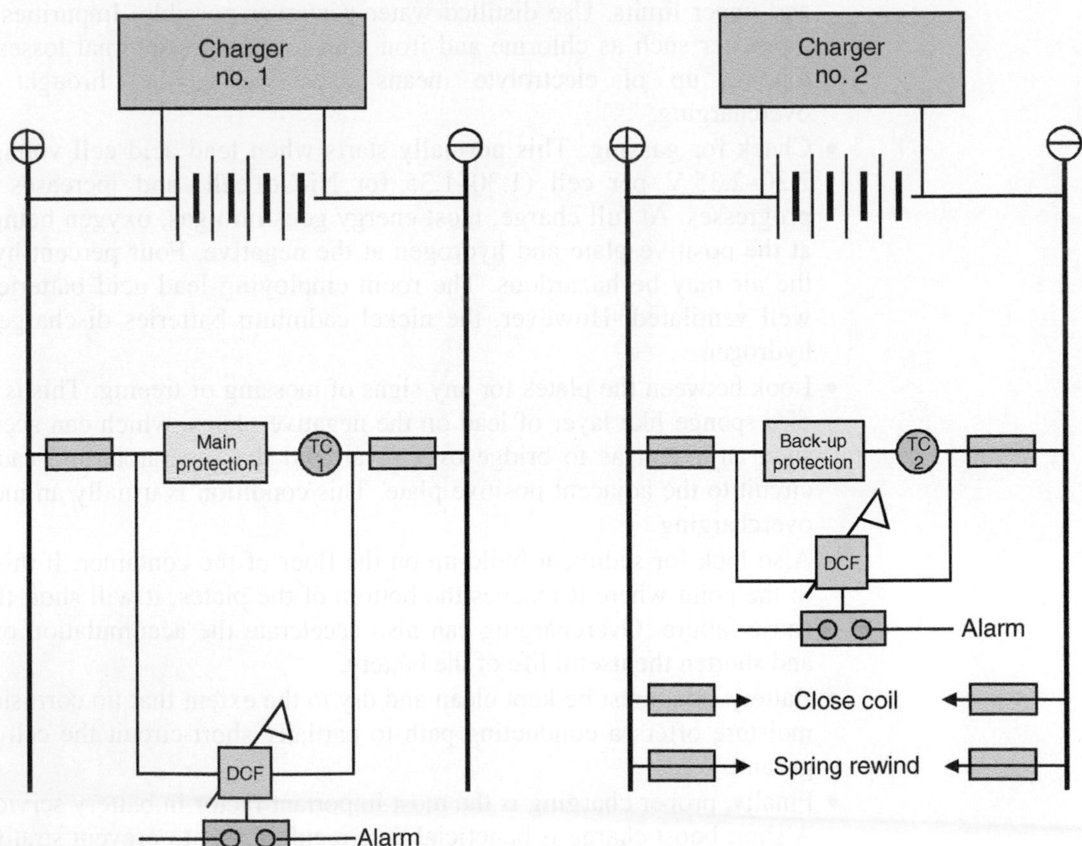

Figure 8.4
Arrangement of DC supplies with two trip coils for each circuit breaker

8.3.2 Earthing of DC supplies

It is a normal practice to earth the tripping battery at one point to prevent it floating all over the place with respect to earth. One popular method is to earth the negative rail (see Figure 8.6). It will be noted that in this case a solid link is used instead of a fuse on the negative side. This ensures that the negative is never lost. If it was, 'sneak circuits' become a distinct possibility causing many vexing problems, etc. The drawback with this system is that the first earth fault on the wiring could possibly create a short on the battery.

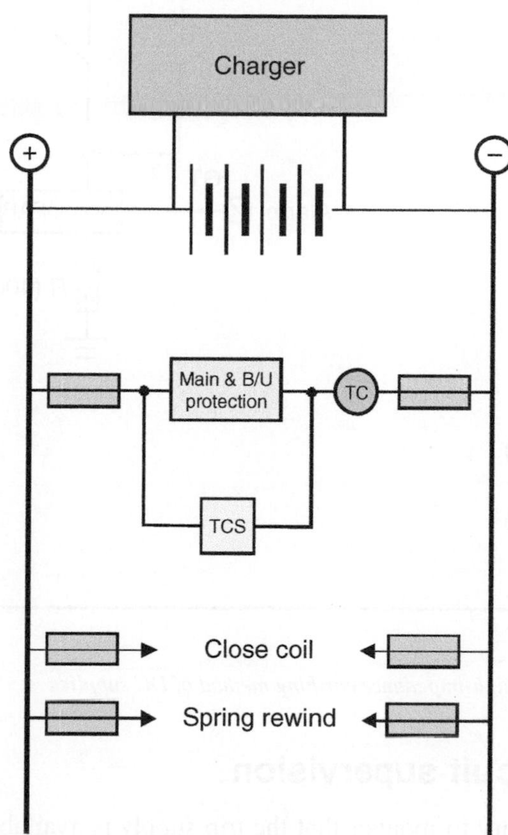

Figure 8.5
Arrangement of DC supplies with one trip coil

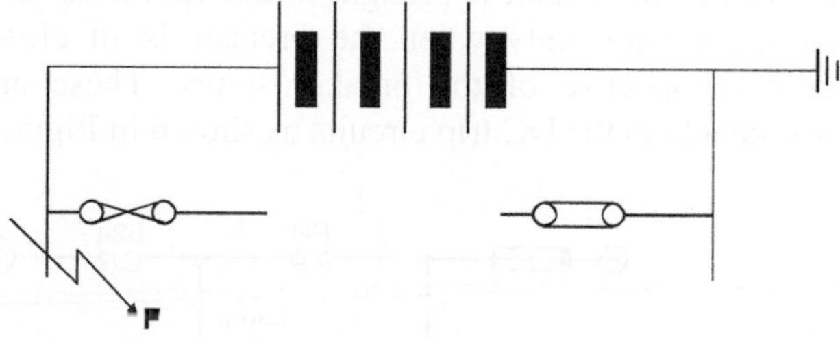

Figure 8.6
Earthing of DC supplies negative rail

A more secure system is the center-point high-impedance earth method, which utilizes a battery earth fault alarm relay (BEFAR) see Figure 8.7. An earth fault on either the positive or negative rail will cause a very small current to flow up the neutral (0 V) connection. This will cause the BEFAR to operate to alarm and indicate which leg has faulted. The supply will however remain on but this condition should be attended to before a second earth fault occurs. A test button is also provided to check the relay is functional at any time, by offsetting this away from the center zero point.

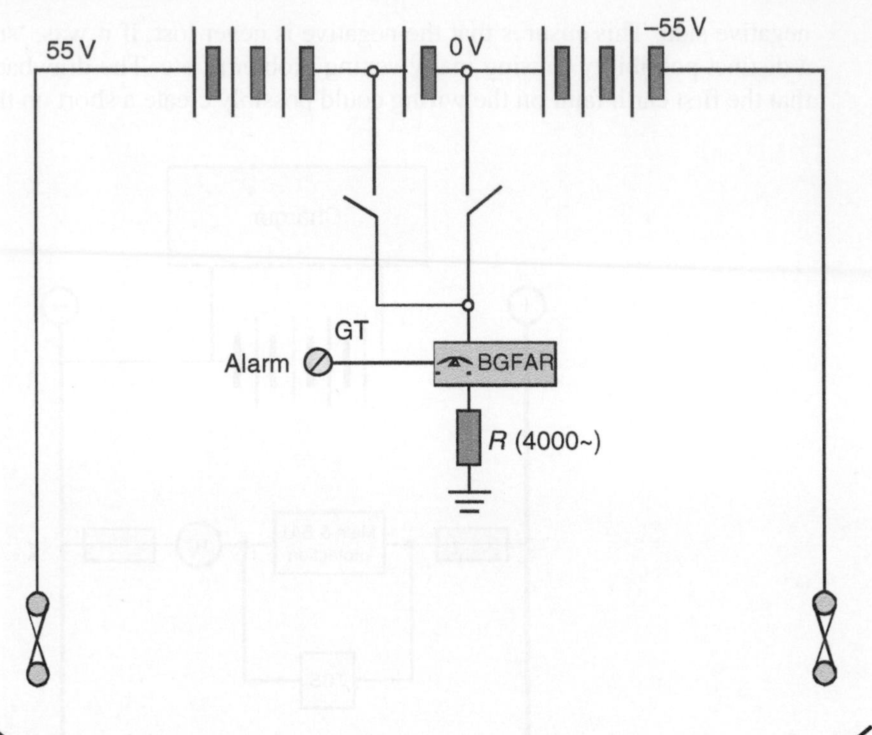

Figure 8.7
Center-point high-impedance earthing method of DC supplies

8.4 Trip circuit supervision

It is necessary to monitor that the trip supply is available in a trip circuit to ensure that the system is continuously being monitored. This is some times achieved by means of 'fail-safe circuits', which ensure that there is no question of operating a system without monitoring.

Trip circuit supervision is a method, where the trip supply is continuously monitored, so that any break in the circuit is brought to the operators' attention. There are two possible types (a) supervision only when the breaker is in closed (in service) condition; (b) supervision irrespective of the breaker status. These are achieved by using breaker auxiliary contacts in the DC trip circuits as shown in Figures 8.8(a)–(c).

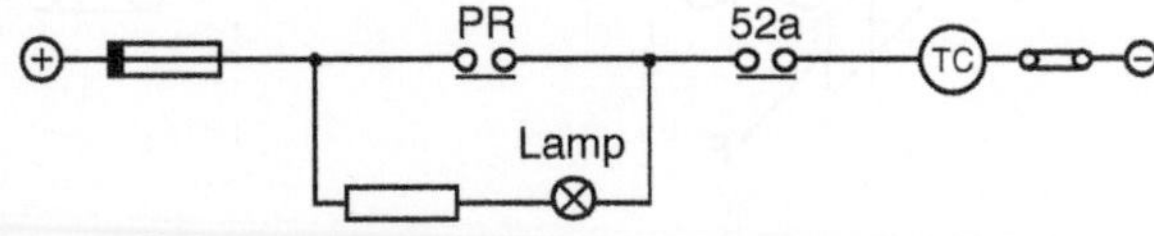

Figure 8.8(a)
Supervision while circuit breaker is closed

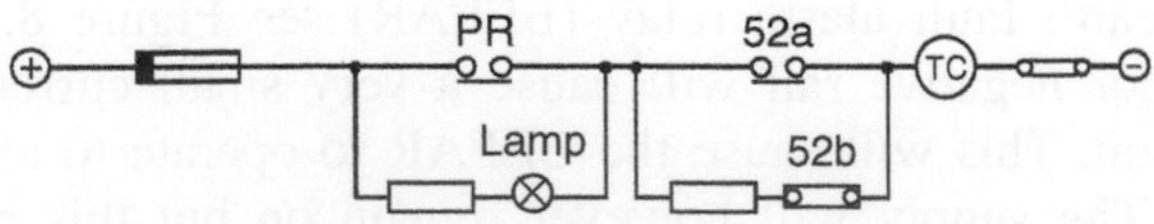

Figure 8.8(b)
Supervision while circuit breaker is open or closed

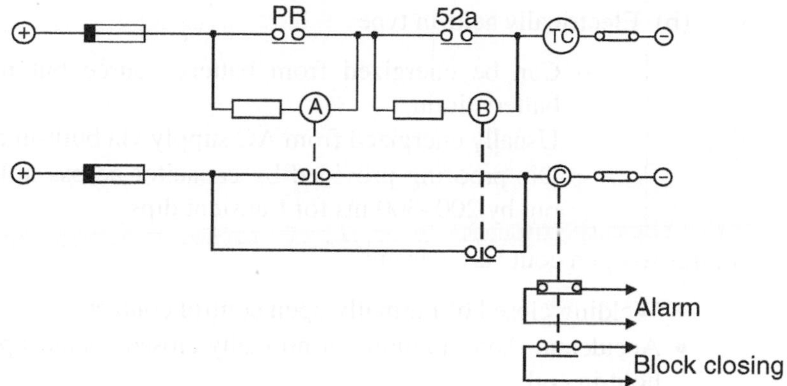

Figure 8.8(c)
Supervision with circuit breaker open or closed with remote alarm

8.5 Reasons why breakers and contactors fail to trip

8.5.1 Breakers

1. Open-circuited DC shunt trip coil
2. Loss of circuit DC trip supply
   - Trip fuse blown or removed or
   - Trip MCB open
3. Loss of station DC trip supply
   - Battery and/or charger failure
   - Battery and/or charger disconnection
4. Burnt out DC shunt trip coil
5. Failure by open circuit of control wiring or defective relay contact
6. Breaker mechanically jammed:
   - Trip bar solid
   - Trip coil mechanically jammed
   - Trip coil loose or displaced
   - Broken mechanism
   - Lack of regular/correct maintenance
   - Main contacts welding
   - Contacts arcing – loss of vacuum or SF6 gas pressure.

Trip circuit supervision (TCS) provides continuous monitoring and gives immediate warning of conditions 1 to 5 before breaker is called upon to trip. Corrective action can then be taken before the event, to prevent breaker failure occurring.

Breaker fail (BF) protection covers all conditions but only highlights problem after a system primary fault occurs and the breaker has failed to clear.

TCS can be regarded as the fence at the top of the cliff whereas BF is the ambulance at the bottom, only operating after the event!

8.5.2 Contactors

(a) Latched type
   - Suffer same failures as breakers

(b) Electrically held-in type

- Can be energized from battery source but not favored because of battery drain
- Usually energized from AC supply via built-in rectifiers
- Dip-proofing provided by capacitor across hold-in coil to delay drop out by 200–300 ms for transient dips.

Failures to open could arise from:

- Welding closed of normally open control contacts
- Accidental short-circuiting of normally closed contacts preventing de-energizing hold-in coil
- Welding closed of main contacts (old vacuum or air break types)
- Mechanism mechanically jammed.

8.6 Capacitor storage trip units

These units are intended for use in substations where no tripping batteries have been provided.

They can also be used as an alternative tripping supply for installations, which employ two-shunt trip coils per circuit breaker, one connected to the tripping battery and the other to the capacitor storage unit.

The unit comprises two capacitors to provide short-time storage of auxiliary energy, one to operate a protective relay whilst the other operates the circuit breaker trip coil. The device has three inputs, which may be fed from DC or AC sources. A test button is provided for checking the degree of capacitor charge, whilst an in-built signaling relay with a normally closed contact offers remote indication of discharged capacitors. Figure 8.9 shows the unit below.

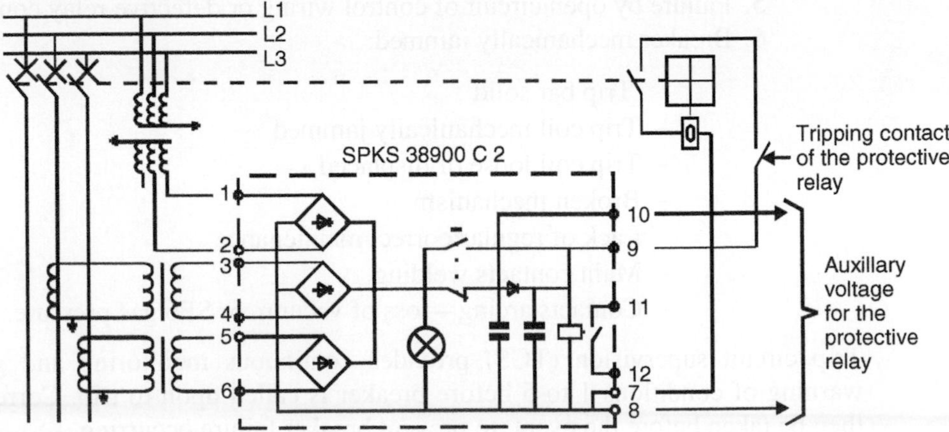

Figure 8.9
Capacitor storage trip unit

8.6.1 Points to watch for

- Always endeavor to feed the unit from a busbar connected VT if possible as the first priority as this is the most secure supply.
- Only feed one relay and circuit breaker from each unit and check that the stored energy is sufficient to do the job required of it.

- When the storage device is fed from current transformers, it is recommended that at least two phases are used.
- A set of special interposing auxiliary CTs should be installed between the main CTs and the storage device, to protect it from overvoltages caused by nearby short-circuits. They should be designed to saturate at about 200 V rms.
- However, they must also ensure that they are able to charge the capacitors from zero to maximum in two cycles to cover the case of 'closing onto fault'.
- Ratios and performance of the line CTs on light loads must also ensure that the capacitors can be charged rapidly.
- Care should be taken when installing these devices into CT circuits that they do not overburden other relays and protection that may be connected to the same CTs. The recommended interposing CTs may well pose a problem in this regard.
- As the protection and tripping will depend entirely on the correct performance of the capacitor storage unit, it would be as well to consider allocating an exclusive dedicated set of CTs for this function.

9

Relays

9.1 Introduction

It had been repeatedly indicated in the earlier chapters that relays are the devices, which monitor the conditions of a circuit and give instructions to open a circuit under unhealthy conditions. The basic parameters of the three-phase electrical system are voltage, current, frequency and power. All these have pre-determined values and/or sequence under healthy conditions. Any shift from this normal behavior could be the result of a fault condition either at the source end or at the load end. The relays are devices, which monitor various parameters in various ways and this chapter gives a brief outline of their principles of operation.

The types of relays can be broadly classified as:

- Electromechanical relays
- Static relays (analog and digital).

The electromechanical relays had been dominating the electrical protection field until the use of silicon semiconductor devices, becoming more common. The use of static relays in the early stages were more due to the advantages like lower weight, non-moving mechanical parts, reduced wear and tear, etc. However, the initial static relays had not been overwhelmingly accepted in the electrical field also due to their 'static' nature. Further, the reliability of electronic components in the initial stages had been unsatisfactory due to the quality issues and their ability (or inability) to withstand source fluctuations and ambient temperature conditions. However, the reliability of electronic components improved subsequently, and the advent of digital electronics technology and microprocessor developments gave a completely different picture to the use of static relays. The earlier analog relays have been slowly replaced with digital relays, and today's protection technology is more inclined towards use of digital relays, though the electromechanical relays are still preferred in certain applications, with cost being one of the main reasons. The use of static analog relays is not so common.

9.2 Principle of the construction and operation of the electromechanical IDMTL relay

As the name implies, it is a relay monitoring the current, and has inverse characteristics with respect to the currents being monitored. This (electromechanical) relay is without doubt one of the most popular relays used on medium- and low-voltage systems for many years, and modern digital relays' characteristics are still mainly based on the torque characteristic of this type of relay. Hence, it is worthwhile studying the operation of this relay in detail to understand the characteristics adopted in the digital relays (see Figure 9.1).

Figure 9.1
Typical mechanical relay

The above relay can be schematically represented as shown in Figure 9.2.

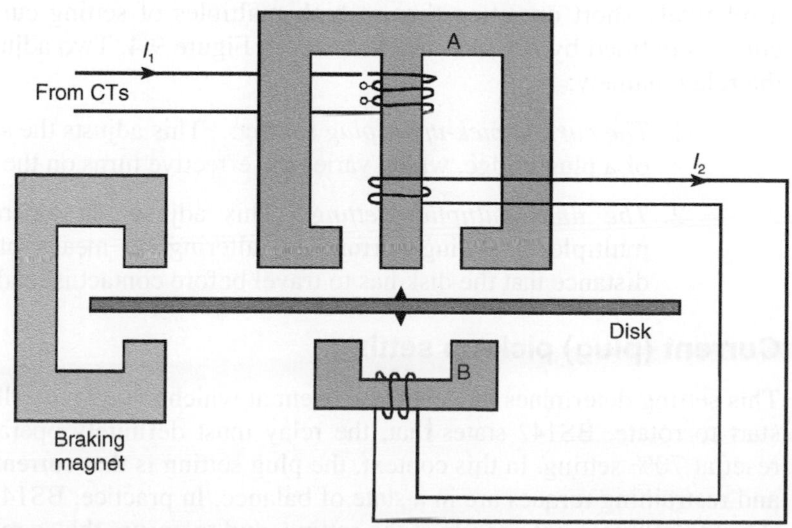

Figure 9.2
The IDMTL relay

The current I_1 from the line CTs, sets up a magnetic flux A and also induces a current I_2 in the secondary winding which in turn sets up a flux in B. Fluxes A and B are out of phase thus producing a torque in the disk causing it to rotate. Now, speed is proportional to braking torque, and is proportional to driving torque. Therefore, speed is proportional to I^2.

But,

$$\text{Speed} = \frac{\text{Distance}}{\text{Time}}$$

Hence,

$$\text{Speed} = \frac{\text{Distance}}{\text{Time}} = \frac{1}{I^2}$$

This therefore gives an inverse characteristic (see Figure 9.3).

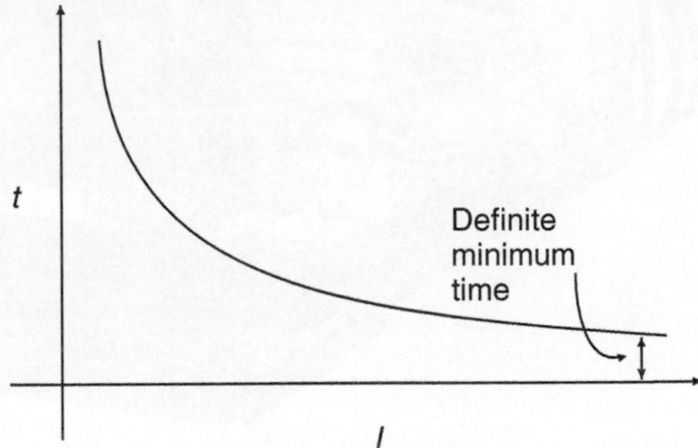

Figure 9.3
Characteristic curve of relay

It can be seen that the operating time of an IDMTL relay is inversely proportional to a function of current, i.e. it has a long operating time at low multiples of setting current and a relatively short operating time at high multiples of setting current. The characteristic curve is defined by BS 142 and is shown in Figure 9.4. Two adjustments are possible on the relay, namely:

1. *The current pick-up or plug setting*: This adjusts the setting current by means of a plug bridge, which varies the effective turns on the upper electromagnet.
2. *The time multiplier setting*: This adjusts the operating time at a given multiple of setting current, by altering by means of the torsion head, the distance that the disk has to travel before contact is made.

9.2.1 Current (plug) pick-up setting

This setting determines the level of current at which the relays will pick-up or its disk will start to rotate. BS142 states that, the relay must definitely operate at 130% setting and reset at 70% setting. In this context, the plug setting is that current at which the operating and restraining torques are in a state of balance. In practice, BS142 requires that the relay should definitely not operate at the setting, and to ensure this, a relay may display a slight tendency to reset at the normal setting.

The relay therefore normally picks up in the range of 105–130% its current plug setting.

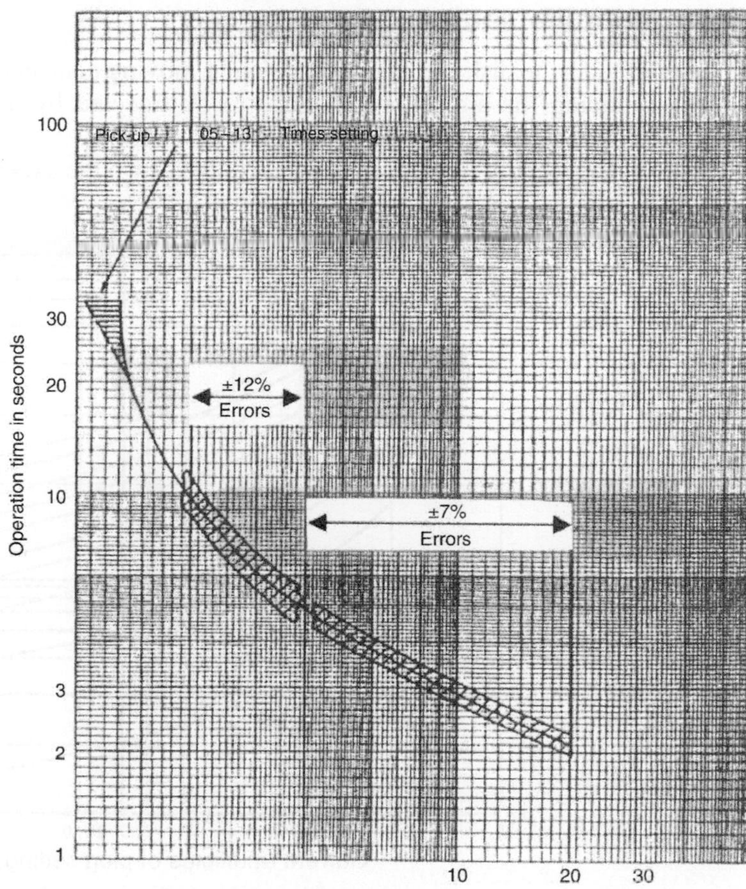

Figure 9.4
Time and pick-up errors at unity time multiplier

Usually the following ranges of nominal current are used, giving a 1:4 ratio in seven steps:

Percentage plug settings (reyrolle)

Overcurrent (%):	50%	75%	100%	125%	150%	175%	200%
Earth fault (%):	20%	30%	40%	50%	60%	70%	80%
Or (%):	10%	15%	20%	25%	30%	35%	40%

Current plug settings (GEC) – for 5 A relay

Overcurrent (A):	1.5	3.5	5.0	6.25	7.5	8.75	10.0
Earth fault (A):	1.0	1.5	2.0	2.5	3.0	3.5	4.0
Or (A):	0.5	0.75	1.0	1.25	1.5	1.75	2.0

Normally, the highest current tap is automatically selected when the plug is removed, so that adjustments can be made on load without open-circuiting the current transformer.

9.2.2 Time multiplier setting

This dial rotates the disk and its accompanying moving contact closer to the fixed contact, thereby reducing the amount of distance to be traveled by the moving contact, hence speeding up the tripping time of the relay.

This has the effect of moving the inverse curve down the axis as shown in Figure 9.5.

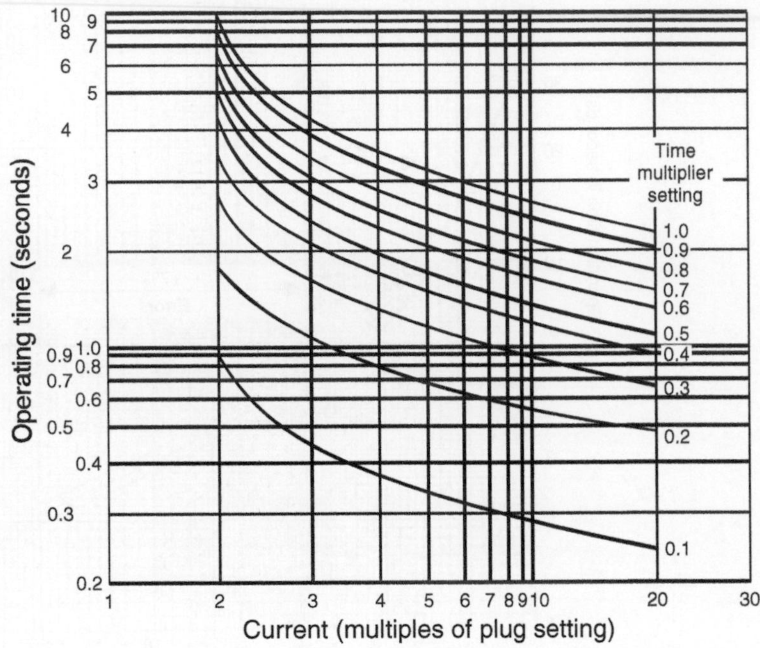

Figure 9.5
Time/current characteristic

The above curve is the most common type used, namely the Normal inverse curve. Its characteristic shows an operating time of 3 s at 10 times the current plug setting i.e. with the plug bridge set at 1 A, when 10 A flows through, the relay will close its contacts after 3 s (i.e. with the time multiplier set at 1.0). The time of operation of the relay is chosen by collectively selecting the current and time plug settings. There is another popular version, which has an operating time of 1.3 s at 10 times the current setting.

It is possible to manufacture relays with different characteristics, but the principle of operation remains the same. Other characteristic curves popular are very and extremely inverse. The different time characteristic curves of an IDMTL relays is shown in Figure 9.6. These are represented in logarithmic graphs due to the exponential nature.

9.2.3 Burden

Burden is the normal continuous load imposed on the current transformers by the relay, normally expressed in VA or some times in ohms. For electromechanical relays, this is normally stated as 3 VA nominal. The modern electronic relays offer a much lower figure, which is one of their virtues.

However, for the electromechanical type, the selection of the plug setting does have an effect on the burden. As stated earlier, the operating coil is wound to give time/current curves of the same shape on each of the seven taps, which are selected on the plug bridge. As there is a required minimum amp-turns of magnetic flux to get the relay to pick-up, the lower the current the more turns are necessary. The lower the setting therefore results in higher the burden on the CTs. Example:

$$VA = I^2 R \quad \text{hence} \quad R = \frac{VA}{I^2}$$

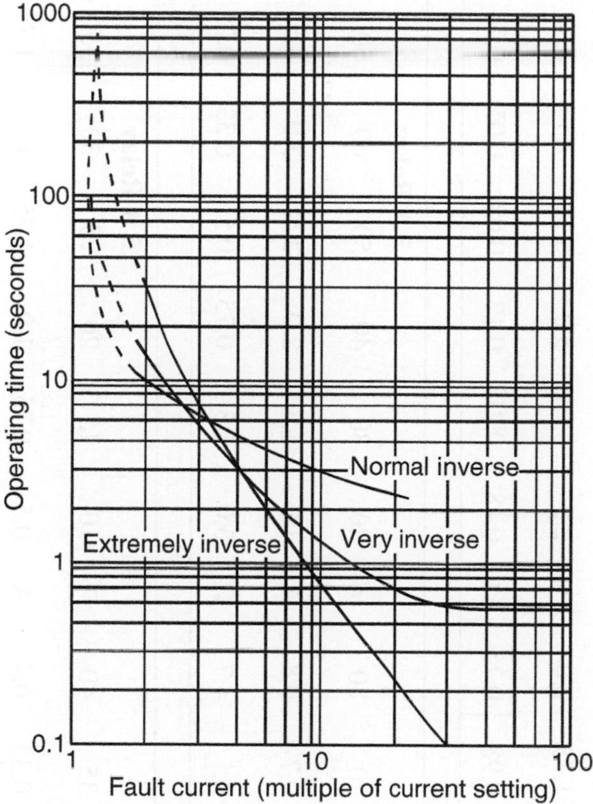

Figure 9.6
Fault current (multiple of current setting)

For 5 A relay on 200% tap,

$$R = \frac{3}{10 \times 10} = 0.03\ \Omega$$

On 10% tap,

$$s R = \frac{3}{0.5 \times 0.5} = 12\ \Omega$$

The lower tap therefore, places a higher burden on the CTs and they must have adequate performance to meet such demands. This is often not the case in low ratio CTs. A mistaken impression is created that the relay is at its most sensitive setting when it is set on its lowest tap. However, the fact is that the CTs may saturate under these conditions due to the higher burden, causing the electromechanical relay to respond more slowly, if at all it picks up. However, the modern digital relays do not exhibit such behavior and have constant burden through out its operating range.

Table 9.1 gives the burden values exhibited by one of the most common type of electromechanical relays used in the electrical systems and the curves that follow show the advantage of static/digital relays over electromechanical relays in this aspect (see Figure 9.7)

	1 A Relay								5 A Relay							
Overcurrent	50	75	100	125	150	175	200	%	50	75	100	125	150	175	200	%
reyrolle – TM 50–200%	0.5	0.75	1.0	1.25	1.5	1.75	2.0	A	2.5	3.75	5.0	6.25	7.5	8.75	10.0	A
GEC-CDG Impedance (ohms)	12.0	5.33	3.0	1.92	1.33	0.98	0.75	Ω	0.48	0.21	0.12	0.08	0.05	0.04	0.03	Ω

	1 A Relay								5 A Relay							
Earth fault	20	30	40	50	60	70	80	%	20	30	40	50	60	60	80	%
reyrolle – TM 20–80%	0.2	0.3	0.4	0.5	0.6	0.7	0.8	A	1.0	1.5	2.0	2.5	3.0	3.5	4.0	A
GEC-CDG Impedance (ohms)	75.0	33.3	18.7	12.0	8.3	6.1	4.7	Ω	3.0	1.3	0.75	0.5	0.33	0.25	0.2	Ω

	1 A Relay								5 A Relay							
Earth fault	10	15	20	25	30	35	40	%	10	15	20	25	30	35	40	%
reyrolle – TM 10–40%	0.1	0.15	0.2	0.25	0.3	0.35	0.4	A	0.5	0.75	1.0	1.25	1.5	1.75	2.0	A
GEC-CDG Impedance (ohms)	300	133	75	48	33.3	25.4	18.75	Ω	12.0	5.33	3.0	1.92	1.33	1.0	0.75	Ω

Table 9.1
Electromechanical relays – coil impedance vs plug setting

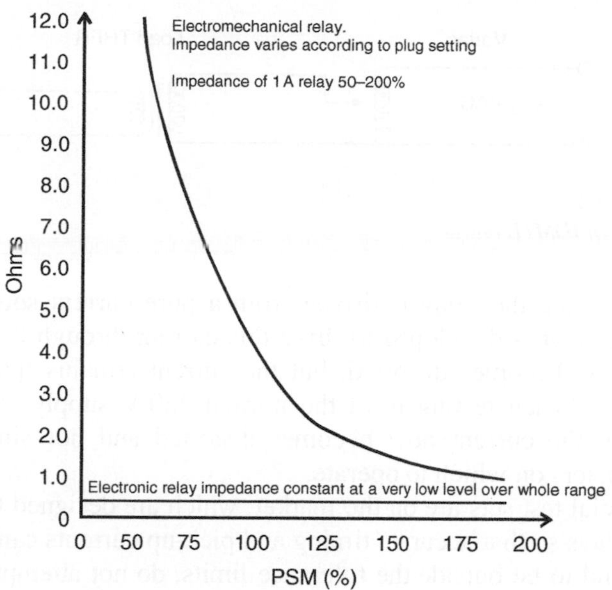

Figure 9.7
Impedance vs plug setting (electromechanical and digital)

9.2.4 General

Since an electrical system employs many relays, mechanical or electrical flag indication is provided in each relay to indicate whether that relay has operated to indicate the type of fault involved. Many modern relays are of the draw out type so that, the relay can be removed from its case even when the CT circuits are alive. This is possible as the associated CT terminals in the case are short circuited just before the relay contacts break whilst the relay is being withdrawn.

Certain models also have catches, which hold the relay in its case. When these catches are unlatched, the tripping circuit is opened so that accidental closing of the trip contacts will not trip the associated circuit breaker. This feature must not always be relied upon to prevent tripping as it does not necessarily isolate all tripping circuits and the feature is not present in all relays.

9.2.5 Testing of IDMTL relays

Modern relays are very reliable and in their dust proof cases, they remain clean. However, dirt and magnetic particles are the biggest cause of problems in electromechanical relays. Hence, when this type of relay is removed for testing, it should be covered, while not being actually tested. It should preferably be kept in a spare case or a plastic bag if stored or transported.

For operational tests, a load transformer and variac can be used to supply current, while a timer will indicate when the tripping contacts close. Typical connections are shown in Figure 9.8. This method can be used to check that the relay operates, that the flag drops correctly just as the contacts are made with slow disk operation and that the contacts do make positively with good pressure.

Caution: This method is not reliable for timing or pick-up tests. A proper relay current test set is necessary for accurate tests as with this simple set up, distorted non-sinusoidal currents result because of the non-linear magnetic circuit of the relay.

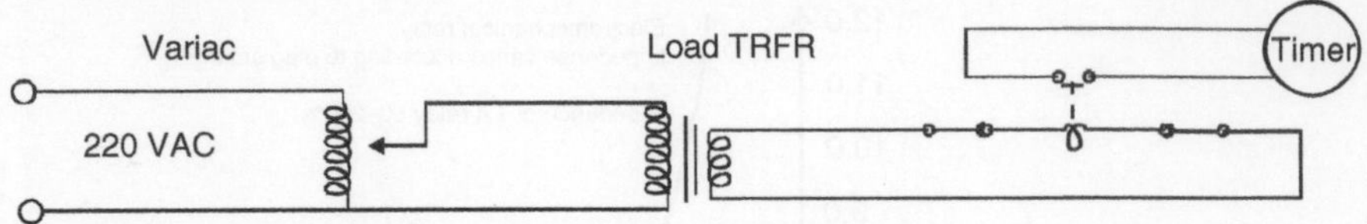

Figure 9.8
Testing of IDMTL relays

In service the relay is driven from a pure current source, namely the line CTs. The voltage that is developed to drive this current through the non-linear magnetic circuit of the relay becomes distorted, but the current remains 'pure' and faithful to the primary current. When testing from the normal 220 V supply, we have a pure voltage source. Hence, the current now becomes distorted and non-sinusoidal, giving the relay false parameters on which to operate.

Special test sets are on the market, which are designed to inject sinusoidal currents into the relays so that accurate timing and pick-up currents can be recorded. If the relay timing is found to be outside the tolerance limits, do not attempt to rectify this by adjusting the spiral hairspring at the top of the disk shaft, as this could upset the whole characteristic. This spring should only be adjusted by trained relay service technicians when checking for 'disc creep' and this together with adjustments of the magnets, determine the accuracy of the timing characteristics (see Figures 9.9– 9.11).

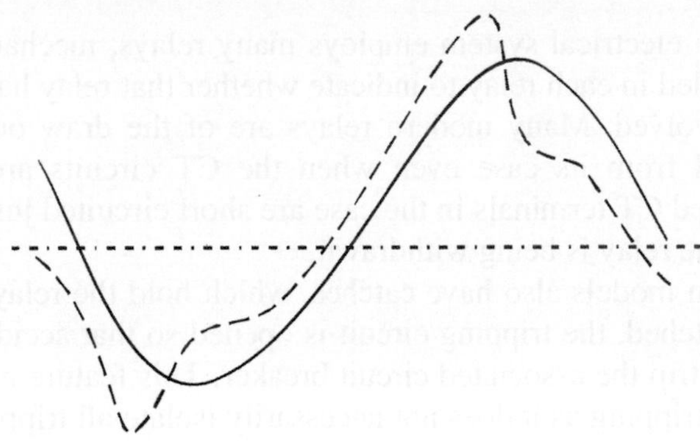

Figure 9.9
Typical waveforms when relay driven from CTs in service

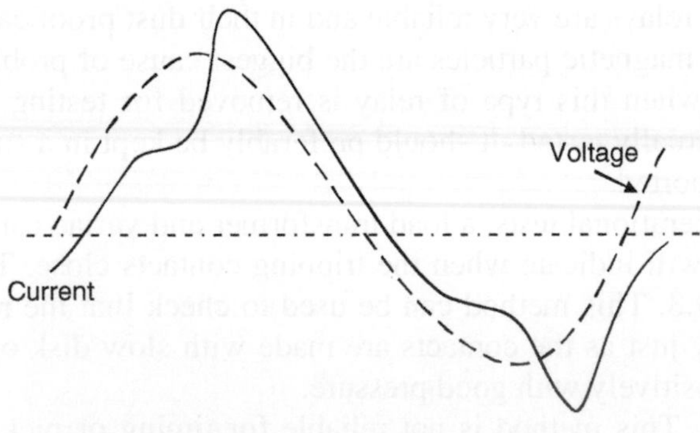

Figure 9.10
Typical waveforms when relay driven from plain 240 V aux. supply and table of errors

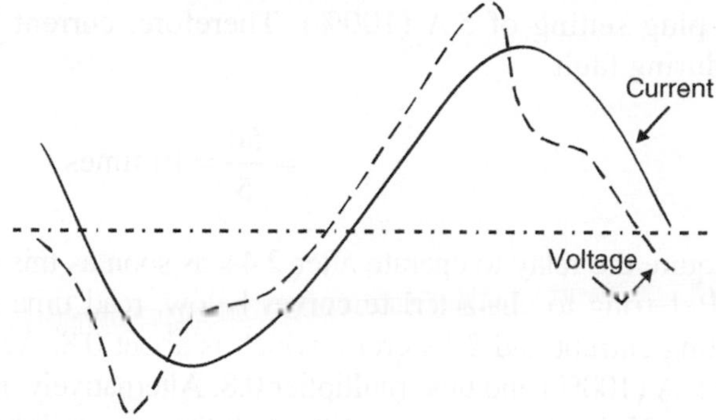

Figure 9.11
A 5 A earth fault relay supplied from 200/5 A current transformers

$I_{\text{relay}} = I$ A, plug setting = 0.5 A. The following table gives the margin of errors in the test results based on the testing source.

Voltage Source vs Current Source Testing of Mechanical Relays									
	5 A Overcurrent			5 A Earth Fault			1 A Earth Fault		
	25 A Plug Setting			0.5 A Plug Setting			0.2 A Plug Setting		
Times Current	Time		Error	Time		Error	Time		Error
	Voltage Source	Current Source		Voltage Source	Current Source		Voltage Source	Current Source	
×2	9.15	11.27	23%	8.99	11.6	29%	10.32	12.15	18%
×3	5.85	7.44	27%	6.04	7.31	21%	6.63	8.1	22%
×5	4.25	4.97	17%	4.15	5.07	22%	4.59	5.5	20%
×10	2.85	3.2	12%	2.79	3.21	15%	3.17	3.58	13%
×20	2.1	2.2	5%	2.01	2.21	10%	2.34	2.46	5%

Setting of an IDMT relay

Example: Calculate the plug setting and time multiplier setting for an IDMTL relay on the following network so that it will trip in 2.4 s (see Figure 9.12).

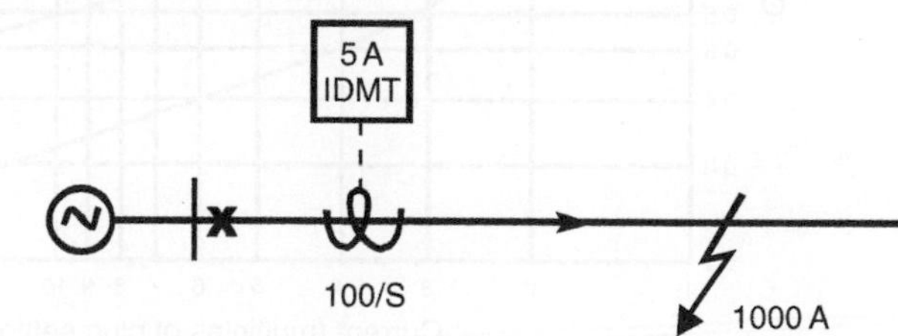

Figure 9.12
Example of calculation of plug and time multiplier setting

Answer:

$$\text{Fault current} = 1000 \text{ A}$$
$$\text{CT ratio} = 100/5 \text{ A}$$

Hence expected current into relay under fault conditions,

$$= 1000 \times \frac{5}{100} = 50 \text{ A}$$

Choose plug setting of 5 A (100%). Therefore, current into relay as a multiple of plug setting during fault:

$$= \frac{50}{5} = 10 \text{ times}$$

We require the relay to operate after 2.4 s as soon as this much current starts flowing in the circuit. Referring to characteristc curves below, read time multiplier setting where 10 times plug setting current and 2.4 s cross, which is about 0.8. Accordingly, relay settings = current plug tap 5 A (100%) and time multiplier 0.8. Alternatively, if the current plug setting is chosen as 125% (6.25 A), the fault current through the relay will be 50/6.25 = 8 A. The graph shows that eight times plug setting to operate in 2.4 s, the time multiplier should be about 0.7. This technique is fine if the required setting falls exactly on the TM curve. However, if the desired setting falls between the curves, it is not easy to estimate the intermediate setting accurately as the scales of the graph are log/log. The following procedure is therefore recommended (see Figure 9.13).

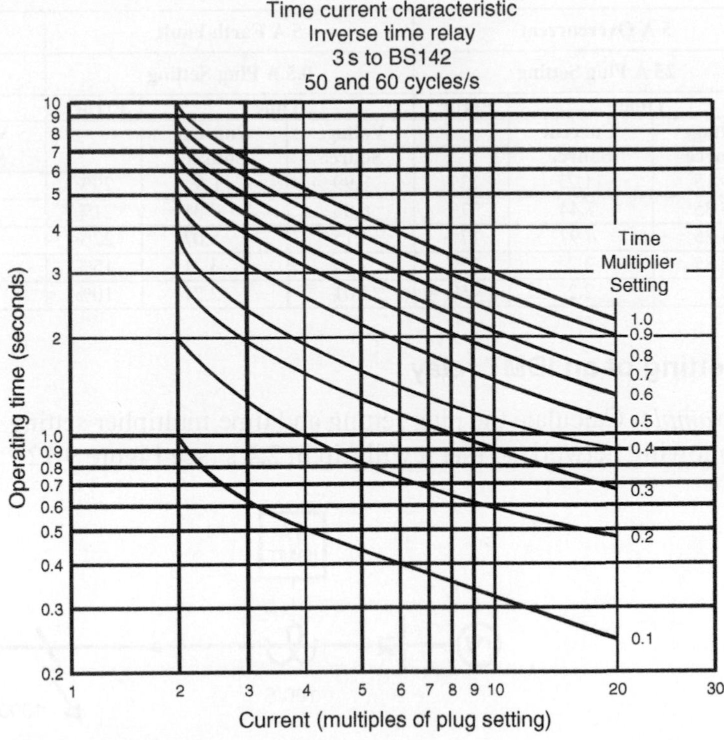

Figure 9.13
Multiples of plug setting current

Go to the multiple of plug setting current and read the seconds value corresponding to the 1.0 time multiplier curve. Then divide the desired time setting by this figure. This will give the exact time multiplier setting:

Seconds value at 10 times = 3 (at 8 times it is about 3.4)
Desired setting = 2.4
Therefore time multiplier = 2.4/3 = 0.8 or 2.4/3.4 = 0.7 in the second case.

9.3 Factors influencing choice of plug setting

1. *Load conditions*: Must not trip for healthy conditions, i.e. full load and permissible overloads, re-energization and starting surges
2. Load current redistribution after tripping
3. *Fault currents*: High fault currents can cause saturation of CTs. Choice of CT ratio is important
4. *CT performance*: Magnetization curve. Its internal resistance
5. *Relay burden*: Increases at lower taps on electromechanical relays
6. *Relay accuracy*: Better at top end of curve. Attempt to use in tight grading applications.

9.4 The new era in protection – microprocessor vs electronic vs traditional

9.4.1 Background

Electromechanical relays of various types have been available from the earliest days of electrical power supply. Some of these early designs have been improved over the years. One of the most successful types of electromechanical protection relays has been the previously discussed inverse definite minimum time (IDMT) overcurrent relay based on the induction disk. With the introduction of electronic devices such as the transistor in the 1950s, electronic protection relays were introduced in the 1960s and 1970s. Since then, the development of relays has been related to the general development of electronics.

By the late 1960s, extensive experience in the use of electronics in simple protection systems enabled the development of many quite advanced protection schemes and the first high-voltage substations were equipped with static protective relays. Over a period, these have been extended to cover other equipments such as transmission lines, motors, capacitors and generators. New measuring techniques have been introduced, measurements that are more accurate can be performed and high overall quality, reliability and performance of the protection system for high-voltage power systems have been reached.

Developments in the 1970s concentrated on improving reliability through improved design of printed circuit boards leading to integrated circuits and general improvements in substation designs, particularly earthing. In general, most static protective relays of that time were designed to match or improve on the basic electromechanical performance features. Improvements introduced included low-current transformer burden, improved setting accuracy and repeatability as well as improved speed. Also, during this period, experiments were conducted in Europe, Japan and the USA to test computer-based protection systems based on the availability of digital electronics.

This is particularly true with IDMT overcurrent relays, where it was both difficult and expensive to provide the inverse time characteristics by means of analog electronic circuits. However with the advent of microprocessors, it is much easier to provide the most commonly used characteristics such as definite time, normal inverse, very inverse, extremely inverse and thermal characteristics using different algorithms stored in the microprocessor's memory.

The overcurrent relay is undoubtedly the most common type of protection relay used by electricity supply authorities for protection on distribution systems. This chapter concentrates on the various features of modern static overcurrent protection relays in relation to the older electromechanical relays, which are still commonly used on distribution systems today. The purpose is to clarify some of the arguments for and against static protection relays, particularly for medium-voltage applications.

What is a static protection relay?

Static relays are those in which the designed response is developed by electronic or magnetic means without mechanical motion. This means, that the designation 'static relay' covers the electronic relays of both the analog and digital designs. The analog relays refer to electronic circuits with discrete devices like transistors, diodes, etc., which were adopted in the initial stages. However, the digital designs incorporate integrated chips, microprocessors, etc., which had been developed subsequently. In recent years, very few relays of the analog type are being developed or introduced for the first time.

Most modern overcurrent relays are of the digital type. There are many reasons for this, the main ones being associated with cost, accuracy, flexibility, reliability, size, auxiliary power drain, etc. Many of these reasons will become evident during the course of this chapter, which will concentrate on relays of the digital type. Microprocessor relays are of the digital type.

The main objective of using static relays is to improve the sensitivity, speed and reliability of a protection system by removing the delicate mechanical parts that can be subject to wear due to vibration, dust and corrosion. During the early development of static relays, the use of static components were particularly attractive for the more complicated relays such as impedance relays, directional relays, voltage regulating relays, etc. On the other hand, the early static IDMT overcurrent relays were expensive because it was difficult to match the inverse time characteristic using analog protection circuits. The battery drain associated with these static IDMT relays was too high and this discouraged the use of this type of relay for medium-voltage applications. The general developments in the field of electronics and the introduction of digital circuits have overcome many of the above problems. Using modern microprocessor relays, almost any characteristic is possible and economical, even for the simplest applications such as, overcurrent relays and motor protection relays.

What is a microprocessor relay?

A microprocessor relay is a digital electronic relay, which derives its characteristics by means of a pre-programed series of instructions and calculations (algorithms), based on the selected settings and the measured current and/or voltage signals. For example the formula used to derive the inverse time characteristics in an overcurrent relay that comply with IEC 255 and BS 142 is mathematically defined as follows:

$$t[s] = \frac{k \times \beta}{(I/I>)^{\alpha} - 1}$$

Where

T = operating time in seconds
K = time multiplier
I = current value
$I>$ = set current value.

The unit includes four BS 142 specified characteristics with different degrees of inverse. The degree of inverse is determined by the values of the constants α and β.

Degree of Inversity of the Characteristic	α	β
Normal inverse	0.02	0.14
Very inverse	1.00	13.50
Extremely inverse	2.00	80.00
Long-time inverse	1.00	120.00

A description of a typical microprocessor (or numerical) relay follows – which includes:

- Introduction
- The simplified block diagram
- Handling of the energizing signal
- The microprocessor circuits
- The output stages
- The self-supervision.

Introduction to the numerical relay

The measurement principle is based on sampling of the energized currents or voltages, analog to digital conversion and numerical handling, where all settings are made in direct numerical form in a non-volatile memory. Setting can be performed either manually on the relay front or by serial communications using either a personal computer or a control/monitoring system. In addition, the operation of the self-supervision is described (see Figures 9.14 and 9.15).

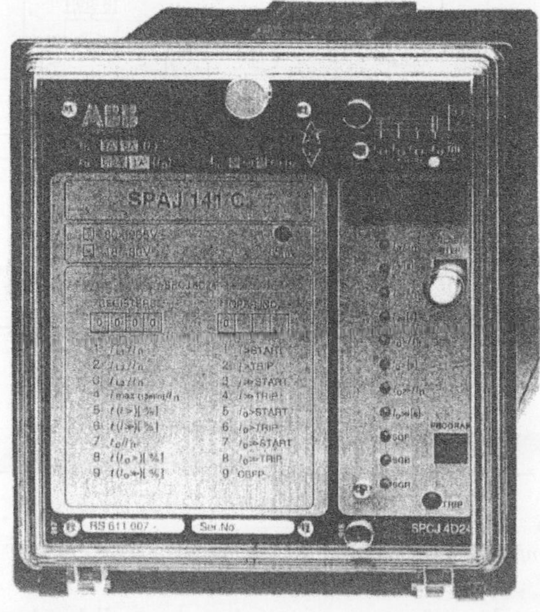

Figure 9.14
A typical microprocessor relay

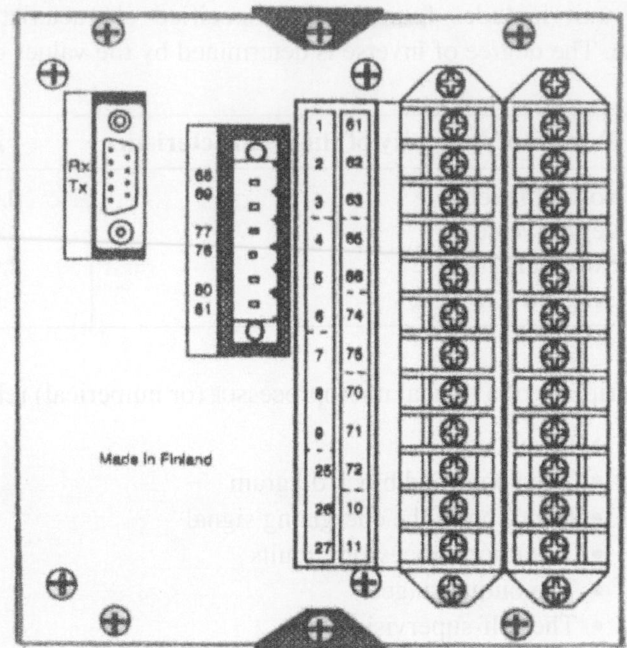

Figure 9.15
Rear view of a microprocessor relay

Simplified block diagram is shown in Figure 9.16.

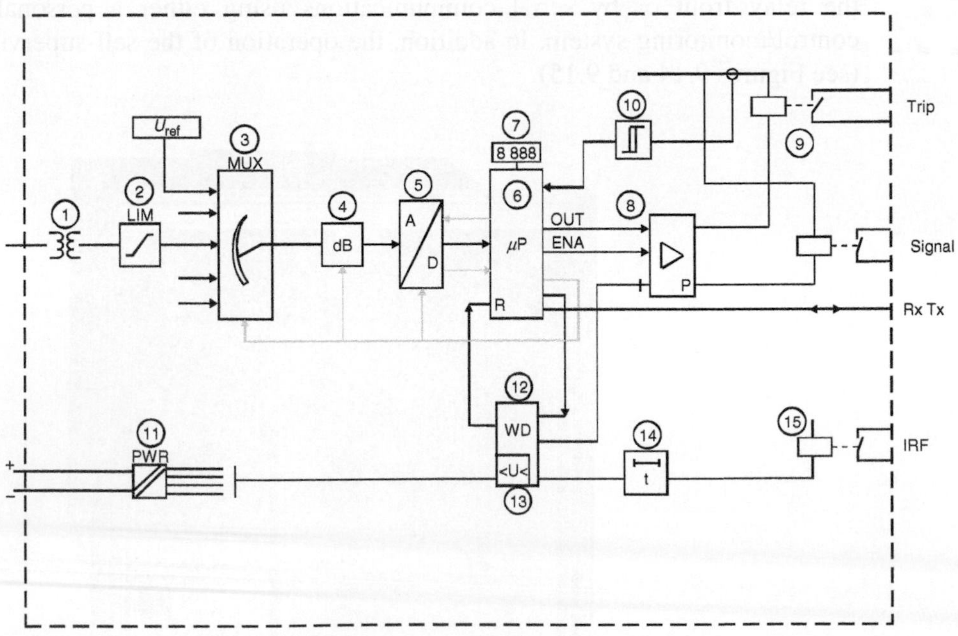

Figure 9.16
A typical simplified block diagram for a relay

The operation of a protective relay can well be described by using a simplified block diagram as shown above. Here we can recognize the input signal path with the signal processing parts, the output circuits for trip and signal, and the self-supervision circuits.

9.4.2 Handling of the energizing signal

The basic function of the relay is to measure the input and assess its condition. A digital relay comprises of sensitive devices and hence it is necessary that they do not fail because of the input changes. This is taken care by the isolating transformer and the limiter used in the relay.

The energizing currents or voltages are brought into the relay using the internal matching transformer (1). After the transformer there is a voltage limiter (2), which cuts the voltages entering the relay at a safe level, should there be an extremely high-current or voltage input. The reasons for this limiting are only to protect the internal circuits of the relay from being destroyed by too high an input voltage. Together with the limiting circuit, a filter can also often be implemented. In this case, the reason is that the harmonics contents of the energizing signal is not wanted and is filtered out. A typical filtering level here is that the third harmonics are attenuated by a factor of 10 and the fifth harmonics or higher are attenuated by factors of 100 and more.

The next stage is to measure the signals, which are to be monitored. In an AC circuit, the voltage, current and power undergo changes in relation to the supply frequency. The multiplexer (3) selects the signals that are to be measured. A sample of each signal is measured once per ms. If the measuring module is of type C, with setting knobs, the setting values are also read through the multiplexer. Finally, the multiplexer also selects a reference channel once per second to have a condition check for the input circuits.

The analog-to-digital-converter (A/D) (5) measures the level of the measured samples and transforms the analog values into a numerical form. The resolution of the A/D converter is 8 bits, representing numerical values 0–255 ($2^0 - 1$ to $2^8 - 1$).

Because the dynamic range of the signal levels to be measured is quite high, the 8-bit conversion is as such not enough to give a good accuracy over the whole current or voltage span. Therefore, a programmable attentuator (4) is needed between the multiplexer and the A/D converter to enable an accurate handling for both low and high current or voltage levels.

For low signals, the amplification is 1, passing the signal directly to the converter. For phase current measurements, e.g. at a current level of 1.00 × In, the numerical value from the A/D is 100. When the signal is too high to be handled directly, i.e. for values above 200–255, an attenuation of ×5 is put on the signal. Furthermore, if the signal is still too high, the attenuation is switched up to ×25. This means that the overall range of measuring capability is 0–6375, corresponding to more than a 12-bit conversion. When the numerical value 100 represents a current of 1.00 × In, the highest measurable current is thus 63 × In.

The sampling rate is typically 1 ms, which means that every half-period of the 50 Hz sine wave is measured by ten samples (see Figure 9.17). This gives a very accurate measure of the peak value during the half-period. For the worst case of the two top samples hitting evenly on both sides of the exact peak, a maximum error could theoretically be about 1%. As all signal handling is based on the mean value of two consecutive half-waves, and the sampling is non-synchronized, the probable theoretical fault is less than 0.5% and is partially compensated in the unit calibration.

As the measured current now is available in a numerical form, several things can be made. For example, the problem with handling of a signal with a DC-component is now very straightforward. When the mean value of two consecutive half-waves is calculated, the DC-component is eliminated to almost 100% without any need for non-linear air gap transformers or similar components (see Figure 9.18). On the other hand, the calculation of the mean value consumes 10 ms, which is not wanted for very high short-circuit current levels where an instantaneous trip is called for. For this case another trip criteria is simply added. If the current in the first half-wave exceeds twice the setting, it is obvious that the mean value of the two half-waves will exceed the set level and therefore a trip can be carried out instantaneously without the need to wait for the next half-wave.

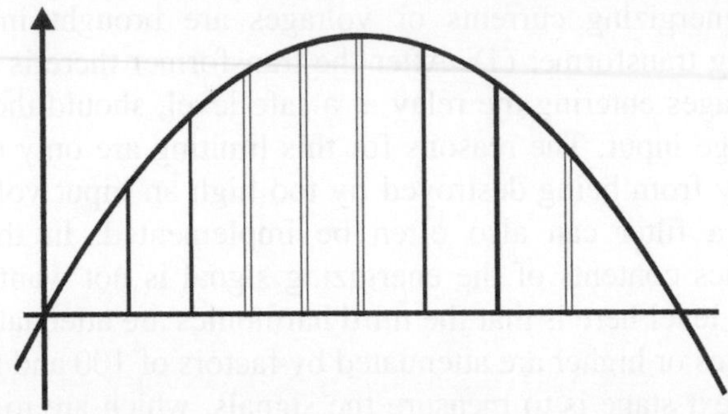

Figure 9.17
Sampling of the sine wave at 1 ms intervals

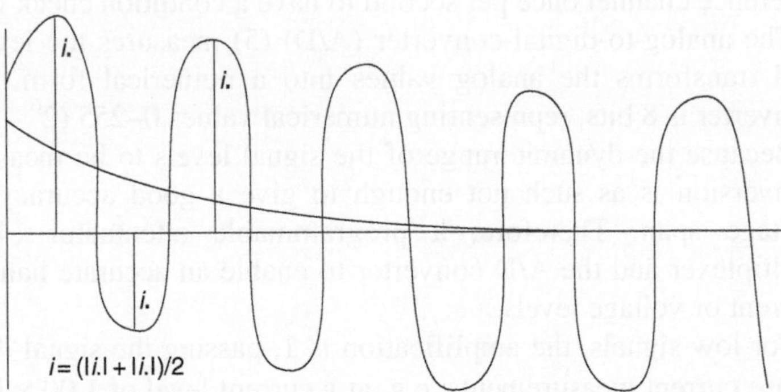

Figure 9.18
Elimination of the DC-component by mean value calculation of two consecutive half-waves

All measured numerical values can of course easily be transferred over the serial communication, be stored in memory banks, etc: for later retrieval e.g. when fault reasons are being investigated. The sampling is also used in another good way i.e. to minimize the transient over-reach. When the operating time for a stage has elapsed and the trip order is to be carried out, the stage will wait for still one single sample exceeding the set level before the trip is linked to the output relay. In this triggered state, the relay will wait for a short time and if no further samples are detected, the relay will reset. This means that the retardation time or the transient over-reach is very short, less than 30 ms.

9.4.3 The microprocessor circuits

In the entire signal handling stages, the microprocessor or more accurately the microcontroller (6) of course takes part as every operation is controlled by this component. Furthermore, all protective decisions are made here, the operating time is counted for every stage and after that, the outputs are linked to the output relays.

The display (7) used for the MMI is also served by the controller.

External components to the controller which are not shown in the figure are also the three different memory components that are used:

1. The random access memory (RAM), which is used as a scratchpad for measuring and calculating results, storing of memorized values, etc.
2. The read-only memory (ROM), which contains the program firmware for the module.
3. The electrically erasable programmable read-only memory (EEPROM), which is being used as parameter storage memory, e.g. for all the setting values.

9.4.4 The output stages

The outputs from the module is linked from the microcontroller, via a power buffer amplifier (8) to the electromechanical output relays (9) on the output card.

To prevent a disturbance condition from causing a false output, a double signal arrangement is used. This means that an enable signal (ENA) must be sent together with the output command before any output relay can be activated. Furthermore, the outputs can also be inhibited by self-supervision. The mechanical operating times of the output relays, typically about 10–15 ms are subtracted from the operate times for the definite time protective stages. Thus, very accurate timings can be achieved for these operations. For IDMT operations, this is however not possible as you cannot decide when you have reached the trip instant – 10 ms for all possible variations of current.

9.4.5 Self-supervision

In order to avoid false operations due to relay internal faults and to maximize the overall availability of the protection, a set of auto-diagnostic circuit arrangements have been implemented in the relay modules. Generally, all tests are performed during a period of 6 min.

The different memory circuits, i.e. the RAM, the ROM and the EEPROM are all continuously tested by different methods at a speed of about one byte per 10 ms. Thus all memory is cyclically checked out.

The microprocessor and the program execution are supervised by a watchdog (12) once every 5 ms. The multiplexer, the switchable attenuator and the A/D converter are tested by measuring of a very accurate reference voltage once a minute and always before tripping. This is to ensure that a measured signal is real, and not caused by a fault or disturbance in some input circuit, all to avoid false outputs.

The settings are tested once a minute by the use of a checksum arrangement. It is obvious that the settings must be ensured under all conditions of power breakdown, etc. Any kind of battery backup must also be avoided as all batteries have too short a lifetime for this purpose. A relay is designed to have a lifetime of 20 years but the best life batteries cannot survive much more than about 5 years.

The solution is to have all settings stored in a non-volatile memory (EEPROM) and in two identical subsets (see Figure 9.19). For both the setting memory areas, the microcontroller automatically keeps track of the exclusive or checksum for the whole block. This checksum is stored in the last memory place of the block and is used as a reference of the correctness of the contents in the memory block. If a fault is detected in one block, the controller will directly check the other one and replace the contents of the faulty block with the contents of the correct one. This means that the settings are self-correcting, for e.g. ageing/refreshment faults due to the properties of the EEPROM. Normally a refresh cycle should be carried out every 10 years, in this way the relay module makes the refreshment when needed.

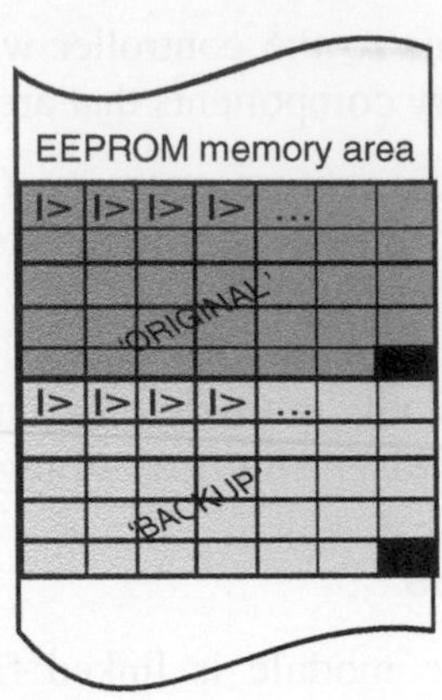

Figure 9.19
The two memory areas allocated for keeping the relay settings secured

The internal supply voltages from the power supply module (11) are all tested once a minute by measurement of the different supply voltages, +8 V, ±12 V and +24 V. This test is performed by the voltage limit detector (13).

The trip output path, the output amplifier and the output relay are checked once a minute by injecting a 40 ms voltage pulse into the circuit and checking (10) that a current flows through the trip circuit.

If the self-supervision detects a fault, the output buffer amplifier (8) is immediately blocked to ensure that no false signals are carried out due to the fault condition. After this, the watch-dog tries to get the microcontroller to work properly again, by resetting the process three times. If this attempt is not successful, a signal about the internal relay fault (IRF) is sent after a time delay (14) linked to the output relay (15). Furthermore information about the fault is sent as an event over the serial communication RxTx and a red LED on the front of the module is activated. If the module is still in an operative condition, also an indication about the character of the fault is shown as a code number in the front display.

Even a full breakdown of the relay, e.g. by loss of power supply will be detected as the IRF relay (15) operates in a fail-safe mode, causing a signal when the relay drops off. In addition, the serial communication will indicate loss of contact to the module and later on the module goes into a reset condition.

9.5 Universal microprocessor overcurrent relay

Electromechanical relays are designed specifically for particular protection applications and they usually have a limited setting range. For example, a different relay is necessary when a 'very inverse' characteristic is required or if a setting is required that is outside the range of the standard relay. This means that at the time when an electric power system is being designed and specified, considerable thought must be given to both the type of protection characteristic that will be required and the likely setting of the relay to ensure that the correct relay is specified.

The concept of many modern microprocessor relays is to provide a protection relay that covers all likely protection requirements in one relay.

This includes wide setting ranges and, in addition, several selectable characteristics and options to cover many protection applications. Microprocessor overcurrent relays are typically selectable for definite time, normal inverse, very inverse, extremely inverse, longtime inverse and sometimes a thermal characteristic as well to cover all likely application requirements. In addition, several output options are often provided to enable the user to select, for example, whether he requires an overcurrent 'starting' output contact or not. From a user's point of view, this delay in decision characteristic and setting range is required to the time of commissioning.

The concept of a universal relay tends to improve the availability of protection relays from the manufacturers by making them 'stock' item. From a manufacturing point of view, this minimizes the number of relay types that have to be manufactured and held in stock and allows to provide a faster and better service to the users of protection relays. This also tends to reduce the cost of protection relays by reducing the number of variations.

Table 9.2 summarizes the available characteristics and setting ranges of a modern microprocessor overcurrent relay in comparison to a typical induction disk IDMT overcurrent relay.

		Static (Digital)		Electromechanical
Characteristics		Selectable		Separate relay
		Definite time	or	Definite time
	and	Normal inverse	or	Normal inverse
	and	Very inverse	or	Very inverse
	and	Extremely inverse	or	Extremely inverse
	and	Long time inverse	or	Long time inverse
Current inputs		1 A and 5 A		1 A or 5 A
Thermal current withstand				
Continuous	:	3 A/15 A		2 × setting current
For 10 s	:	25 A/100 A		–
For 3 s	:	–		20 A/100 A
For 1 s	:	100 A/300 A		–
Overcurrent setting		Continuous 50–500%		Plug setting 50–200% in 7 steps
Earth fault setting		Continuous 10–80%		Plug setting 10–40% in 7 steps
			or	20–80% in 7 steps
Time multiplier		Continuous 0.05–1.0		Continuous 0.1–1.0
High-set overcurrent		Included 0.5–40 times		Extra add-on
High-set time delay		Included 0.05–300 s		Extra add-on

Table 9.2
Comparison of microprocessor vs electromechanical relays

9.6 Technical features of a modern microprocessor relay

9.6.1 Current transformer burden

One of the disadvantages of the IDMT relays of the induction disk type is that they have relatively high CT burdens when compared to static IDMT relays. The ohmic value of these burdens varies with the setting as shown in Table 9.3. As the setting is reduced, the burden on the CT is increased. Induction disk relays have a burden typically specified as 3 VA. Modern static relays, on the other hand, have a very low burden of less than 0.02 Ω for 5 A input and 0.10 Ω for 1 A input, which is independent of the setting. The table below shows the calculated ohmic burden of a 1 A induction relay at the various settings compared to a microprocessor overcurrent relay. (Also refer Table 9.1 in the beginning of this chapter.)

The main consequence of the high burden is the poor performance of the CT/relay combination under high fault current conditions, particularly when low CT ratios are used. The high burdens can affect the actual primary setting achieved by the CT/relay combination. The example below shows that, with an electromechanical relay, the actual primary setting increases even though the plug setting is reduced on the relay.

Setting (%)	Induction Disk Relay Burden (Ω)	Microprocessor Relay Burden (Ω)
10	300	0.10
15	133	0.10
20	75	0.10
25	48	0.10
30	33	0.10
35	24	0.10
40	19	0.10

Table 9.3
Comparison between the CT burdens in Ω of equivalent 1 A relays of the induction disk and microprocessor types

With the static relay, almost any primary setting is possible. This means that on a distribution network using static relays, relay coordination is still possible at high fault levels even for a very low relay current setting and low CT ratios (see Figure 9.20(a)).

9.6.2 Accuracy of settings

The current and time-multiplier settings on a microprocessor relay are done with the aid of a digital display, which is part of the measuring unit. The accuracy and repeatability of the settings on this type of relay is far greater than that for electromechanical relays. Setting accuracys of ±1% and operating accuracys of ±3% of set value for the static relay compare very favorable with the ±7.5% accuracy of the electromechanical device. The accuracy of the electromechanical relay is also dependent on the frequency, and the presence of harmonics further affects accuracy.

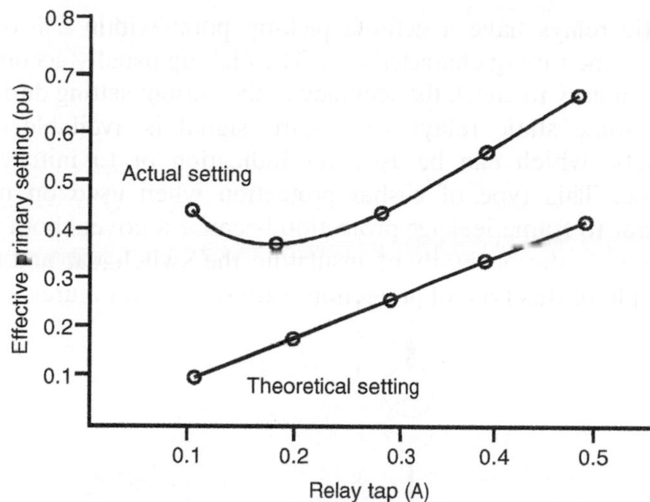

Figure 9.20(a)
Primary to relay tap setting

This greater accuracy and repeatability of the static relay, generally independent of harmonics, combined with negligible 'overshoot' means that reduced grading intervals are now possible, especially when these relays are used in combination with the faster operating SF6 and vacuum switchgear. This is clear when one recalls that the grading times are dependent on the following:

- Errors in CTs
- Errors in the relay operating time
- Relay 'overshoot' time
- Circuit breaker operating time
- Safety margin.

It is practical to consider grading intervals of as low as 0.2 s when using microprocessor relays in combination with SF6 or vacuum breakers as compared to 0.4 ~ 0.5 s needed with electromechanical ones.

9.6.3 Reset times

Electromechanical IDMT relays have reset times of up to 10 s at time multiplier settings = 1, which means that during auto-reclose sequences an integration effect can take place and coordination can be lost. This situation can occur when the disk has turned some distance in response to a fault in the network cleared possibly by some other breaker with an auto reclose feature. If the fault is still present when the breaker recloses and if the disk has not fully returned to its reset position, the relay would take less time than calculated to trip. Uncoordinated tripping is then possible. The reset times of static relays are negligible.

9.6.4 Starting characteristics

An IDMT relay of the induction disk type is an electromechanical device, which includes mechanical parts such as a disk, bearings, springs, contacts, etc., which are subject to some mechanical inertia. When the current exceeds the setting, the disk only starts to move somewhere between 103 and 110% of the setting and closes for currents between 115 and 120% of the setting.

Static relays have a definite pick-up point within 5% of the current setting and this initiates the timing characteristics. The pick-up usually accompanied by an LED indication, makes it easy to check the accuracy of the current setting during testing of the relay.

On some static relays, this 'start' signal is available on a separate pair of output contacts, which can be used for indication or to initiate a simple busbar protection scheme. This type of busbar protection when used on metal clad MC switchgear is superior to frame leakage protection because it covers both phase-faults in the switchgear and avoids the necessity of insulating the switchgear and cable glands from earth. The principle of this type of protection is illustrated in Figures 9.20(b), (c).

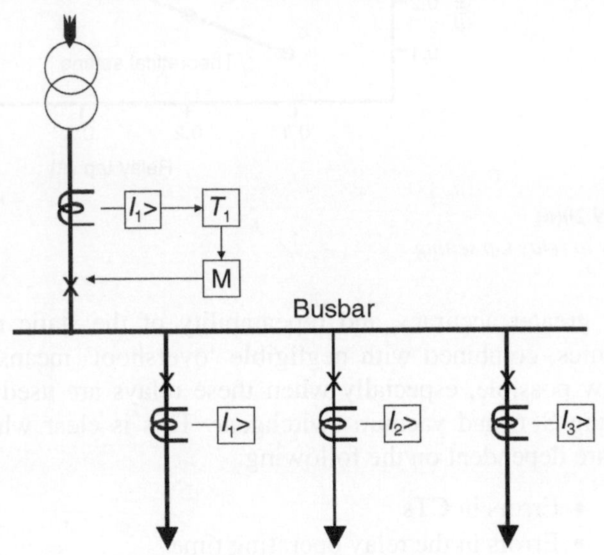

Figure 9.20(b)
Busbar protection scheme using starting contact of the static overcurrent relays

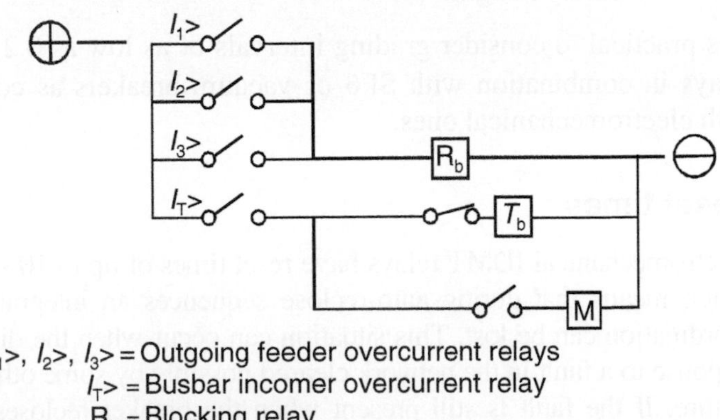

Figure 9.20(c)
Busbar protection scheme using starting contact of the static overcurrent relays

9.6.5 Dual setting banks

Some digital relays are now designed to provide a dual settings bank, which provides a complete duplication of all the settings and operating switch positions. Setting 1 or setting 2 can be selected at the relay, via the serial communications system or a remote switch, which can be an output contact of another relay or a circuit breaker auxiliary switch.

In many instances, when setting relays, such as the example shown in Figures 9.21(a), (b), we are faced with having to set a relay for the lowest or an average value of two possible settings. Now we can have both settings, calculated exactly, and switch from setting 1 to setting 2 at will.

This dual setting bank can also be useful in a ring main circuit, which can be opened at different places, necessitating differing settings when a relay can be in two different places in two radial feeders.

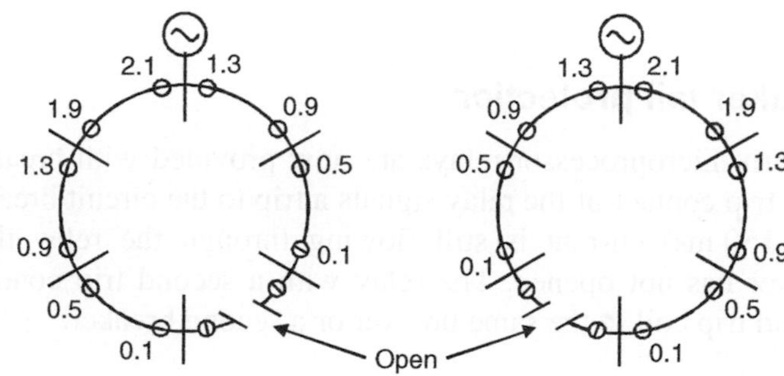

Figure 9.21(a)
Open ring protection

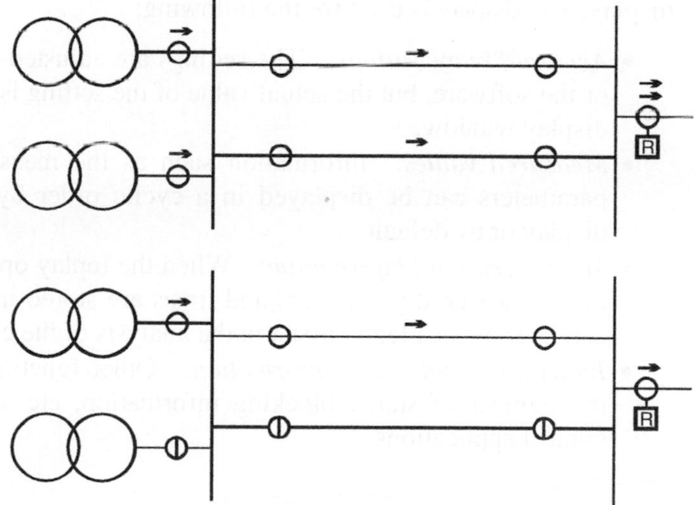

Figure 9.21(b)
Parallel feeder protection

9.6.6 High-set instantaneous overcurrent element

In microprocessor overcurrent and earth fault relays, a high-set overcurrent element is provided as a standard feature and often has a timer associated with it to provide a time delay. If not required, it can be set to be 'out-of-service'. Because of the measurement method, the transient overreach is very low and the instantaneous overcurrent setting can be set much closer to the maximum fault current for a fault at the remote end of the feeder. The transient overreach is the tendency of the relay to respond to the DC-offset, which is commonly present in most fault current waveforms. To avoid this problem on electromechanical relays, the setting of the high-set element has to be at least twice the calculated maximum fault current, making the protection less effective.

High-set overcurrent protection is particularly useful on the higher voltage side of a transformer, where it provides fast protection for most faults on the HV side while the time-delayed overcurrent relay provides protection for the faults on the lower-voltage side of the transformer.

9.6.7 Breaker fail protection

Modern microprocessor relays are now provided with breaker fail protection. When the main trip contact of the relay signals a trip to the circuit breaker and if after pre-set delays (say 150 ms) current is still flowing through the relay this indicates that the circuit breaker has not opened. The relay with a second trip contact sends another signal to a second trip coil in the same breaker or a second breaker.

9.6.8 Digital display

Most of the present day microprocessor relays are now provided with alpha-numerical data display. The display is used for the following:

- *Accurate relay settings*: The settings are adjusted by means of potentiometers or the software, but the actual value of the setting is accurately displayed on the display window.
- *Measured values*: Information such as the measured values of the various parameters can be displayed in a cyclic order by selecting the sequence of display or by default.
- *Memorized fault information*: When the replay operates for a fault, the values of the measured parameters and times are stored in memory. This information can later be recalled to assist in the analysis of the cause of the fault.
- *Indications and status information*: Other functions and information such as the number of starts, blocking information, etc. can be displayed for motor control applications.

9.6.9 Memorized fault information

Microprocessor relays, which provide memorized fault information, have been available for some years but this information had been initially limited to the maximum value of the measured current or the most recent fault. The advancement in digital technology nowadays enable much more comprehensive fault analysis with up to quite a number of memorized values of the three-phase currents, zero sequence current, maximum demand current (15 min) duration of the start of the low-set and high-set overcurrent. There are possibilities to have these data stored in hard discs connected on a continuous manner for later retrieval (see Figure 9.22).

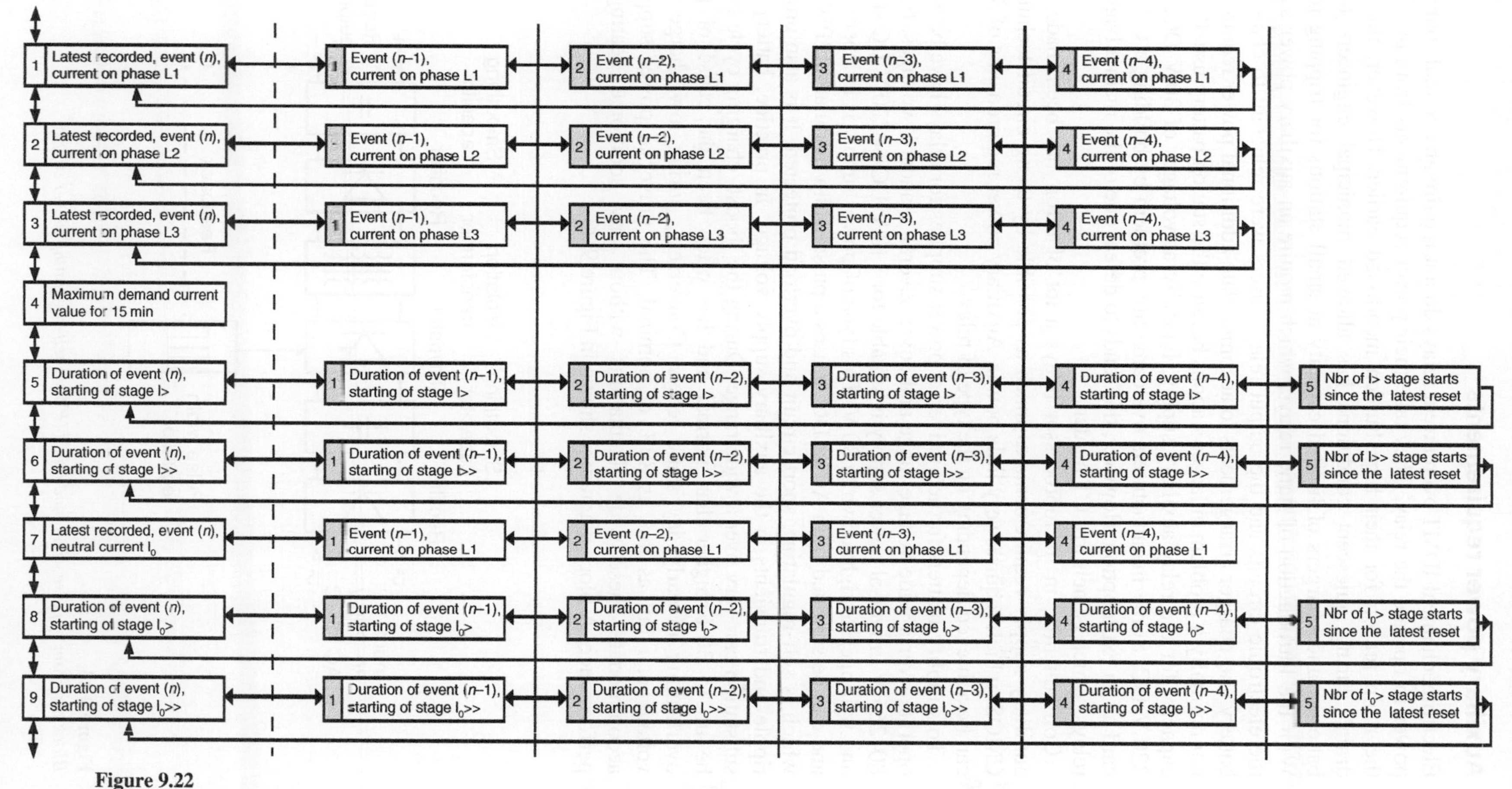

Figure 9.22

Example of memorized fault information

9.6.10 Auxiliary power requirements

Electromechanical IDMT overcurrent relays do not require an external source of auxiliary power to operate the relay. They take their power requirements from the CT and this is the main reason for their high burden mentioned earlier. However, this 'zero battery drain' during quiescent conditions has allowed municipal engineers to fit tripping batteries and chargers of limited capacity at small stations for tripping purposes only. With the introduction of static relays, which require an auxiliary power supply to drive the electronic circuits and the output relays, users were reluctant to change these small battery and charger arrangements to accommodate additional power requirements. This is not normally a problem in larger stations because the station battery usually has sufficient capacity for the relay auxiliary supply, typically at voltages of 30 V DC, 110 V DC or 220 V DC. Some manufacturers overcame this problem by building a CT power supply card as an extra option. However, this tends to defeat one of the main advantages of static relays, which is their low CT burden.

Consequently, in microprocessor relays a lot of effort has been made to reduce the auxiliary supply requirements as much as possible by using circuit techniques, such as CMOS, which requires very little power. Auxiliary power requirements of 3 W and lower can be achieved depending on the type of relay.

To simplify matters further, universal power supplies for relays have been developed to operate over a wide voltage range and cover several 'standard' voltages. For example an 80–265 V universal power supply is suitable for 110 V DC or 220 V DC station batteries and will operate right down to 80 V. This type of power supply is independent of polarity and can be supplied from AC or DC. It uses a pulse width modulation (PWM) technique, which is self-regulating, short circuit and overload protected. It is also protected against ripple and transients in the auxiliary supply voltage. In practice, battery voltages in a substation can vary over a wide range. During the 'boost' charging cycle the voltage can be up to 30% higher than normal and has often been the cause of power supply overheating in early day static relays. Conversely, during low charge situations the voltage can fall as low as 80% of nominal. The universal power supply can easily accommodate these wide fluctuations without any additional heating or loss of performance. A block diagram is shown in Figure 9.23.

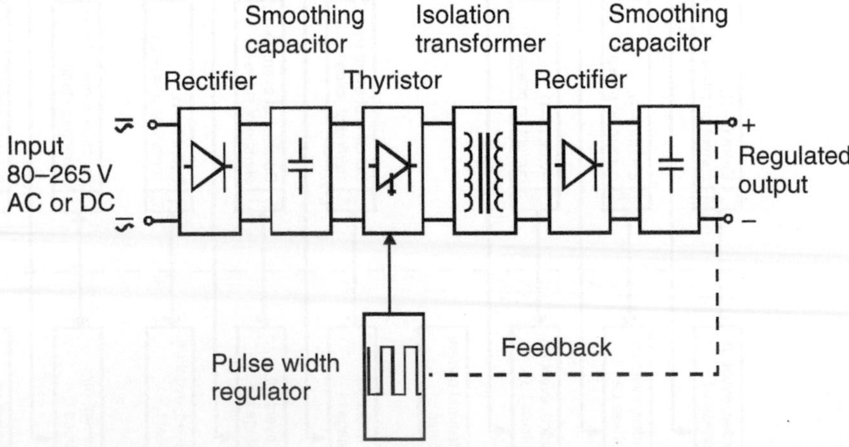

Figure 9.23
Block diagram of pulse width modulated (PWM) self-regulating power supply

Some of the benefits of PWM self-regulated universal power supply units are as follows:

- The same relay can be used in several applications for a wide range of battery voltages resulting in one power supply for all standard battery voltages from 30 to 220 V.
- Battery fluctuations due to the charger do not affect the replay performance.
- Low battery voltage, within reasonable limits, does not affect the relay performance.

In small stations where no station battery is available or economic, auxiliary supply can be arranged from a capacitor storage unit fed from both the CT and the PTs. This unit will provide auxiliary power to the relay even when no current is flowing in the primary circuit. Because of the relays, universal power supply, fluctuations of voltage due to the variations in the supply do not affect relay performance. A typical connection of a capacitor storage unit supplying a relay using a PWM self-regulating power supply is shown in Figure 9.24. The capacitor storage unit also provides the energy to trip the circuit breaker where there is no tripping battery.

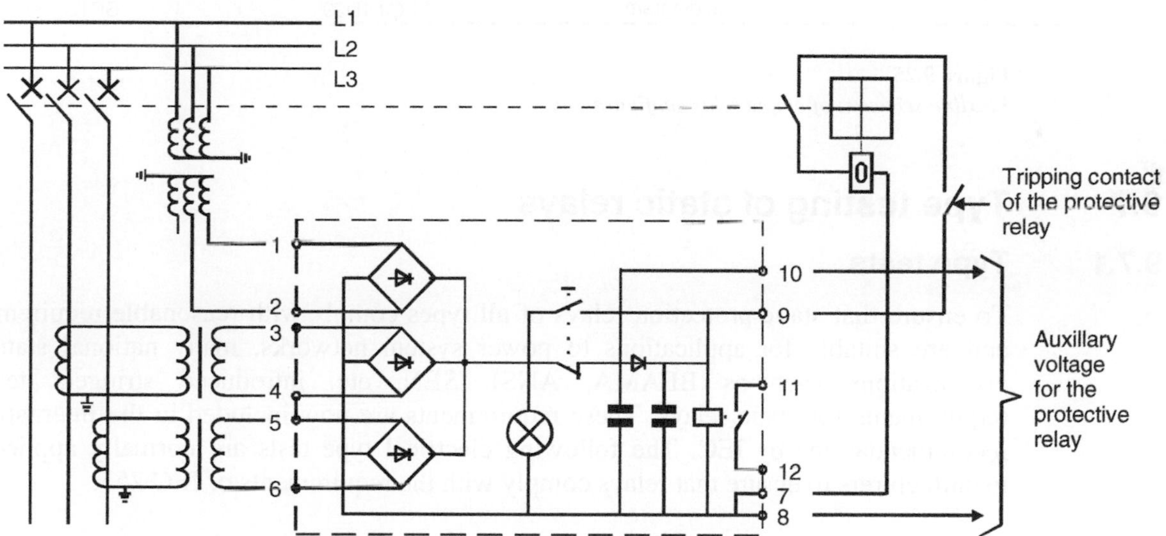

Figure 9.24
Capacitor storage unit is supplied from both the CTs and a CT to provide auxiliary power for the relay and for tripping the circuit breaker

9.6.11 Flexible selection of output relay configuration

With the help of six output relays (two heavy duty and three medium duty) and a completely flexible software-switching program, we can choose to have any function to operate any combination of output relays – including the various 'start' operations, see Figure 9.25.

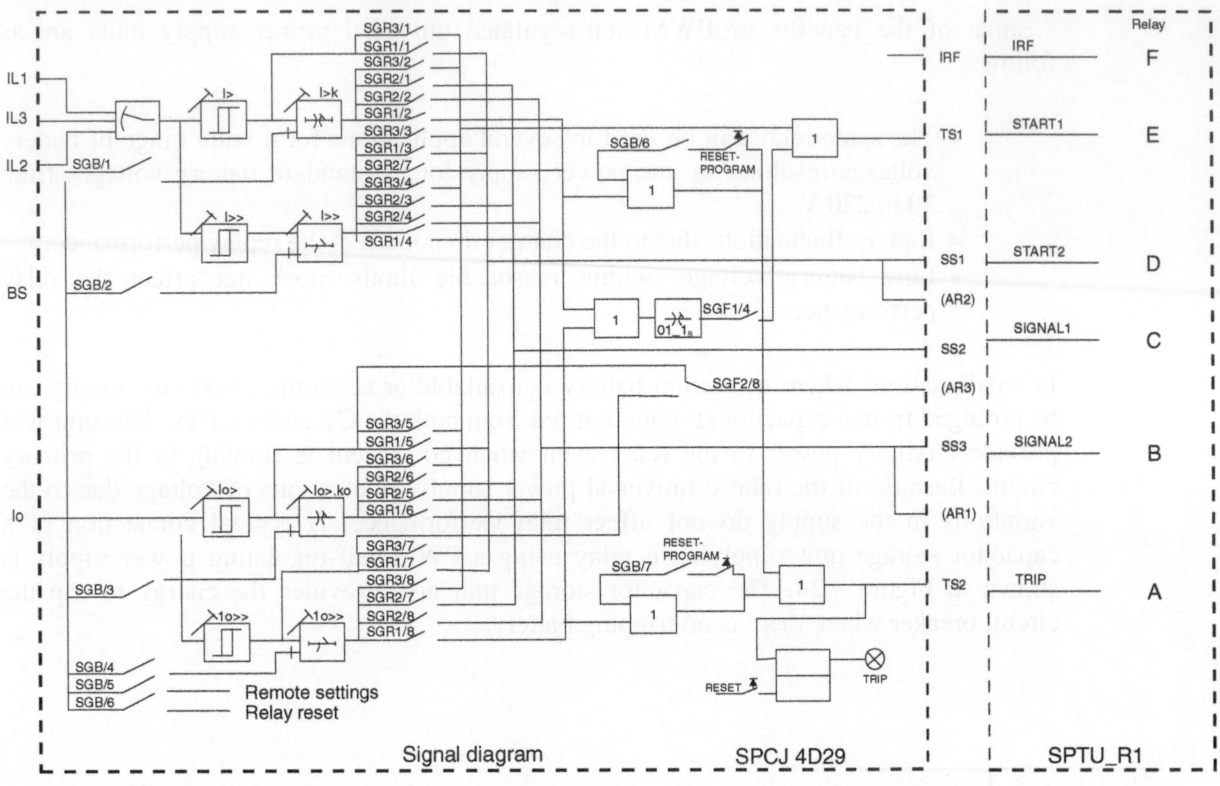

Figure 9.25
Flexible selection of output relay configurations

9.7 Type testing of static relays

9.7.1 Type tests

To ensure that static protection relays of all types comply with reasonable requirements and are suitable for applications to power system networks, many national standard organizations such as BEAMA, ANSI, SEN, etc. introduced stringent testing requirements for static relays. These requirements are now included in the international recommendations by IEC. The following electrical type tests are normally applied by manufacturers to ensure that relays comply with the requirements of IEC 255:

- *Insulation test voltage*: 2 kV, 50 Hz, 1 min IEC 255-5
- *Impulse test voltage*: 5 kV, 1.2/50 micro s, 0.5 J IEC 255-5
- *High-frequency interface test*: 2.5 kV, 1 MHz 255-6
- *Spark interference test*: 4–8 kV SS 436 15 03.

9.7.2 Self-supervision

Perhaps the most important feature introduced by microprocessor relays is that of continuous self-supervision. One of the classical problems of the older protection relays lies in the absence of any ready means to identify a fault in the relay. As protective relays are, for most of their lives, in a quiescent state, regular secondary injection tests are necessary to prove that the relays are operational.

The microprocessor relays, on the other hand, utilize their capacity during quiescent periods to continuously monitor their circuits and will provide an alarm if a failure occurs. The digital readout can be used to diagnose the problem. This enhances protection system reliability on a continuous basis and intervals between manual inspections can be prolonged.

Digital devices tend to work either 100% or not at all. Consequently, it is very easy to check a microprocessor relay on a regular basis and achieve a very high certainty that the relay is operational. By pressing the button requesting the display of the phase current, a reading that matches the ammeter on the panel confirms the following:

- The CTs are healthy
- The wiring from the CT to the relay is okay
- The relay is working.

If necessary, a trip-test can be done from the relay to ensure that the relay output trip contacts are working and that the breaker trip coil and mechanism are okay.

9.8 The future of protection for distribution systems

With the second generation of microprocessor relays now available, the emphasis is on the broader use of the protection relays as data acquisition units and for the remote control of the primary switchgear. Protection relays continuously monitor the primary system parameters such as current, voltage, frequency, etc. as part of the protection function of detecting faults. Since faults seldom occur, protection relays are expected to fulfill the protection requirements for a very small portion of their lifetime.

By utilizing the protection relays for other duties during the periods when the power system is normal, it permits integration of the various systems such as protection, supervisory control and data acquisition and results in savings on other interface components such as measuring transducers for current and voltage, meters, circuit breaker control interface, etc.

Improvements in digital communications by means of optical fibers allow the information available at the relay to be transferred without interference to the substation control level for information or event recording.

The following information is typically available from the relay:

- Measurement data of current and voltage
- Information stored by the relay after a fault situation
- Relay setting values
- Status information on the circuit breakers and isolators
- Event information.

The communication link to the relay can also be used for control purposes:

- Circuit breaker open/close commands
- Remote reset of the relay or auto-reclose module
- Changes to the protective relay settings.

Figure 9.26 shows the components of an integrated protection and control system that could be implemented in distribution substations.

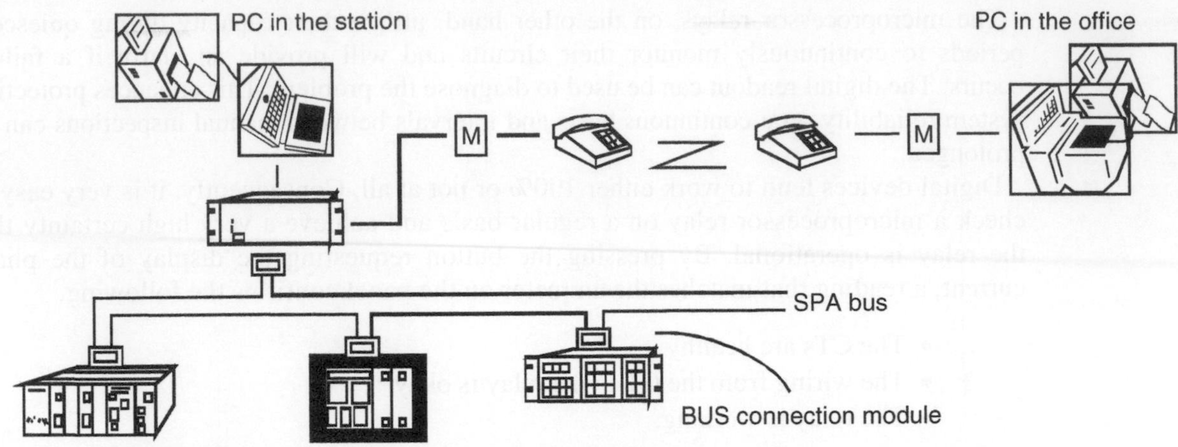

Figure 9.26
Components of an integrated protection and control system

9.9 The era of the IED

As already discussed earlier, protection relays became more advanced, versatile and flexible with the introduction of microprocessor-based relays. The initial communication capabilities of relays were intended mainly to facilitate commissioning. Protection engineers realized the advantages of remotely programing relays, the need developed for data retrieval, and so the communication aspects of relays became steadily more advanced.

PLC functionality became incorporated into relays, and with the development of small RTUs, it was soon realized that relays could be much more than only protection devices. Why not equip protection relays with advanced control functions? Why shouldn't protection functions be added to a bay controller? Both of these approaches have been followed, with different manufacturers (and sometimes different divisions within the same manufacturing group) adopting different approaches to the question of protection, control and communications. This resulted in an extensive range of devices on the market, some stronger on protection, some stronger on control, and the term protection relay became too restrictive to describe these devices. This resulted in the term 'intelligent electronic device' (IED).

9.9.1 Definition

The term 'intelligent electronic device' (IED) is not a clear-cut definition, as for example the term 'protection relay' is. Broadly speaking, any electronic device that possesses some kind of local intelligence can be called an IED. However, concerning specifically the protection and electrical industry, the term really came into existence to describe a device that has versatile electrical protection functions, advanced local control intelligence, monitoring abilities and the capability of extensive communications directly to a SCADA system.

9.9.2 Functions of an IED

The functions of a typical IED can be classified into five main areas, namely protection, control, monitoring, metering and communications. Some IEDs may be more advanced than the others, and some may emphasize certain functional aspects over others, but these main functionalities should be incorporated to a greater or lesser degree.

9.9.3 Protection

The protection functions of the IED evolved from the basic overcurrent and earth fault protection functions of the feeder protection relay (hence certain manufacturers named their IEDs 'feeder terminals'). This is because a feeder protection relay is used on almost all cubicles of a typical distribution switchboard, and that more demanding protection functions are not required to enable the relay's microprocessor to be used for control functions. The IED is also meant to be as versatile as possible, and is not intended to be a specialized protection relay, for example generator protection. This also makes the IED affordable.

The following is a guideline of protection related functions that may be expected from the most advanced IEDs (the list is not all-inclusive, and some IEDs may not have all the functions). The protection functions are typically provided in discrete function blocks, which are activated and programed independently.

- Non-directional three-phase overcurrent (low-set, high-set and instantaneous function blocks, with low-set selectable as long time-, normal-, very-, or extremely inverse, or definite time)
- Non-directional earth fault protection (low-set, high-set and instantaneous function blocks)
- Directional three-phase overcurrent (low-set, high-set and instantaneous function blocks, with low-set selectable as long time-, normal-, very-, or extremely inverse, or definite time)
- Directional earth fault protection (low-set, high-set and instantaneous function blocks)
- Phase discontinuity protection
- Three-phase overvoltage protection
- Residual overvoltage protection
- Three-phase undervoltage protection
- Three-phase transformer inrush/motor start-up current detector
- Auto-reclosure function
- Under frequency protection
- Over frequency protection
- Synchro-check function
- Three-phase thermal overload protection.

9.9.4 Control

Control function includes local and remote control, and are fully programmable.

- Local and remote control of up to twelve switching objects (open/close commands for circuit breakers, isolators, etc.)
- Control sequencing
- Bay level interlocking of the controlled devices
  - Status information
  - Information of alarm channels
- HMI panel on device.

9.9.5 Monitoring

Monitoring includes the following functions:

- Circuit-breaker condition monitoring, including operation time counter, electric wear, breaker travel time, scheduled maintenance
- Trip circuit supervision
- Internal self-supervision
- Gas density monitoring (for SF6 switchgear)
- Event recording
- Other monitoring functions, like auxiliary power, relay temperature, etc.

9.9.6 Metering

Metering functions include:

- Three-phase currents
- Neutral current
- Three-phase voltages
- Residual voltage
- Frequency
- Active power
- Reactive power
- Power factor
- Energy
- Harmonics
- Transient disturbance recorder (up to 16 analog channels)
- Up to 12 analog channels.

9.9.7 Communications

Communication capability of an IED is one of the most important aspects of modern electrical and protection systems, and is the one aspect that clearly separates the different manufacturers' devices from one another regarding their level of functionality (see Figure 9.27).

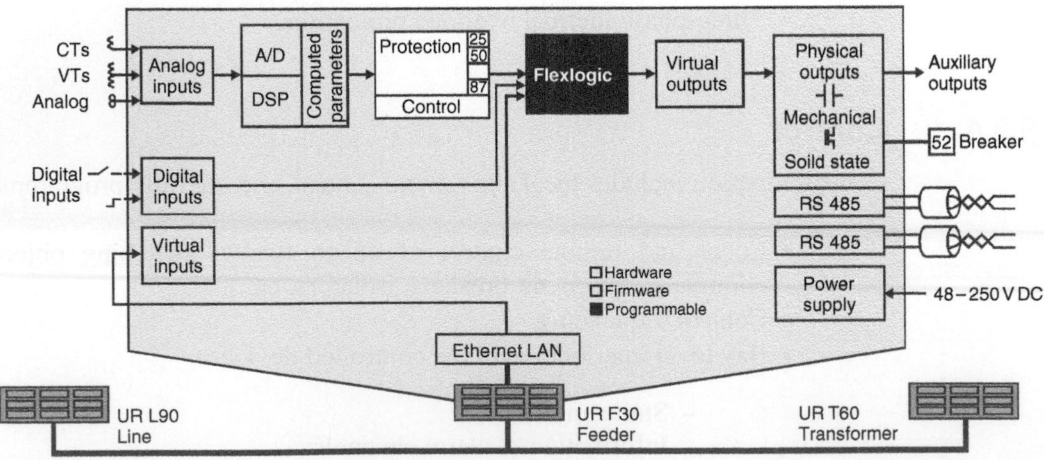

Figure 9.27
Typical IED internal configuration (source: GE universal relay)

9.10 Substation automation

Substation automation is not an easy thing to be achieved in existing substations. Automation in a substation is considered as provision of new generation intelligent electronic devices (IEDs), programmable logic controllers (PLCs) and computers to monitor and communicate. It is always simple to incorporate these components in new substations at the design stage itself. But in an existing substation, which is already using some old types relays, automation is a question of what can be done to reduce the operating expenses and improve customer service, from practical and economical viewpoint. One way to improve customer service is giving the control over the feeder breakers to the operator of distribution substations. This effectively reduces the number of trips that have to be made to substations and allows service to be restored more rapidly after an outage. The question is also, how the operator is able to decide the tripping or otherwise of certain feeders, which are achieved by the capability of the substation to record the data over a period.

9.10.1 Existing substations

A typical scheme is shown in Figure 9.28, for providing just those functions needed to reduce expense and improve customer service. It helps planners to reduce capital expenditures and helps operators to minimize trips to substations and reduce out-age time. Existing relays are retained and feeder automation units are added on each feeder panel as shown in the diagram below. This is referred to as a distributed architecture whereby the feeder automation units are installed close to the input/output wiring sources.

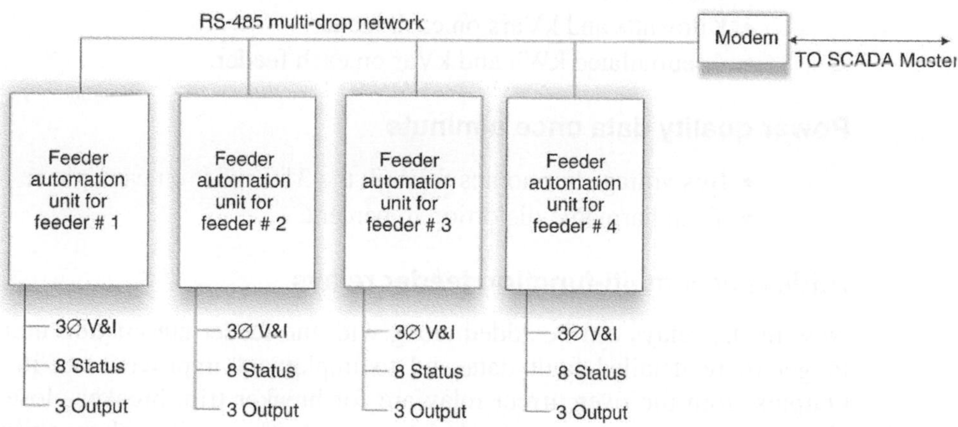

Figure 9.28
Typical incorporation of feeder automation units in a substation

The feeder automation units are available as add-on units to the existing feeders. Feeder current and bus voltage are given as inputs for each feeder automation unit, in addition to the status inputs like reclose status, breaker status and output from trip current relays. Some outputs are needed from the automation unit for trip, close, and enable/disable reclose. The feeder automation units generate relay target information from the trip current relay inputs. Trip current relays are available that can be mounted directly on the studs of relay cases. This allows them to be installed without changing the trip circuit wiring.

The provision of simple feeder automation units provides the following capabilities in automation of existing substations.

Control functions

- Trip/close for each feeder breaker
- Disable/enable reclosing relay on each feeder.

Status information

- Breaker position for each feeder (open/closed)
- Reclose status of reclosing relay on each feeder (enabled/disabled).

Target data

- Sequential listing of all time–overcurrent trips on each feeder
- Sequential listing of all instantaneous-overcurrent trips on each feeder.

Revenue accuracy real-time metering

- Bus voltage (average three-phase measurement)
- Current on each feeder
- Kilowatts and kVars on each feeder.

Planning data on 30 day long at 15/30 min intervals

- Bus voltage (average three-phase measurement)
- Current on each feeder
- Kilowatts and kVars on each feeder
- Accumulated kWh and kVar on each feeder.

Power quality data once a minute

- Bus voltage harmonics through the 31st order on each phase
- Total harmonic distortion in percent.

Adding new multi-function feeder relays

New feeder relays can be added along with the feeder automation units described above to get more detailed fault data and to implement improvements in feeder protection. Outputs from the overcurrent relay are for breaker trip, breaker close, and breaker fail. The relay requires an input for breaker status. The additional capabilities provided over and above the previous scheme are as follows:

Fault data

- Fault summary reports
- Oscillographic records of fault clearings
- Sequence of events data.

Additional protective functions

- Negative sequence overcurrent for more sensitive phase-to-phase protection
- Four relay setting groups for adapting protection to system conditions

- Coordination with bus relay to achieve feeder relay backup
- Automatic reclosing
- Breaker failure detection.

Maintenance data

- Breaker contact wear data
- Breaker operations counter
- Breaker trip speed monitoring.

Operating information

- Yesterday's peak demand current
- Today's peak demand current
- Peak demand current since last reset.

Use of computer in the substation

Ultimately a computer is the most useful addition to record all the information about the various systems and feeders. A substation computer is a key system element in that it allows intelligence to be moved downward to the substation. That intelligence reduces the amount of data that must be communicated between substations and the master station. For example, information can be retrieved as and when needed from databases maintained at the substation. It also provides a way to overcome the need for a standard protocol.

If the optional monitor is employed along with a keyboard, the computer can also serve as the human–machine interface for information and control in the substation.

The additional capabilities added by including a computer in the substation are as follows:

Maintains databases at substation

- One or more years of demand data
- One or more years of chart data.

Mathematical operations on the data

- Watts and VARs can be derived from the three-phase current and voltage data.

Flexibility

- Easy to add new functions and for expansions.

Human–machine interface in the substation

- Elimination of manual control switches
- Displays reports
- Alarm panel
- Event log
- Load control schedules
- Daily summary
- Volts and amps
- Monthly peaks report.

General system requirements

There are some basic requirements that should be considered when selecting an automation system for a substation. These are:

- An open system
- Distributed architecture to minimize wiring
- Flexible and easily set up by the user
- Inclusion of a computer to store data and pre-process information
- Setting up the computer software without requiring a programer
- Ability to communicate with the existing SCADA master
- Ability to handle various communications protocols simultaneously.

9.11 Communication capability

IEDs are able to communicate directly to a SCADA system. Figure 9.29 shows the present day concept of SCADA using IEDs connected through LAN and other network configurations.

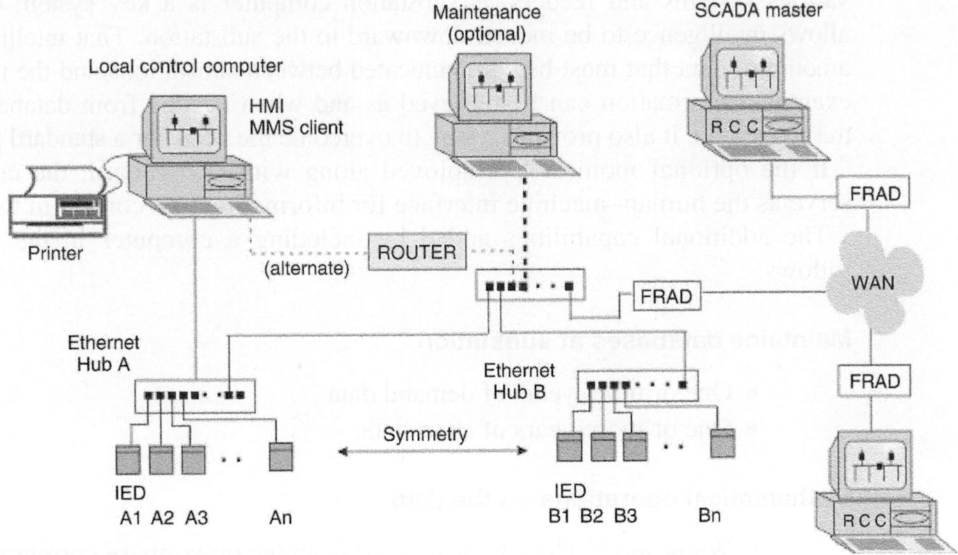

Figure 9.29
Typical structure of IED communication

10

Coordination by time grading

10.1 Protection design parameters on medium- and low-voltage networks

Although not appreciated by many engineers, the widespread use of inverse definite minimum time overcurrent and earth fault (IDMT OCEF) relays as the virtual sole protection on medium- and low-voltage networks requires as much detailed study and applications knowledge as does the more sophisticated protection systems used on higher-voltage networks.

10.1.1 Introduction

Traditionally, design engineers have regarded medium- and low-voltage networks to be of lower importance from a protection view, requiring only the so called simpler type of IDMT overcurrent and earth fault relays on every circuit. In many instances, current transformer ratios were chosen mainly based on load requirements, whilst relay settings were invariably left to the commissioning engineer to determine. Most of the times, the relay settings had been chosen considering the downstream load being protected without an effort to coordinate with the upstream relays. However, experience has shown that there has been a total lack of appreciation of the fundamentals applicable to these devices. Numerous incidents have been reported where breakers have tripped in an uncoordinated manner leading to extensive network disruption causing longer down times or failed to trip causing excessive damage, extended restoration time and in some cases loss of life.

This chapter reviews some of the fundamental points for the design engineer to watch for in planning the application of IDMTL OCEF protection to medium-voltage switchboards and networks.

10.1.2 Why IDMT?

Though it may be possible to grade the relay settings based on the fault currents, it is noted that the fault currents in a series network differs marginally when the sections are connected by cables without any major equipment like transformers in between the two ends. In such types, if networks grading the settings based on current values do not serve the purpose. It is required to go for time grading between successive relays in most of the networks.

To achieve selectivity and coordination by time grading two philosophies are available, namely:

1. Definite time lag (DTL), or
2. Inverse definite minimum time (IDMT).

For the first option, the relays are graded using a definite time interval of approximately 0.5 s. The relay R_3 at the extremity of the network is set to operate in the fastest possible time, whilst its upstream relay R_2 is set 0.5 s higher. Relay operating times increase sequentially at 0.5 s intervals on each section moving back towards the source as shown in Figure 10.1.

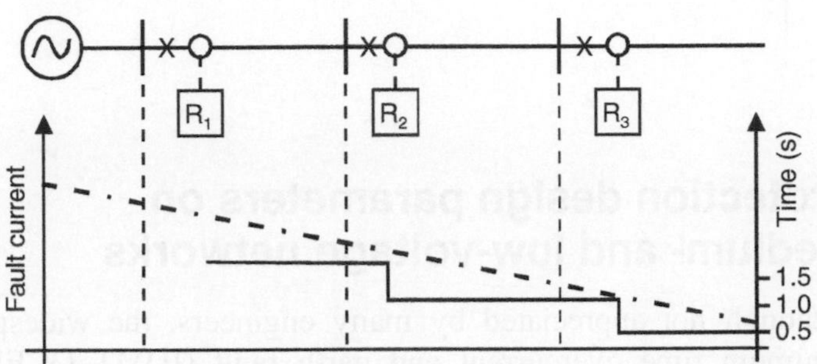

Figure 10.1
Definite time philosophy

The problem with this philosophy is, the closer the fault to the source the higher the fault current, the slower the clearing time – exactly the opposite to what we should be trying to achieve. On the other hand, inverse curves as shown in Figure 10.2 operate faster at higher fault currents and slower at the lower fault currents, thereby offering us the features that we desire. This explains why the IDMT philosophy has become standard practice throughout many countries over the years.

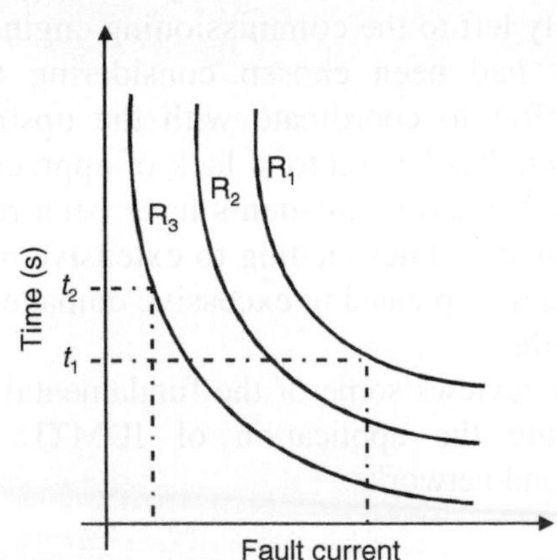

Figure 10.2
Inverse definite minimum time

10.1.3 Types of relays

Until the eighties in the last century, electromechanical disk-type relays were the standard choice, the most popular being GEC's type CDG and the TJM manufactured by Reyrolle.

They both follow the BS 142 specification for the normal inverse curve as highlighted in Figure 10.3, with an acceptable error margins as identified in the graph.

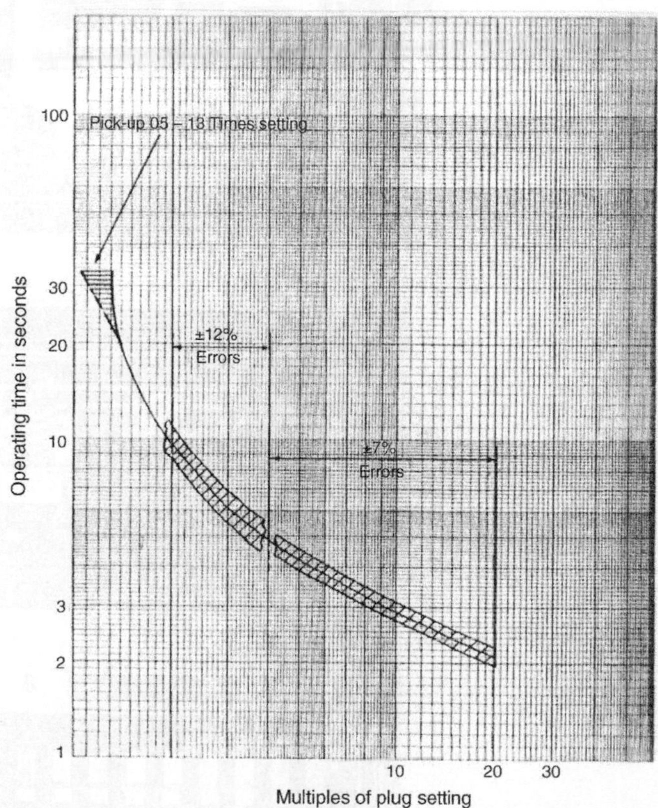

Figure 10.3
BS 142 tolerance limits

Being of the moving disk principle, the disk has a tendency to 'overshoot' before resetting after the fault current is removed by a downstream breaker. This phenomenon has to be considered, together with the tolerance on the tripping characteristic coupled with the breaker clearing time when determining the optimum time grading interval of 0.4 s as developed in Figures 10.4 and 10.5. However, on the modern electronic digital versions there is no overshoot to worry. In addition, they offer better tolerance over the whole curve – better than 5% is claimed – so the combination of these two factors means the time-grading interval can be reduced to 0.3 s (see Figure 10.6). Another point often overlooked in the use of electromechanical relays is that the burden of the relay varies inversely with the plug setting. The lower the plug taps setting, the higher the burden. This had been illustrated in the previous chapter where it had been noticed that for a 1 A relay the range varies from 0.75 Ω on the 200% tap to 300 Ω on the 10% tap. Similarly for the 5 A relay the range varies from 0.03 Ω on the 200% tap to 12 Ω on the 10% tap. The choice of plug tap could therefore have a significant effect on the performance of the current transformers to which the relay is connected.

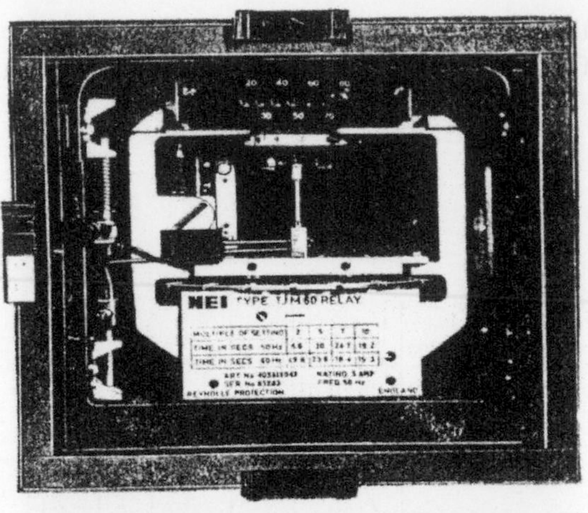

Figure 10.4
Typical disk-type relays

This shortcoming has been addressed on modern electronic relays, the burden remaining constant over the whole setting range and at a very low value, typically 0.02 Ω as seen in the previous chapter, without any major implication on the settings adopted.

10.1.4 Network application

When deciding to apply IDMT relays to a network, a number of important points have to be considered.

Firstly, it must be appreciated that IDMT relays cannot be considered in isolation. They have to be set to coordinate with both upstream and downstream relays. Their very purpose and being is to form part of an integrated whole system. Therefore, whoever specifies this type of relay should also provide the settings and coordination curves as part of the design package to show that he knew what he was doing when selecting their use. This very important task should not be left to others and once set, the settings must not be tampered (even by the operating staff) as otherwise coordination is lost.

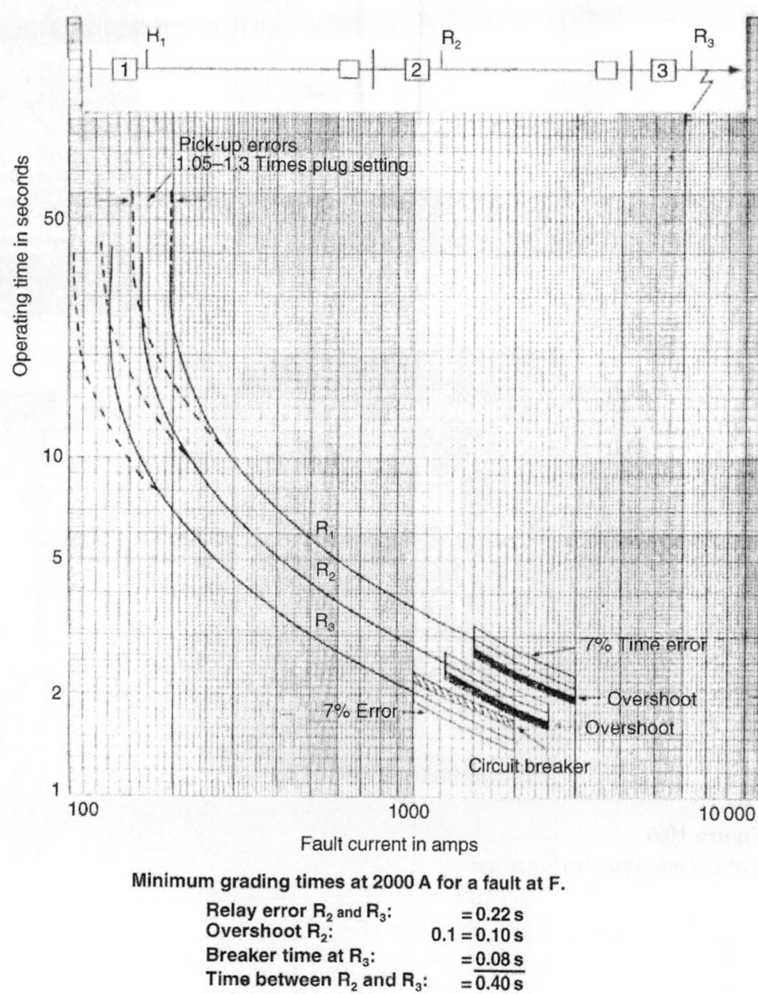

Figure 10.5
Minimum grading intervals

Overcurrent grading

When assessing the feasibility of applying such protection, one must be aware of certain constraints that will be applied by the supply authority at the one end and the coordination requirements of the low-voltage network at the other. These factors can often place severe limitations on the number of grading steps that can be achieved. It will be seen from Figure 10.7 that one could soon run out of time on overcurrent relays at the downstream side and it may be impossible to provide the settings for the downstream end relays.

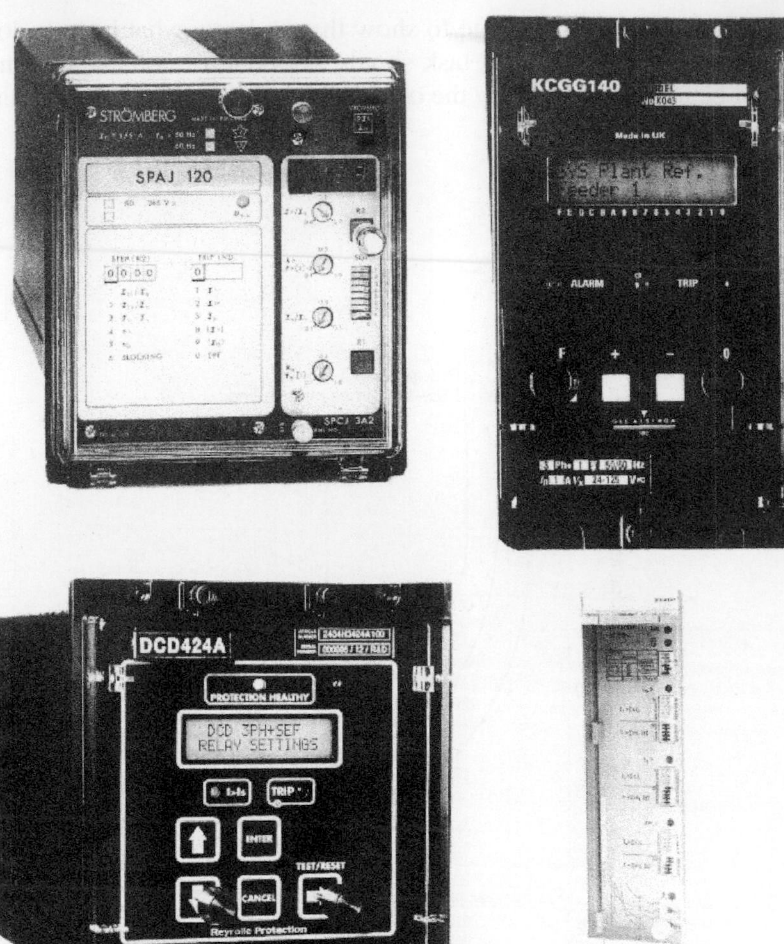

Figure 10.6
Typical microprocessor relays

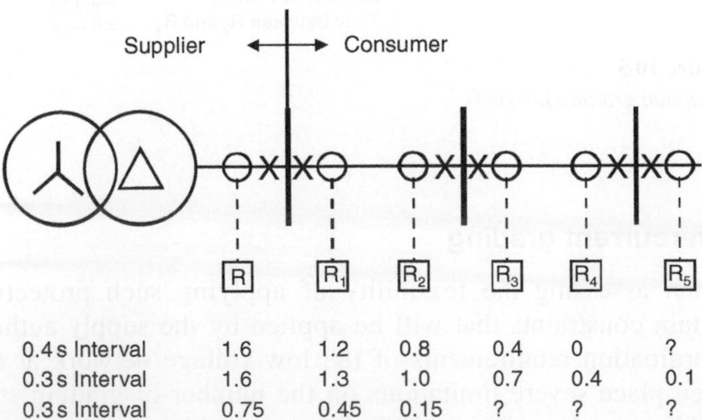

Figure 10.7
Overcurrent–time grading intervals

Using modern electronic relays does help in that with a 0.3 s time interval a couple of extra grading steps can be gained. The message here is therefore very clear. When designing medium-voltage network, one must aim for the minimum number of grading levels. In other words, medium-voltage networks should be designed 'short and fat' rather than 'long and thin' as shown in Figures 10.8(a), (b).

If this recommendation is not followed, then one may be faced with running a network radially by opening a ring at a specific point in order to achieve some moderately

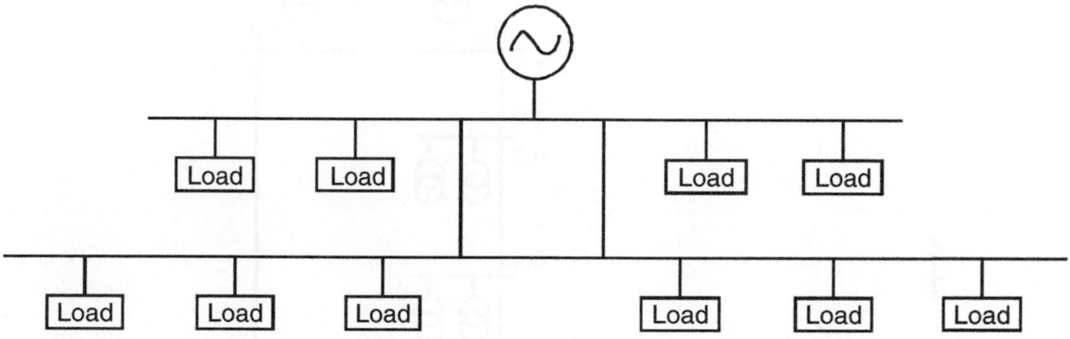

Figure 10.8(a)
Influence on network design (short and fat)

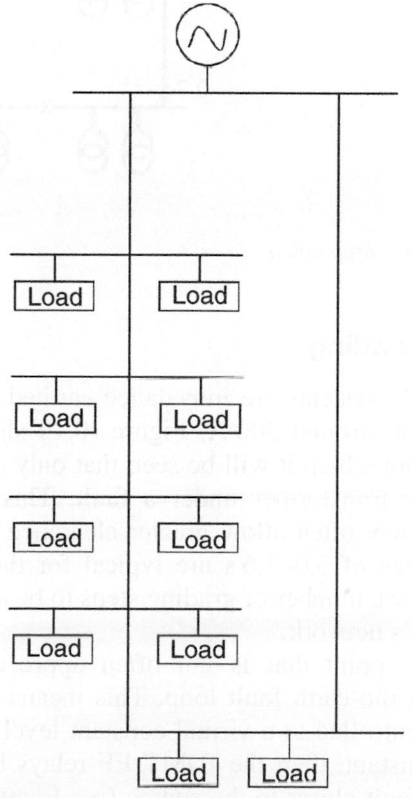

Figure 10.8(b)
Influence on network design (long and thin)

acceptable form of grading as seen in Figure 10.9. The relay settings would thus be dictating at which point the ring must be open. This is a bad design practice as protection relays should never place any limitations on which way the network should be operated, as maximum flexibility is essential at all times. Furthermore, running radially means continuity of supply is lost under fault conditions and coordination would be lost when the network is re-arranged to restore supply from an alternative source.

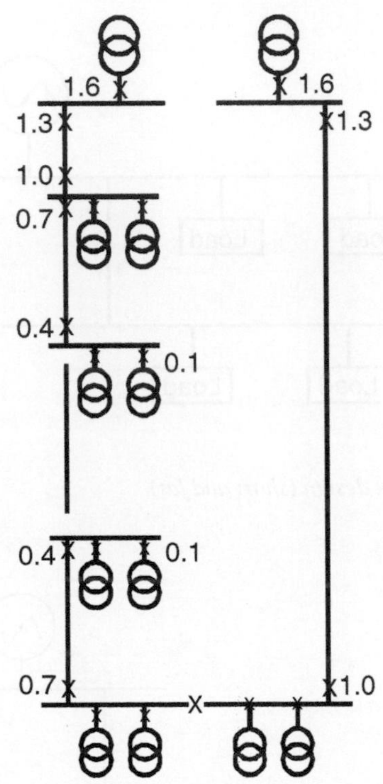

Figure 10.9
Impact on system configuration

Earth fault grading

Generally, MV systems are impedance earthed at its neutral end, which limits the earth fault current to around 300 A. Figure 10.10 shows the current distribution for an MV earth fault from which it will be seen that only a small overcurrent flows on the HV side of the supply transformer under a fault. This is of no embarrassment to the supply authority so they often allow greater clearance times for earth faults on the consumer's network. Times of 3.0–3.6 s are typical for the supply authority's back-up protection. This allows for a number of grading steps to be achieved for the earth fault protection into the consumer's network.

However, a point that is not often appreciated is that the NEC is the dominant impedance in the earth fault loop. This means that an earth fault anywhere on the MV network is controlled to a virtual constant level of current by the NEC. If the earth fault current is constant, then the IDMT EF relays behave as definite time relays, operating longer for a fault closer to the source (see Figure 10.1). An earth fault near to the source can therefore easily develop into high-current phase fault before the earth fault relay has timed out leading to enormous damage.

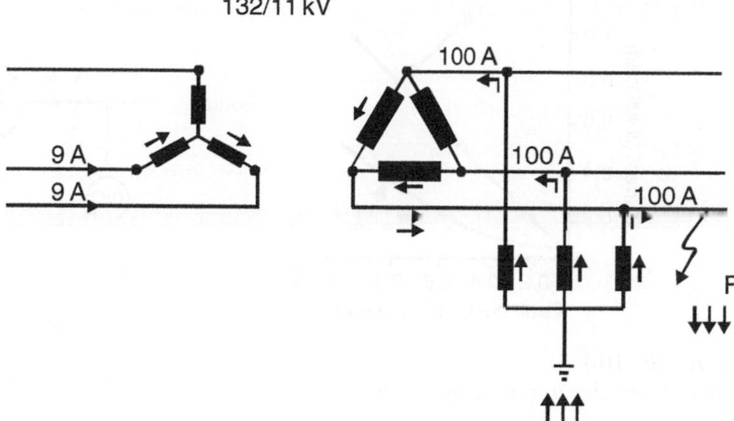

* LV earth fault means very small hv overcurrent

* Earth fault level tends to remain constant

Figure 10.10
Earth fault

Transformer protection

IDMT relays have often been used as the main HV protection on distribution transformers, without due regard for certain limitations.

It is often very difficult, in fact impossible, to set an HV IDMT relay to detect an earth fault on the LV winding of a transformer. As will be seen from Figure 10.11 the equivalent HV current for an earth fault on the LV winding (especially towards the neutral end) can be below the primary full load current. If the LV winding is earthed via a resistor the earth fault current will not exceed 57% of the full load rating of the transformer. The transformer should therefore be fitted with an additional protection such as Buchholz or restricted earth fault (REF) on the LV side to cover this condition.

It has been common practice not to fit Buchholz relays to small transformers because it is too expensive relative to the cost of the transformer. However, the size of the transformer is not that important but its strategic location in the network counts. Its loss may have a major impact on production downtime, etc., irrespective of its size, so it is worth protecting it properly and the Buchholz alarm does give early warning of impending trouble (see Figures 10.11(a), (b)).

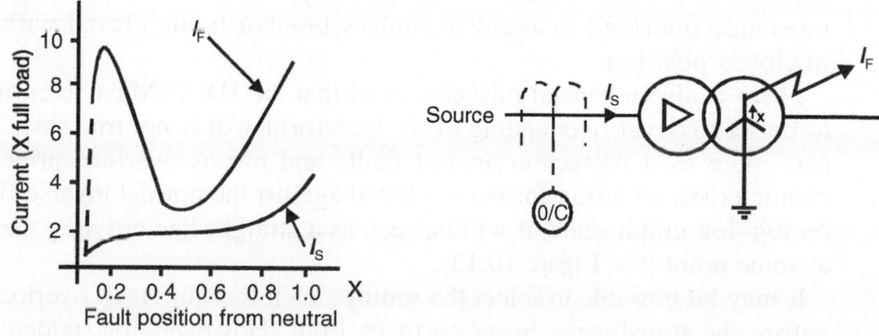

Figure 10.11(a)
HV overcurrent on transformers

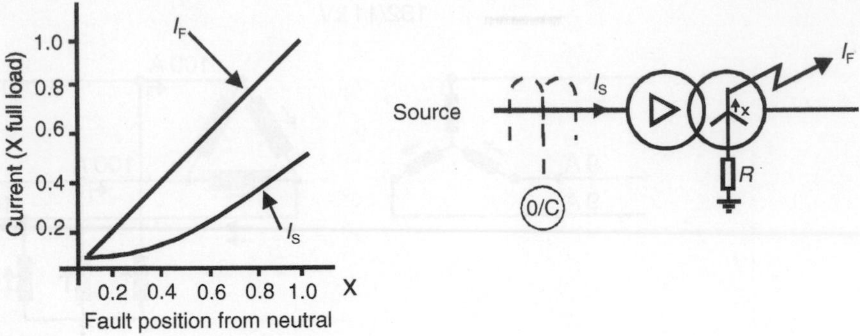

Figure 10.11(b)
Fault current due to grounding resistor

Another poor design practice is for the HV IDMT relay only to trip the HV breaker. For an HV winding fault the breaker is tripped but the fault can continue to be fed via the low-voltage side, the back-feed coming from the adjacent transformer(s) where the LV protection is set high to coordinate with downstream requirements. Transformer protection should always trip both HV and LV circuit breakers wherever possible as shown in Figure 10.12.

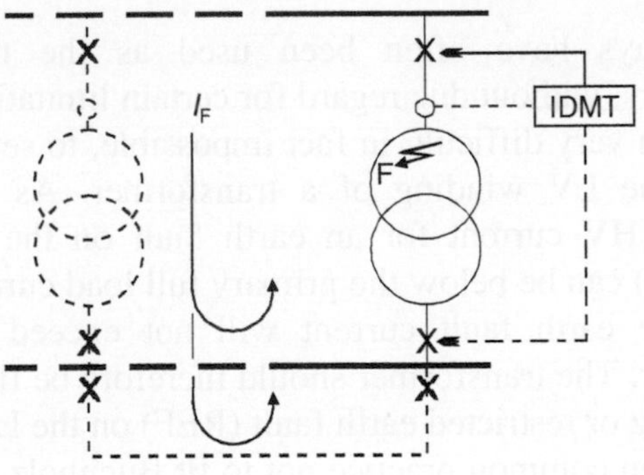

Figure 10.12
Protection should trip HV and LV

This is very much important when a low-voltage bus is connected from more than one source. In some cases, it is a common practice to ensure that the secondary breaker is tripped whenever the primary breaker is tripped/open (intertrip) and to introduce interlock to avoid secondary breaker being closed without primary breaker in closed position.

There is also a general misconception that the HV IDMT overcurrent relay is there and to be set to cover overloading of the transformer. It is not true since the overcurrent relay acts more as a protection against faults and not as overload protection. If the thermal characteristic of a transformer is plotted against the normal inverse IDMT curve (which is on log–log graph scale) it will be seen as a straight line crossing the IDMT characteristic at some point (see Figure 10.13).

It may be possible to select the settings such that for small overloads, the relay will trip before the transformer heats up to its limit. However, for sudden heavy overloads the transformer will cook before the relay trips. The normal inverse IDMT relay is therefore NOT suitable for overload duty – it is a fault protection. If overload protection is desired then a relay with a suitable thermal characteristic should be applied or alternatively select the inverse characteristic in the IDMT range, the time being approximately inversely proportional to the square of the current. This characteristic is much closer to the thermal characteristic of a transformer and any fuses downstream in the LV network.

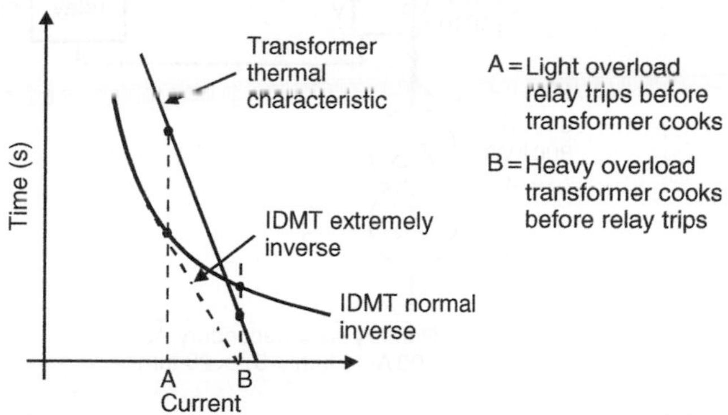

Figure 10.13
IDMT normal inverse not for overload

However, why trip the transformer out for overload – why not just remove the load by tripping the LV breaker? The same current that flows through the transformer flows through the LV breaker and nearly all have thermal characteristics as an integral part of their protection. This approach would certainly save operators having to test and check out a transformer before returning it to service, thereby reducing downtime.

10.1.5 Current transformers

There are numerous installations where the performance of the current transformers has been overlooked or misunderstood.

The most widely mistaken error has undoubtedly been the choice of CT ratio which was invariably selected on the basis of full-load current without due regard to performance under fault conditions. Despite its name, the performance parameter that we are most interested in is the current transformer's secondary voltage. In other words, the voltage developed across the CT secondary shall be sufficient to drive the transformed current through its own internal impedance plus the impedance of the relay and any other equipment to which it is connected (see Figure 10.14). It is necessary to fix the typical form of a magnetization curve for the CT. The knee-point voltage and the CT's internal resistance should be specified/checked to ensure that they are adequate for the application.

A classical example of how-not-to-do-it was seen recently on a feeder to a mini-sub from a main 11 kV switchboard whose fault level was approximately 9000 A. Refer to Figure 10.15.

CT ratio: 25/5
Burden: 5 VA
Accuracy: 10P10
Overcurrent relay: CDG set at 50%

In the above example, for a 9000 A fault on the feeder cable, the CT secondary current would be 1800 A. The burden of the relay at the chosen setting is 0.48 Ω, therefore the secondary voltage required to drive 1800 A through an impedance of 0.48 Ω = 1800 × 0.48 = 864 V. With five turns on the CT the best that can be expected is 1 V per turn which equals 5 V.

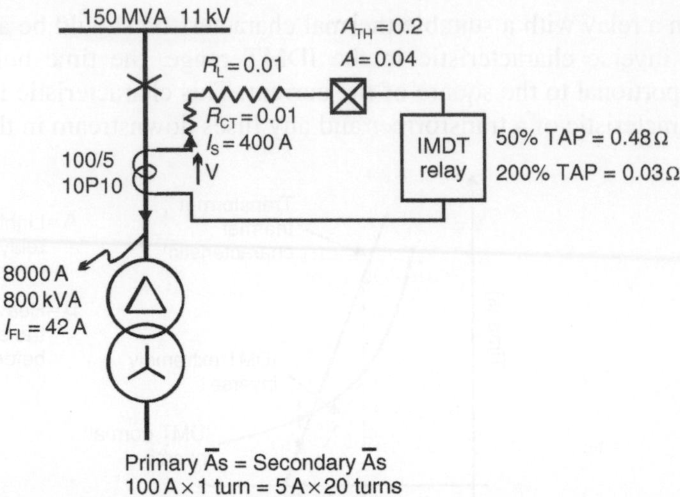

Figure 10.14
CT performance important

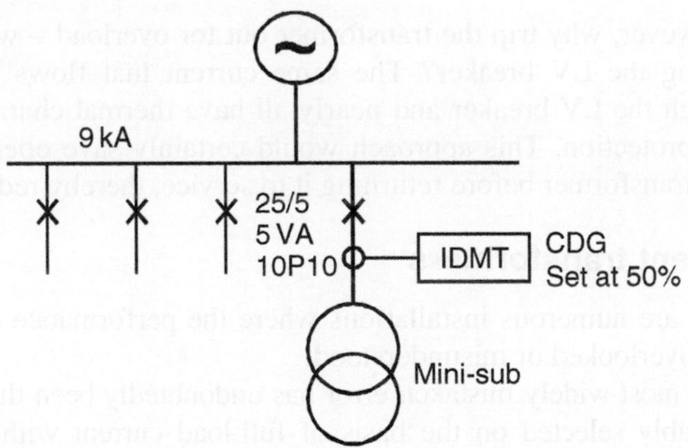

Figure 10.15
Typical selection of CT

It will now be evident why the magnetization curve and internal resistance of the CT are so important, especially when electromechanical relays are used for protection. With such a low ratio, this CT would undoubtedly be of the wound primary type in order to get the required number of amp-turns to magnetize the core. Unless the CT manufacturer was told, the high fault current would cause this CT to burst due to the high thermal and magnetic stresses set up in the primary winding under fault conditions which are a function of the fault current squared. CT ratios should therefore be chosen based on fault current NOT load current.

In such applications, the use of 1 A secondary is recommended as there would be one-fifth of the current and five times the voltage available to drive this smaller current around its connected load. The accuracy stated in the above example of 10P10 means that the CT will remain 10% accurate up to 10 times its rating (i.e. up to 250 A) – not much use for a fault current of 9000 A. Ideally, the chosen ratio should have been 900/1 10P10 or 450/1 10P20.

The correct technique in choosing a CT ratio is to calculate the fault current and divide this by 10 for a 10P10 accuracy rating or by 20 for a 10P20 specification. Accuracy is important, especially in tight grading applications as highlighted above.

It is also a common practice by some designers to add an ammeter into the protection circuit. This is invariably done for reasons of economy to save fitting another set of CTs. However, this is not a good practice because:

1. It adds more burden into the CT circuit – typically 1 VA for a normal ammeter and 5 VA for a thermal maximum demand ammeter – thereby giving the CT more work to do, thereby pushing into saturation.
2. Ammeters are designed to operate under healthy conditions over the range 0–1.2 times full load. Ideally, they should be connected to metering cores, which saturate at this upper level to protect the instrument under fault conditions.
3. Protection cores on the other hand are designed to operate under fault conditions at 10 to 20 times full-load current. Connecting ammeters to protection cores therefore subjects the instruments to enormous shocks under fault conditions, which initially affects the accuracy but ultimately destroys the bearings, and mechanisms, etc.

It is evident that this common wrongful practice of adding meters onto protection cores is the main influencing factor in deciding the CT ratio. A classic case of the tail wagging the dog!

Figure 10.16 indicates the CT ratios and the settings adopted in a typical network comprising of various transformers. It can be noted that the CT knee-point voltages as per the actual fault current figures are quite different from the actual knee-point voltage of the available CTs (refer lines marked '?'). It is a very clear case, wherein the wrongful selection of current transformers could lead to relays not operating at fault conditions. This shall be avoided, which can be done by doing proper calculations before deciding the component details.

Take the settings for a transformer in line 3. The fault current is around 6200 A and with a CT ratio of 100/5 A, the relay, current under fault will be about 310 A. This requires a knee-point voltage of not less than 164.5 for the provided CT but the actual knee-point is only around 46 V. The relay is not going to serve any purpose under fault conditions. Just consider the CT ratio of 100/1 A. The relay current under fault will reduce by one-fifth, and the knee-point voltage required will be only around 33 V and the available CT will do its job without any problem. This kind of analysis is necessary while choosing CT ratio and knee-point voltages while designing a network, which requires coordination for relay grading.

10.1.6 Importance of settings and coordination curves

One can have the finest protection in the world, correctly designed, installed, commissioned and maintained but if it is not set correctly then it is not of much use. Careful attention should therefore be paid to the settings of IDMT relays in particular, as they have to coordinate with upstream and downstream relays. To repeat what has been stated previously, the person who selects and specifies this type of relay should therefore also provide the settings and coordination curves to prove that the relay can fit correctly into the new or existing network. Many instances have arisen where it has not been possible to achieve any sort of adequate settings (as has been highlighted in the examples above) thereby proving the designer did not know what he was doing, or misunderstood the fundamentals.

In setting the IDMT relay it must be appreciated that moving the plug bridge moves the curve left or right in the horizontal direction and selects the current pick-up value. Adjusting the time multiplier dial moves the curve up or down in the vertical direction to select the time of operation (see Figure 10.17).

SUBSTATION	CIRCUIT	FAULT CURRENT	C.T RATIO	OVERCURRENT RELAY CURRENT	SETTING BURDEN	C.T. INT RES	TOTAL BURDEN	VOLTS REQUIRED	C.T. V knee	REMARKS
CENTRAL WORKSHOPS	INCOMER	6756	800/5	42.23	0.21	0.5	0.71	30.12	60	OK
	M3 FEEDER	6756	200/5	168.90	0.03	0.4	0.43	72.63	25	?
	M4 FEEDER	6756	200/5	168.90	0.03	0.4	0.43	72.63	25	?
	W FEEDER	6756	200/5	168.90	0.03	0.4	0.43	72.63	25	?
	C FEEDER	6756	200/5	168.90	0.03	0.5	0.53	89.52	25	?
	A FEEDER	6756	200/5	168.90	0.03	0.4	0.43	72.63	25	?
	E4 FEEDER	6756	200/5	168.90	0.03	0.5	0.53	89.52	25	?
	P FEEDER	6756	200/5	168.90	0.03	0.6	0.63	106.41	50	?
P SUBSTATION	TRANSFORMERS	6278	800/1	7.85	12.00	3.8	15.80	123.99	230	OK
M4 SUBSTATION	TRANSFORMERS	6208	100/5	310.40	0.03	0.5	0.53	164.51	46	?
M1 SUBSTATION	TRANSFORMERS	6177	100/5	308.85	0.03	0.4	0.43	132.81	25	?
COMP.HOUSE	FEEDER	6122	?					0.00		
W SUBSTATION	TRANSFORMERS	6253	100/5	312.65	0.03	0.5	0.53	165.70	25	?
C SUBSTATION	TRANSFORMERS	5972	100/5	298.60	0.03	0.4	0.43	128.40	25	?
A SUBSTATION	TRANSFORMERS	6305	100/5	315.25	0.03	0.4	0.43	135.56	25	?
A1 SUBSTATION	TRANSFORMERS	6294	800/1	7.87	12.00	3.9	15.90	125.09	220	OK
E4 SUBSTATION	TRANSFORMERS	6069	100/5	303.45	0.03	0.5	0.53	160.83	25	?
E3 SUBSTATION	TRANSFORMERS	5943	100/5	297.15	0.03	0.5	0.53	157.49	25	?
E2 SUBSTATION	TRANSFORMERS	5845	100/5	292.25	0.03	0.5	0.53	154.89	25	?
E1 SUBSTATION	TRANSFORMERS	5770	100/5	288.50	0.03	0.5	0.53	152.91	25	?
M3 SUBSTATION	TRANSFORMERS	5722	100/5	286.10	0.03	0.7	0.73	208.85	25	?
M5 SUBSTATION	INCOMER FROM M3	5716	1000/1	5.72	12.00	4.9	16.90	96.60	300	OK
	TRANSFORMERS	5716	1000/1	5.72	12.00	4.9	16.90	96.60	300	OK
	INTERCONNECTOR TO E3	5716	1000/1	5.72	12.00	4.9	16.90	96.60	300	OK
	COMPRESSOR HSE FEEDER	5716	1000/1	5.72	12.00	4.9	16.90	96.60	300	OK

Figure 10.16
Typical case of CT selection in a large network

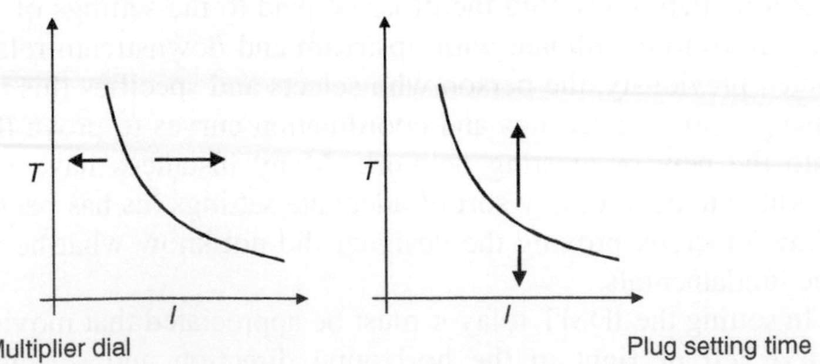

Figure 10.17
Effect of settings

It is therefore possible to achieve the same setting using two different combinations of plug setting and time multiplier as illustrated in Figure 10.18. Notice how the curves cross. Even with a time interval between them, it is still possible to choose plug/time dial settings such that the curves cross. This means that for low fault currents relay A operates before relay B but for high fault currents relay B operates before relay A. Coordination is therefore lost. It is vitally important that after selecting settings for the relays, the coordination curves are drawn to ensure that they do not cross and that they stack up nicely on top of one another as shown in Figure 10.19.

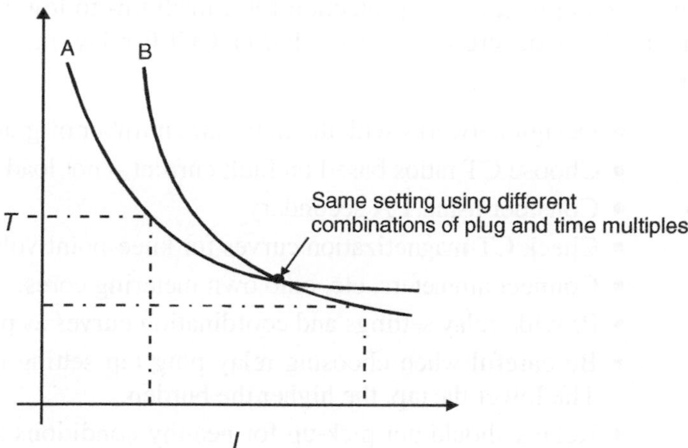

Figure 10.18
Curves must not cross

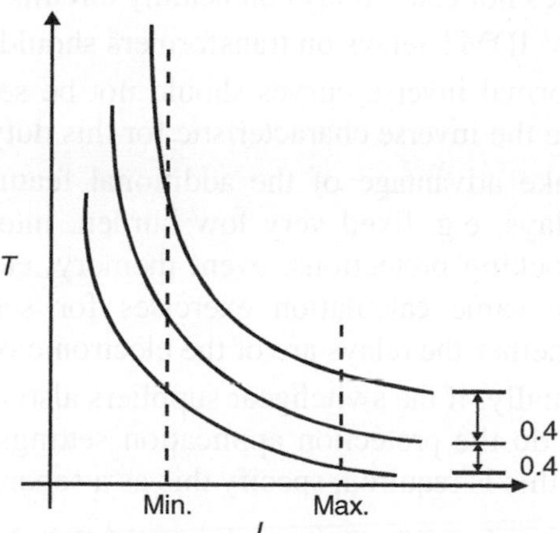

Figure 10.19
Ideal coordination of setting curves

Two basic rules are that they must pick-up for the lowest fault level (minimum plant) and must coordinate for the highest fault level (maximum plant).

One final word on a point not often appreciated. The electrical network is in fact a living thing. It grows and changes over time. Generation is added, load centers develop, old plant's decommissioned, new plant's extended and so on. Fault levels therefore change and invariably increase. IDMT relay settings should therefore be reviewed on a regular basis, especially if there are extensions or changes planned. The opportunity should be taken to up-grade the protection at the same time. The old electromechanical disk relays have done a wonderful job over the last century and can still continue to do so in certain applications provided one is aware of their limitations. However, we are now starting to ask more than they are capable of delivering. Nothing wrong with the relay – it's just the application. So one should take advantage of the additional facilities offered by the new range of numerical relay equivalents.

10.1.7 Conclusion

To engineers planning the protection for a medium- to low-voltage network and wishing to adopt the widespread use of the IDMT OCEF relay, the above can be summarized as follows:

- Design networks with the minimum number of grading levels possible.
- Choose CT ratios based on fault current – not load current.
- Consider using 1 A secondary.
- Check CT magnetization curves for knee-point voltage and internal resistance.
- Connect ammeters, etc. onto own metering cores.
- Provide relay settings and coordination curves as part of the design package.
- Be careful when choosing relay plug tap setting on electromechanical relays. The lower the tap, the higher the burden.
- Relays should not pick-up for healthy conditions such as permissible transient overloads, starting surges and reconnection of loads, which have remained connected after a prolonged outage.
- Care should also be taken that the redistribution of load current after tripping does not cause relays on healthy circuits to pick-up and trip.
- HV IDMT relays on transformers should trip both HV and LV breakers.
- Normal inverse curves should not be selected for overload protection. Rather use the inverse characteristic for this duty.
- Take advantage of the additional features offered by the modern electronic relays, e.g. fixed very low burden, integral high-set, breaker fail and busbar blocking protections, event memory, etc. However, remember, one has to do the same calculation exercises for settings and draw coordination curves whether the relays are of the electronic or electromechanical design.
- Finally, if the switchgear suppliers also manufacture relays, do not expect them to do the protection application settings free of charge as part of the service. If this is required, specify this as a separate cost item in the specification.

Many problems down the line can be avoided and the performance, efficiency and safety of the plant improved if a protection engineer is included in the design team, if not full time, but at least to do an audit on the proposals.

Finally, remember – whilst IDMT relays are the most well known and the cheapest, they are in fact the most difficult relays to set.

10.2 Sensitive earth fault protection

A number of instances arise where the load current demands high-ratio line current transformers, and the neutral current has been limited to a low value by the use of a remote high-impedance neutral earthing device for safety reasons.

In this situation, the use of a conventional IDMTL earth fault relay with a minimum setting of 10% would be unable to detect an earth fault condition. For such an application, sensitive earth fault protection should be considered (see Figure 10.20). In addition, on overhead rural distribution systems it is possible for high-resistance earth faults to occur, especially if a conductor breaks and falls on very dry ground having say high silica content. Instances have been recorded where the initial rush of fault current has caused the silica to form a glass envelope around the end of the broken conductor; so a live conductor then remained undetected following the auto-reclose shot.

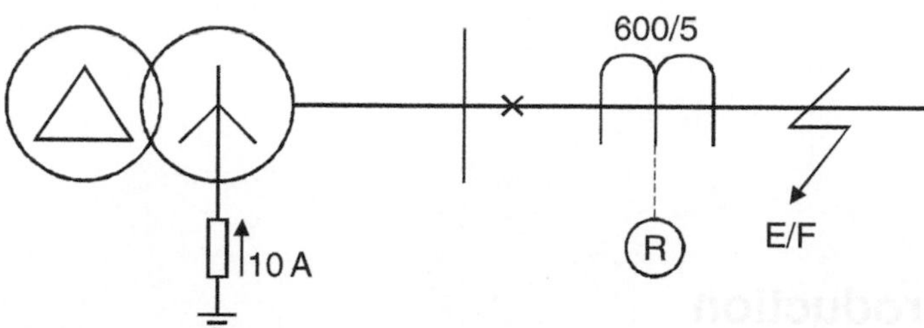

Figure 10.20
Earth fault current limited by NER

Besides this type of unusual incident, it is not uncommon for broken conductors to fall on dry ground, and this presents a hazard to human life and livestock if left undetected for any length of time, particularly if the line does not have earth wires installed. It is therefore necessary to apply sensitive earth fault relays to cover these and other demanding situations.

Another factor is the possible predominance of a third harmonic component in the residual current even under quiescent conditions. Third harmonics appear as zero sequence currents and could cause mal-operation of the relay. It is therefore wise to select a design of relay, which has been tuned to reject currents of this frequency. Also, ensure that the response decreases at higher frequencies to render the relay immune to harmonic resonant conditions, which may vary according to the power system configuration. There are a number of sensitive earth fault relays on the market, essentially of the definite-time variety, having pick-up setting ranges typically of the order of 0.4–40% of 5 A.

One model offers a useful digital read-out of the standing neutral current when interrogated via a push-button. If this is recorded on a regular basis, trends can be established which would assist in determining the deterioration of the line insulators and help to plan preventative maintenance programs.

Time setting ranges vary from 0.1 to 99 s. Adjustable time delays ensure stability during switching and other transient disturbances and allow for adequate grading with other protection systems.

On rural networks, it is as well necessary to ensure that the relay contacts are self-reset (i.e. they do not latch) so that auto-reclosing can take place.

11

Low-voltage networks

11.1 Introduction

The low-voltage network is a very important component of a power system as it is at this level that much of the power is distributed and utilized by the end consumer. Essential loads such as lighting, heating, ventilation, refrigeration, air-conditioning and so on are generally fed at voltages such as 380, 400, 415, 480, 500, 525 V, three-phase 3 wire and three-phase 4 wire.

In mining industry, heavy motor loads often require voltages as high as 1000 V. Because of the diverse nature of the loads coupled with the large number of items requiring power, it is usual to find a bulk in-feed to an LV switchboard, followed by numerous outgoing circuits of varying current ratings, in contrast to the limited number of circuits at the medium-voltage level. Large frame (high current rated) air circuit breakers are therefore specified as incomers from the supply transformer and moulded case circuit breakers (MCCBs) for all outgoing feeders. The downstream network generally consists of MCCBs of varying current ratings and as the current levels drop miniature circuit breakers (MCBs) are used for compactness and cost saving.

11.2 Air circuit breakers

These breakers are available in frame sizes ranging from 630 to 6000 A in 3 and 4 pole versions and are generally insulated for 1000 V. Rated breaking capacities of up to 100 kA rms symmetric to IEC947-2 are claimed at rated voltage of 660 V.

Fixed and draw-out models are available and each unit invariably comes complete with a protection device, which, in keeping with modern trends, is generally of the electronic type and will be discussed in more detail later.

Typical total breaking times are of the order of 40–50 ms for short circuit faults. Their operating speed is important as air circuit breakers (ACBs) are applied as the main incoming devices to the low-voltage network where they are subject to the highest fault levels determined by the supply transformer.

A typical construction of an ACB is shown on the next page. The spring charging can be manual and electrical. The operating and tripping mechanisms are similar to the ones used in high-voltage oil/vacuum/SF6 circuit breakers.

It is common to note that most of the present day ACBs are fitted with solid-state built-in overcurrent and earth fault relay. The same can be set for various current and time characteristics without the need for using external relays. These relays are provided with ample current and time setting ranges to achieve discrimination for various types of LV distribution systems (see Figure 11.1).

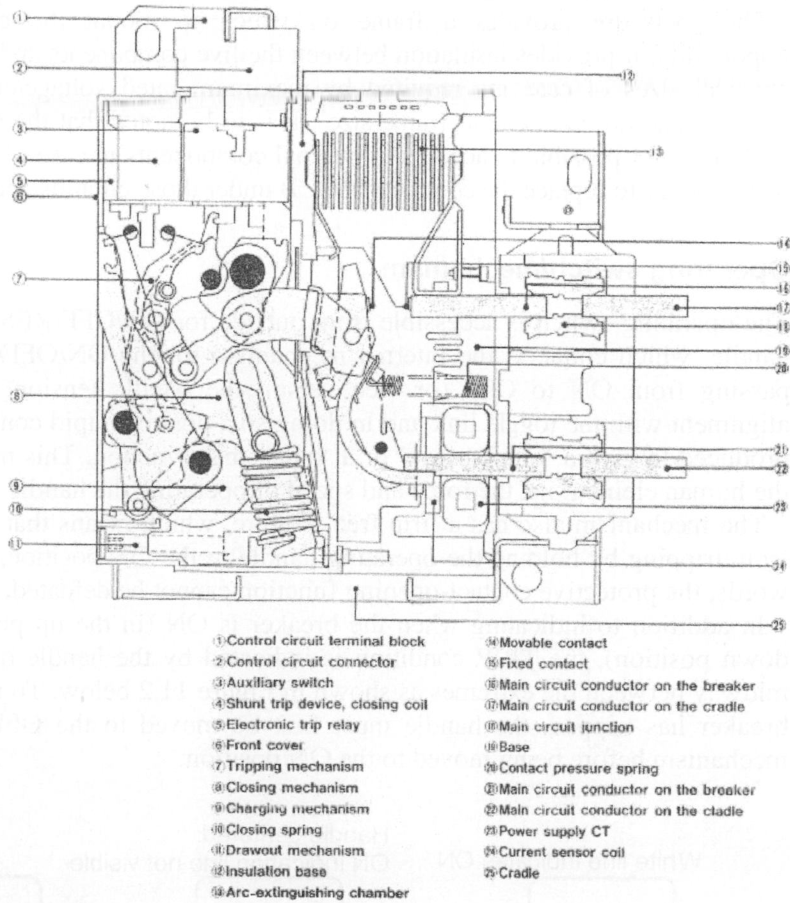

Figure 11.1
Typical internal construction of an air circuit breaker

11.3 Moulded case circuit breakers

MCCBs are power switches with built-in protective functions used on circuits requiring lower current ratings. They include the following features:

- Normal load current open and close switching functions
- Protection functions to automatically disconnect excessive overloads and to interrupt short circuit currents as quickly as possible
- Provide indication status of the MCCB either open, closed or tripped.

Although many different types are manufactured, they all consist of five main parts:

1. Moulded case (frame)
2. Operating mechanism
3. Contacts and extinguishers
4. Tripping elements
5. Terminal connections.

11.3.1 Moulded case

This is the external cover of the MCCB, which houses all the sensing and operating components. This is moulded from resin/glass-fiber materials, which combine ruggedness with high dielectric strength.

The enclosure provides a frame on which to mount the components, but more importantly, it provides insulation between the live components and the operator. Different physical sizes of case are required by maximum rated voltage/current and interrupting capacity and are assigned a 'frame size'. It is to be noted that the case is moulded and as such, it is not possible to access the internal components in case of any failures and it will be necessary to replace the complete MCCB under those circumstances.

Operating switch/mechanism

The operating switch is accessible from outside for ON/OFF/RESET purposes. This is a handle, which connects the internal mechanism for the ON/OFF/RESET operations. In passing from ON to OFF (or vice versa), the handle tension spring passes through alignment with the toggle link and in doing so a positive rapid contact-operating action is produced to give a 'quick break' or a 'quick make' action. This makes it independent of the human element i.e. the force and speed of operating the handle.

The mechanism also has a 'trip free' feature, which means that it cannot be prevented from tripping by holding the operating handle in the ON position, during faults. In other words, the protective contact-opening function cannot be defeated.

In addition to indicating when the breaker is ON (in the up position) or OFF (in the down position), the TRIP condition is indicated by the handle occupying the positions midway between the extremes as shown in Figure 11.2 below. To restore service after the breaker has tripped, the handle must first be moved to the OFF position to reset the mechanism before being moved to the ON position.

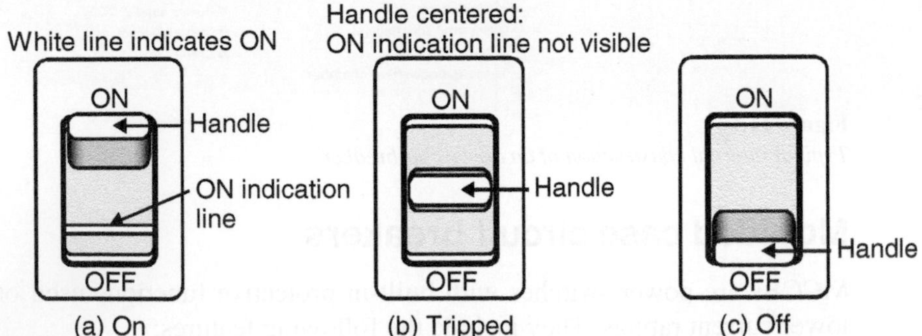

Figure 11.2
Handle positions

Contacts and extinguishers

A pair of contacts comprises a moving contact and a fixed contact. The instants of opening and closing impose the most severe duty. Contact materials must therefore be selected with consideration to three criteria:

1. Minimum contact resistance
2. Maximum resistance to wear
3. Maximum resistance to welding.

Silver or silver-alloy contacts are low in resistance but wear rather easily. Tungsten or tungsten-alloys are strong against wear due to arcing but rather high in contact resistance. Contacts are thus designed to have a rolling action, containing mostly silver at the closing current-carrying points, and mostly tungsten at the opening (arcing) point (see Figure 11.3).

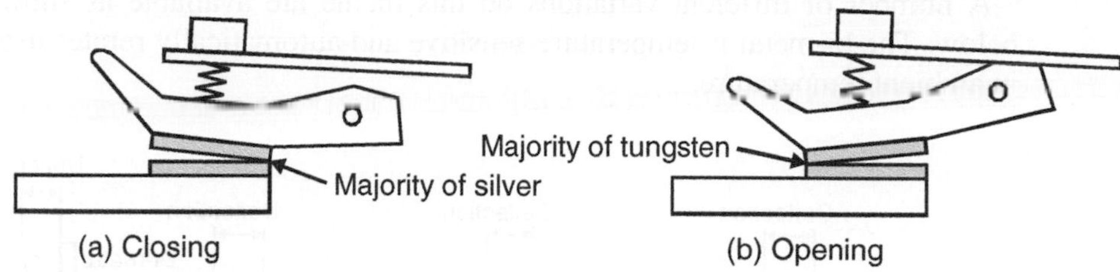

Figure 11.3
Dual function contacts

In order to interrupt high short-circuit currents, large amounts of energy must be dissipated within the moulded case. This is achieved by using an arc-chute, which comprises a set of specially shaped steel grids, isolated from each other and supported by an insulated housing. When the contacts are opened and an arc is drawn, a magnetic field is induced in the grids, which draws the arc into the grids. The arc is thus lengthened and chopped up into a series of smaller arcs, which are cooled by the grids heat conduction. Being longer it requires far more voltage to sustain it and being cooler tends to lose ionization and extinguish at first current zero (see Figure 11.4).

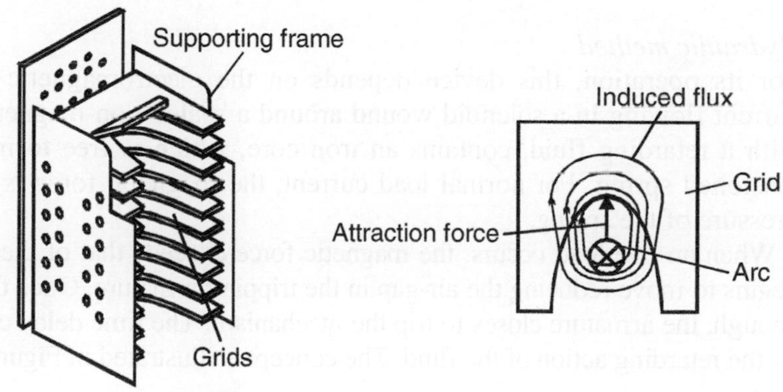

Figure 11.4
The arc chute

Tripping elements

The function of the trip elements is to detect the overload or short-circuit condition and trip the operating mechanism.

Thermal overload

The thermal trip characteristic is required to be as close as possible to the thermal characteristics of cables, transformers, etc. To cover this overload condition two types of tripping methods are available, namely Bi-metallic and hydraulic.

Bi-metallic method

The thermal trip action is achieved by using a bi-metallic element heated by the load current. The bi-metal consists of two strips of dissimilar metals bonded together. Heat due to excessive current will cause the bi-metal to bend because of the difference in the rate of expansion of the two metals. The bi-metal must deflect far enough to physically operate the trip bar. These thermal elements are factory-adjusted and are not adjustable in the field. A specific thermal element must be provided for each current rating.

A number of different variations on this theme are available as shown in Figure 11.5 below. The bi-metal is temperature-sensitive and automatically rerates itself with variations in ambient temperature.

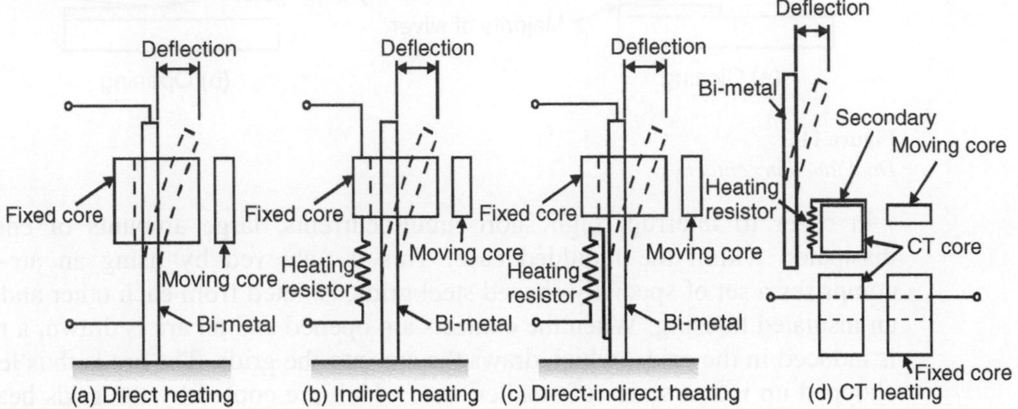

Figure 11.5
Thermal tripping methods

Hydraulic method

For its operation, this device depends on the electromagnetic force produced by the current flowing in a solenoid wound around a sealed non-magnetic tube. The tube, filled with a retarding fluid, contains an iron core, which is free to move against a carefully tensioned spring. For normal load current, the magnetic force is in equilibrium with the pressure of the spring.

When an overload occurs, the magnetic force exceeds that of the spring and the iron core begins to move reducing the air-gap in the tripping armature. Once the magnetic field is large enough, the armature closes to trip the mechanism. The time delay characteristic is controlled by the retarding action of the fluid. The concept is illustrated in Figure 11.6 below.

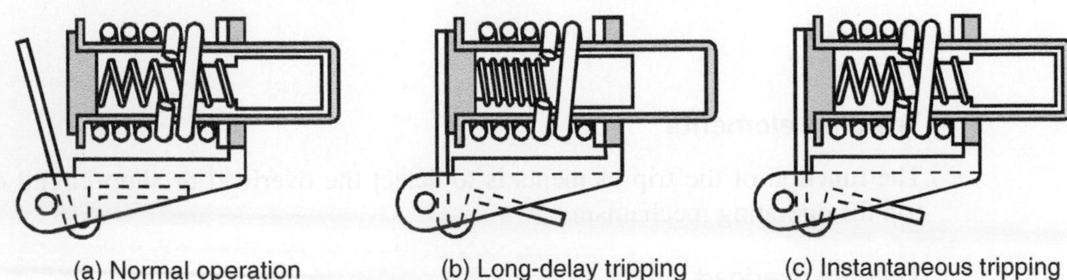

Figure 11.6
Hydraulic tripping method

Short circuits

For short-circuit conditions, the response time of the thermal element is too slow and a faster type of protection is required to reduce damage, etc. For this reason, a magnetic trip action is used in addition to the thermal element. When a fault occurs, the short circuit current causes the electromagnet to attract an armature, which unlatches the trip mechanism. This is a fast action and the only delay is the time it takes for the contacts to physically open and extinguish the arc. This is normally of the order of 20 ms – typically 1 cycle.

In the hydraulic method, the current through the solenoid will be large enough to attract the armature instantaneously, irrespective of the position of the iron core. The interruption speeds for this type of breaker for short circuit currents are also less than 1 cycle (20 ms) similar to the bi-metallic type.

In both of the above methods, the thermal magnetic and the hydraulic magnetic, the tripping characteristics generally follow the same format as shown in Figure 11.7.

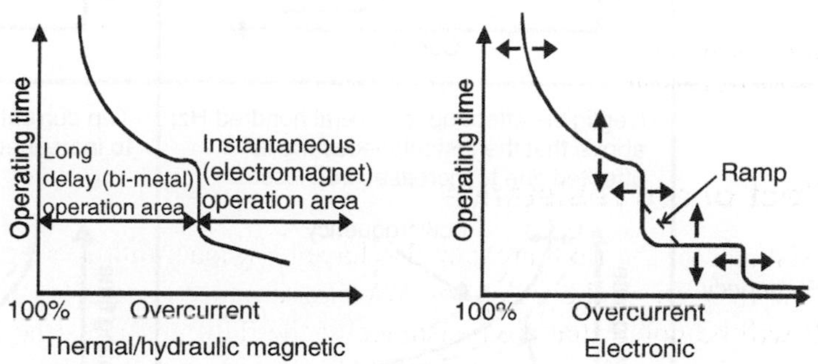

Figure 11.7
Typical tripping characteristics

Electronic protection MCCBs

Moulded case circuit breakers of the conventional types mentioned above are increasingly being replaced by electronic trip units and current transformers, which are an integral part of the breaker frame. This modern trend in the technology results in increased accuracy, reliability and repeatability. However, the main advantage is the adoptability of the tripping characteristics compared to the above-mentioned electromechanical devices, which are generally factory pre-set and fixed for each current rating. Discrimination can then be improved. Furthermore, semi-conductor-controlled power equipment can be a source of harmonics which may cause mal-operations.

Electronic protective devices detect the true rms value of the current, thereby remaining unaffected by harmonics. A comparison of the thermal-magnetic and hydraulic-magnetic types is given in Figure 11.8.

Terminal connections

These connect the MCCB to a power source and a load. There are several methods of connection such as busbars, straps, studs, plug-in adapters, etc. Up to 250–300 A whenever cables are used, compression type terminals are used to connect the conductor to the breaker. Above 300 A, stubs, busbars or straps are recommended to ensure reliable connections, particularly when using aluminum cables. Please refer to the subsequent pages regarding the cautions to be taken while connecting the MCCB in a power circuit.

Miniature circuit breakers

The miniature circuit breakers are similar to moulded case circuit breakers but as their name implies, these are smaller in size and are mostly used for current ratings below 100 A. These are normally available in single pole (SP), single pole neutral (SPN), double pole (DP), triple pole (TP), triple pole neutral (TPN) and four pole (FP) versions.

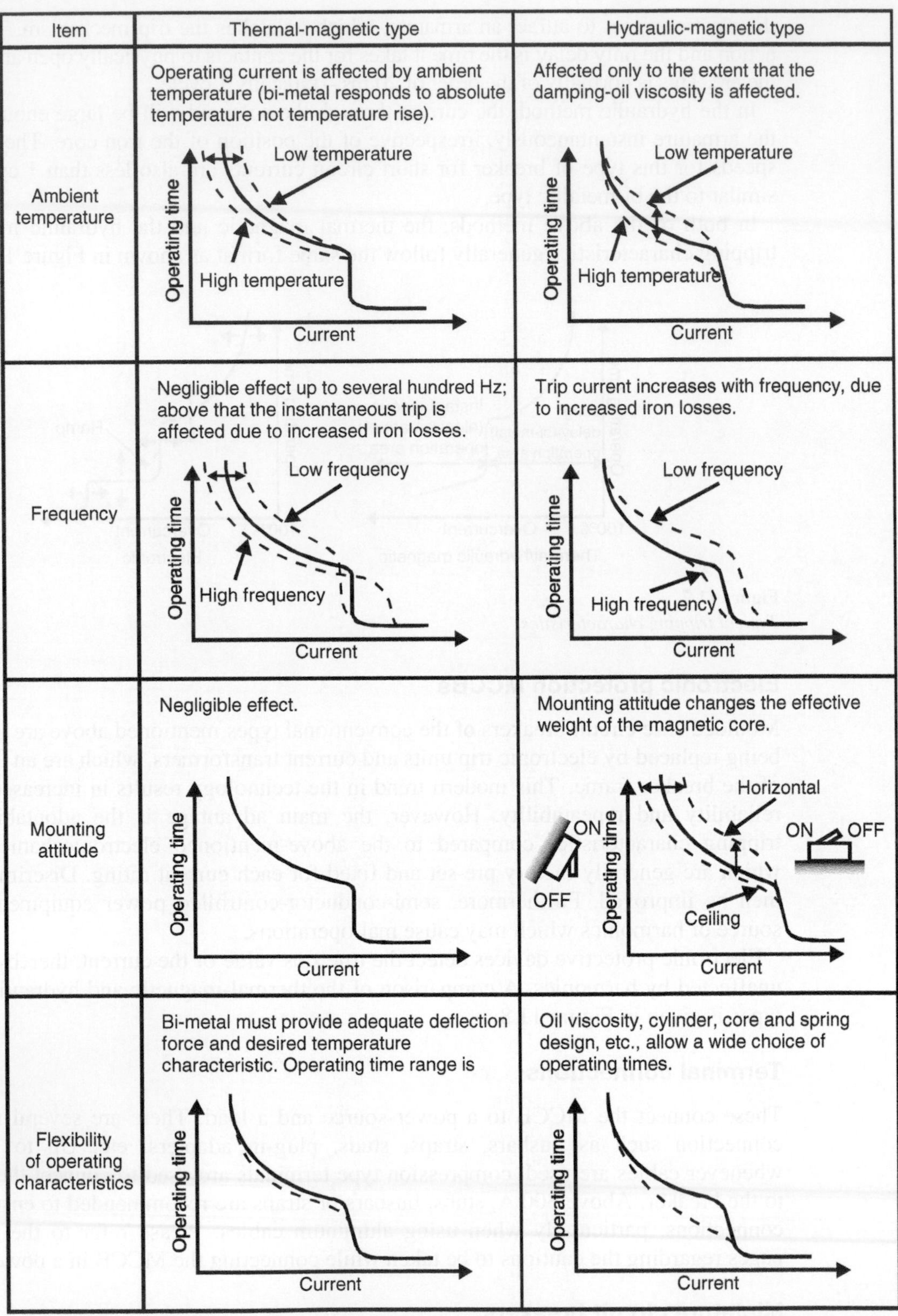

Item	Thermal-magnetic type	Hydraulic-magnetic type
Ambient temperature	Operating current is affected by ambient temperature (bi-metal responds to absolute temperature not temperature rise). Low temperature High temperature Operating time Current	Affected only to the extent that the damping-oil viscosity is affected. Low temperature High temperature Operating time Current
Frequency	Negligible effect up to several hundred Hz; above that the instantaneous trip is affected due to increased iron losses. Low frequency High frequency Operating time Current	Trip current increases with frequency, due to increased iron losses. Low frequency High frequency Operating time Current
Mounting attitude	Negligible effect. Operating time Current	Mounting attitude changes the effective weight of the magnetic core. Horizontal ON OFF Ceiling ON OFF Operating time Current
Flexibility of operating characteristics	Bi-metal must provide adequate deflection force and desired temperature characteristic. Operating time range is Operating time Current	Oil viscosity, cylinder, core and spring design, etc., allow a wide choice of operating times. Operating time Current

Figure 11.8
Comparison of thermal-magnetic and hydraulic-magnetic types

DC circuits

The MCCBs and MCBs are available for both AC and DC ratings for the various standard voltages. However, it must be remembered that a 220 V AC MCB may not be suitable for a 110 V DC application, unless it is tested and approved by the manufacturer. Hence, proper care must be taken while using MCBs and MCCBs in DC circuits.

11.3.2 Current-limiting MCCBs

Current-limiting MCCBs are essentially extremely fast-acting breakers that interrupt the short circuit fault current before it reaches the first peak, so reducing the current or energy let-through in the same manner as a fuse. They are therefore required to operate in the first quarter of a cycle, i.e. 5 ms or less and limit the peak short circuit current to a much lower value, after which can be switched on again, if necessary, without replacement of any parts or elements (see Figure 11.9).

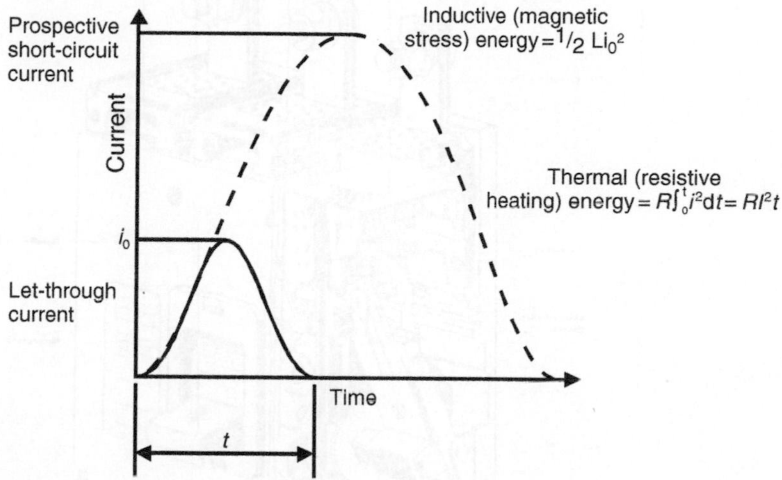

Figure 11.9
Limited short circuit let-through current

This high contact speed of separation is achieved by using a reverse loop stationary contact. When a fault develops, the current flowing in the specially designed contacts causes an electrodynamic repulsion between them. The forces between the contact arms increase exponentially rather than linearly. As the contact gap widens, the arc is quickly extinguished by a high-performance arc-chute.

By limiting the let-through current, the thermal and magnetic stresses on protected equipment such as cables and busbars is reduced in case of a short circuit. Provided combination series tests have been done and certified, this also permits the use of MCCBs with lower short-circuit capacities to be used at downstream locations from the current limiting MCCB. This is known as cascading and results in a more economical system, but additional care must be taken to preserve discrimination between breakers. This will be discussed in more detail in the next Section 11.4.

Accessories

The following accessories are available with all different makes of MCCB:

- Shunt trip coils
- Under-voltage release coils

- Auxiliary switches
- Mechanical interlocks
- Residual current devices (earth fault protection).

A typical moulded case circuit breaker (MCCB) is shown in Figure 11.10. When selecting an MCCB for an application it is important to ensure that the following ratings are correct:

1. Voltage rating (AC/DC)
2. Current rating
3. Breaking capacity rating.

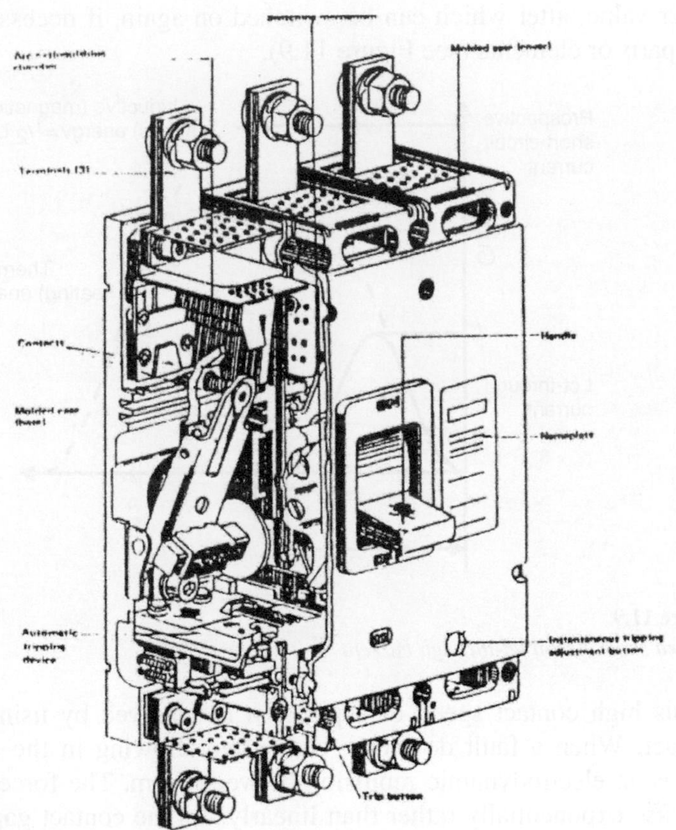

Figure 11.10
Typical moulded case circuit breaker

Installation

Figure 11.11 indicates the standard precautions to be followed while installing the MCCB.

Connection

Figure 11.12 indicates the standard precautions to be followed while connecting the MCCB.

Cautions for installation

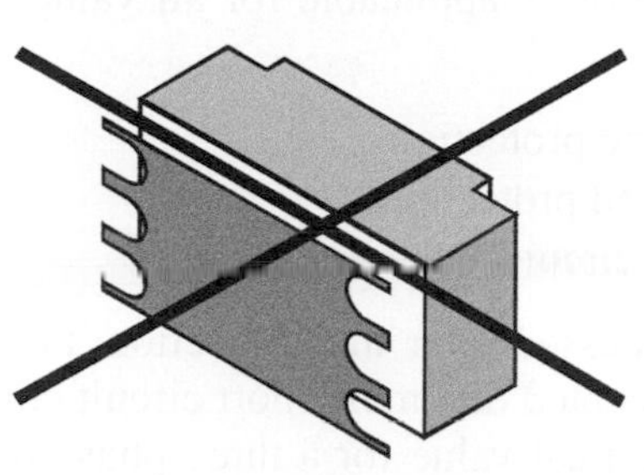

Do not remove the compound inserted into the screw part of the base rear or the rear cover

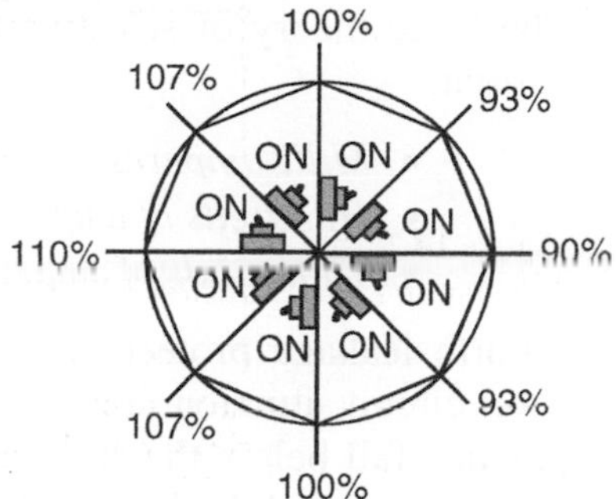

Change ratio of the rated current value according to the installation angle

Figure 11.11
Cautions for installation

1. Take sufficient insulation distance

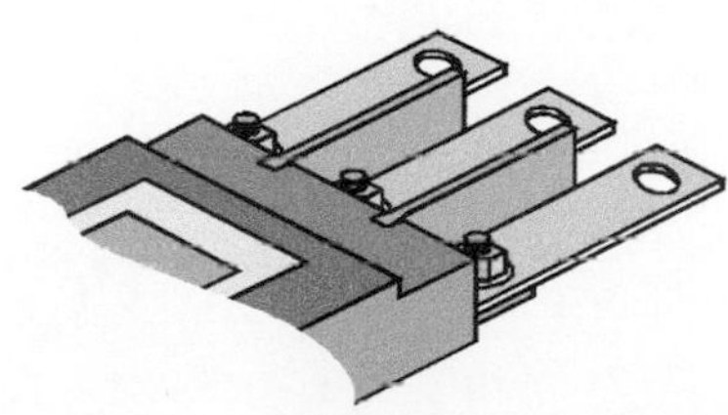

Take care, as the insulation distance may be insufficient according to the installation position of the connection conductor.
As some type are provided with insulation barriers, they should be used under reference top appendix table 6

2. Do not apply oil to the threaded parts

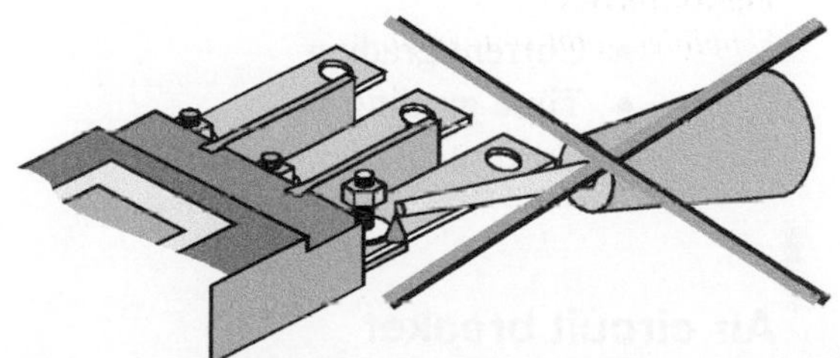

Do not apply lubrication oil to the threaded parts. Application of lubrication oil reduces the friction of the threaded part, so that loosening and overheat can be caused. In case of lubrication, even the standard tightening torque can produce excessive stress in the threaded part and thus breaking of the screw

1. Parallel conductors for all poles

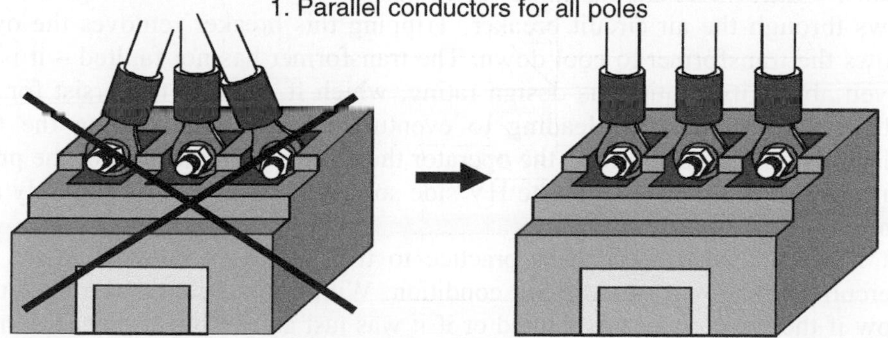

Install the connection conductors In parallel for all poles

Figure 11.12
Cautions for connections

11.4 Application and selective coordination

The basic theory of selective coordination is applicable for all values of electrical fault current.

- *Milli-amperes*: Earth leakage protection
- *Hundreds of amps*: Overload protection
- *Thousands of amps*: Short circuit protection.

Earth leakage protection will be discussed later under Section 11.5. Considering the short circuit situation, it is generally accepted that most short circuit currents that occur in practice fall below the calculated theoretical value for a three-phase bolted fault. This is because not all faults occur close to the MCCB (except when the supply cable is connected to the bottom of the MCCB!). The resistance of the cable between the MCCB and the fault reduces the fault current; also, most faults are not bolted faults – the arc resistance helping to reduce the fault current even further.

For economic and practical reasons it is not feasible to apply the same sophisticated relay technology as used on the medium-voltage to low-voltage networks, as this would result in a very complicated and expensive system. The present system therefore of using air and moulded case circuit breakers is a successful compromise developed over many years.

These devices, however, are current operated as described previously, so it is possible to achieve varying degrees of coordination by the use of:

- Current grading
- Time grading
- Current and time grading.

11.4.1 Air circuit breaker

Let us now consider the protection provided by the air circuit breaker on the LV side of the main in-feed transformer.

Transformer overload condition

The thermal element on the air circuit breaker can be set to protect the transformer against excessive overloading, as the same current that flows through the transformer flows through the air circuit breaker. Tripping this breaker removes the overload and allows the transformer to cool down. The transformer has not faulted – it is only being driven above its continuous design rating, which if allowed to persist for some time, will cook the insulation leading to eventual failure. By checking the temperature indicators on the transformer, the operator then has clear indication of the problem. The transformer is still alive from the HV side so it has not faulted. It is purely an overload condition (see Figure 11.13).

It has often been a common practice to trip the transformer from the HV IDMT overcurrent relay for an overload condition. With this approach, the operator does not know if the transformer has faulted or if it was just an overheating condition. He is now faced with a decision and if he is conscientious he may decide to test the unit before switching in again. This could lead to excessive downtime. In addition, the HV IDMT overcurrent relay (normal inverse) does not have the correct characteristic for overload protection as pointed out in Chapter 10.

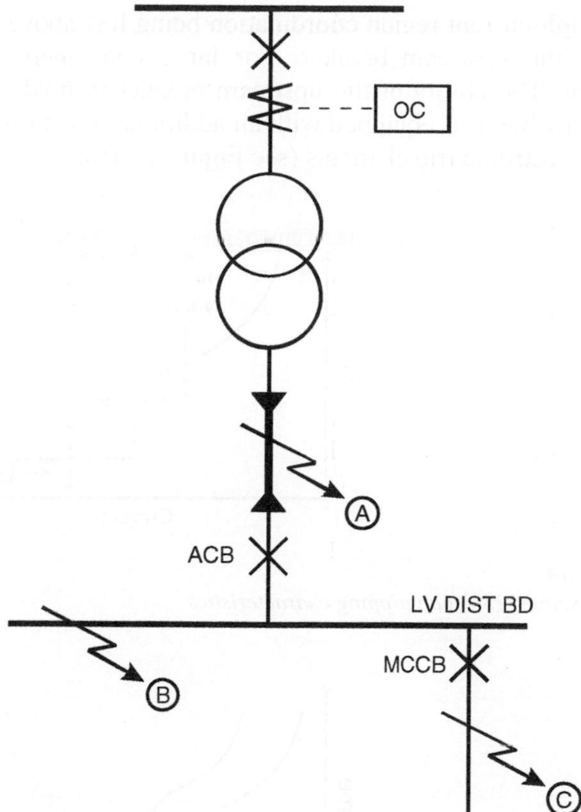

Figure 11.13
LV air circuit breaker on transformer

Short circuit protection

Short circuits at points A, B and C must now be considered. The fault currents will bethe same as there is virtually no impedance between them. The short circuit protection on the air circuit breaker (ACB) should therefore be set with a short time delay to allow the downstream MCCB to clear fault C. However, if the fault is on the busbar the time delay should be short enough to effect relatively fast clearance to minimize damage and downtime.

Fault A will have to be cleared by the HV overcurrent relay in order to protect the cable from the transformer to the LV switchboard. This in turn should have a longer time delay to coordinate with the LV ACB and provide discrimination for faults B and C.

These requirements show the value of specifying adjustable current pick-ups and time delays for the protection devices on air circuit breakers, most of which are available in electronic form. In addition, they also come equipped with a very high-set instantaneous overcurrent feature having a fast fixed time setting of 20 ms to cover 'closing-onto-fault' conditions (see Figure 11.14).

Moulded case circuit breakers

A reasonable degree of current grading can be achieved between two series-connected MCCBs by simply applying a higher-rated breaker upstream of a given unit. The extent of the coordination is shown on the following time–current characteristic curves (see Figure 11.15). It will be noted that selectivity is obtained in the thermal overload and partial high-current region coordination being lost above the short circuit pick-up current level of the upstream breaker. For large consumers, the integrity of the supply is important. The ability of the upstream breaker to hold in under such fault conditions is enhanced when it is equipped with an additional short time delay facility, provided by the modern electronic trip elements (see Figure 11.16).

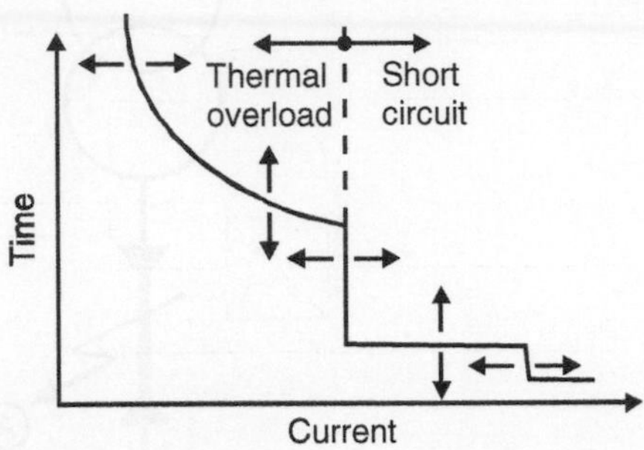

Figure 11.14
ACB adjustable protection tripping characteristics

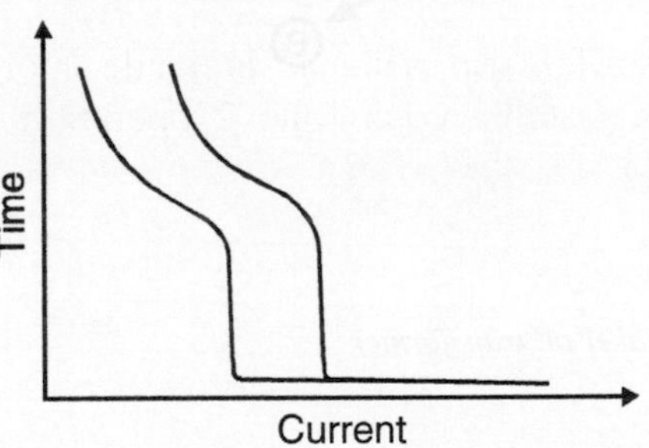

Figure 11.15
Current coordination in MCCB

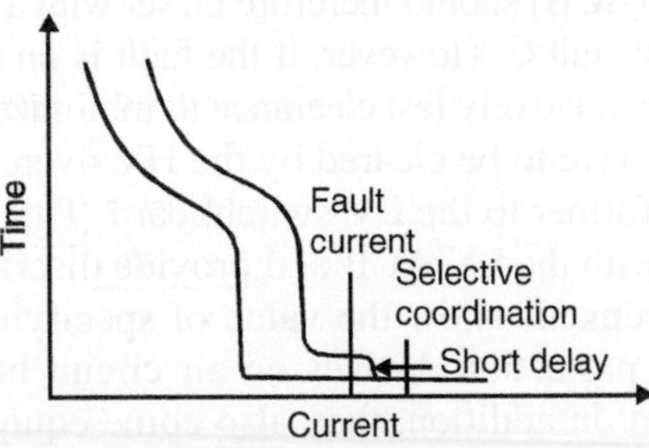

Figure 11.16
Current–time coordination

MCCB unlatching times

Once triggered, MCCBs have an unlatching time, which is dictated by the physical size and inertia of the mechanism. It stands to reason that the physically smaller, lower-rated breakers will have a shorter unlatching time than the higher-rated, larger upstream breakers, thereby enhancing their clearing time.

Experience in practical installations of fully rated breakers has shown that unexpected degrees of discrimination have been achieved because of this. For current-limiting circuit breakers, where contact parting occurs independent of the mechanism, the unlatching times do not have such an impact on their clearance times.

Fully rated systems

When time delayed MCCBs are used to achieve extended coordination, all downstream circuit breakers must be rated to withstand and clear the full prospective short-circuit current at the load side terminals.

Cascading systems

This approach can be used if saving on the initial capital cost is the overriding factor. This necessitates using a current-limiting breaker to contain the let-through energy so allowing lower-rated (hence less costly) breakers to be used downstream. To achieve successful coordination it will be appreciated that careful engineering is required especially with regard to clearance and unlatching times, in addition to size and length of interconnecting cables together with accurate calculation of fault levels.

If the let-through energy is sufficient to cause the downstream breaker to unlatch, then the faulty circuit will be identified, although the upstream current-limiting MCCB will also have tripped to drop the whole portion of the network being served by this main breaker. However, if the downstream breaker does not unlatch then extended outage time is inevitable to trace the fault location. It is vital that the complete system be tested and approved to ensure the delicate balance of the system is not disturbed. There are number of factors that need careful consideration.

Sluggish mechanisms

It is well known that any electromechanical assembly of links, levers, springs, pivots, etc. which remain under tension or compression for a long period, tend to 'bed in'. Also, dust and corrosion contribute further to retarding the operation after long periods of inactivity. The combined effect could add a delay of 1–3 ms when eventually called into operation.

This additional delay has little effect on fully rated breakers which generally operate after one cycle (20 ms), but on current-limiting MCCBs, which are required to operate in 5 ms, the additional 1–3 ms will have a significant impact on their performance. The increased energy let-through could have disastrous results for both itself and in particular the downstream breaker.

Point-on-wave switching

Most specifications and literature show current/energy-limitation based on fault initiation occurring at a point-on-wave corresponding to current zero. Should the fault occur at some other point on the wave, the d*i*/d*t* of the fault current would be much greater than that shown, resulting in higher-energy let-through (see Figure 11.17).

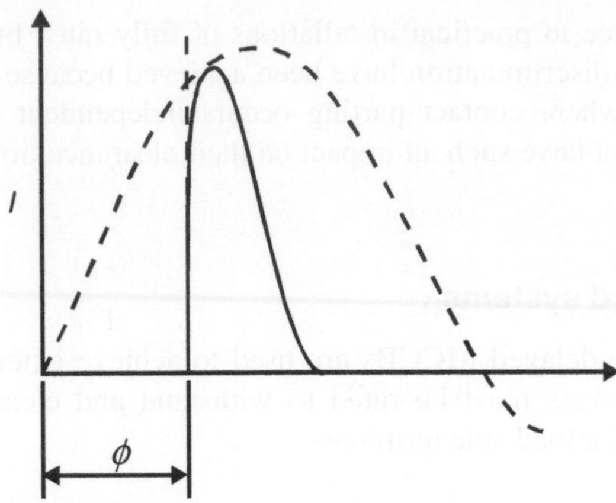

Figure 11.17
Effect of point-on-wave fault occurrence

Service deterioration

Qualification-type tests in most international specifications require the MCCB to successfully perform one breaking operation and one or two make–break operations. In practice, it is rare that the number of operations by a breaker under short-circuit conditions is monitored. This shortcoming is not critical on fully rated systems as the protection of the downstream breakers is of no consequence.

However, in a series-connected cascading system, where the downstream breakers rely for their survival on the energy-limiting capabilities of the upstream current-limiting breaker, there is always the danger that replacement of the upstream device is overlooked. There is therefore a strong case for monitoring the number of operations.

Maintenance

For reasons stated above, any upstream or downstream breakers in a cascade system must be replaced with identical breakers from the same manufacturer in accordance with the original test approvals. This also applies to any system extensions. Any deviation could prove disastrous. Incorrect replacement of the upstream breaker could result in higher-energy let-through and longer operating times, whilst incorrect replacement of downstream breakers may lead to lower energy-handling capability coupled with shorter operating times.

These conflicting requirements are such that even experienced or well-trained technicians may be confused unless they are fully conversant with the principle of the cascade system. There could be an even greater problem for the maintenance electrician and his artisan, in selecting a replacement device, which may often be dictated by availability.

Identification

In view of the problems of staff turnover and the possibility of decreasing skills, it therefore becomes a stringent requirement that all switchboards carry a prominent identifying label together with all relevant technical information to ensure the satisfactory operation and maintainability of cascaded or series-connected systems.

General

Although cascaded systems may offer an attractive saving in initial capital expenditure, it requires a higher level of engineering for the initial design and extensions. Maintenance can be difficult, as total knowledge and understanding of the system and all its components is required by all operating personnel. The consultant, contractor or user is thus faced with the decision of choosing between two quite different systems:

1. A fully rated properly coordinated system
2. A system based on cascaded ratings.

The first choice may have a slightly higher initial cost. The alternative, offers some initial cost savings whilst sacrificing some system integrity, selectivity and flexibility.

11.5 Earth leakage protection

In the industrial and mining environment the possibility of persons, making direct contact with live conductors is very remote. This is because the conductors are housed in specially designed enclosures, which are lockable and where only trained qualified electricians are allowed access.

The danger lies, however, when an earth fault occurs on a machine and because of poor earth bonding, the frame of the machine becomes elevated to an unsafe touch potential as illustrated in Figure 11.18 below. This is an 'indirect' contact situation, which must be protected.

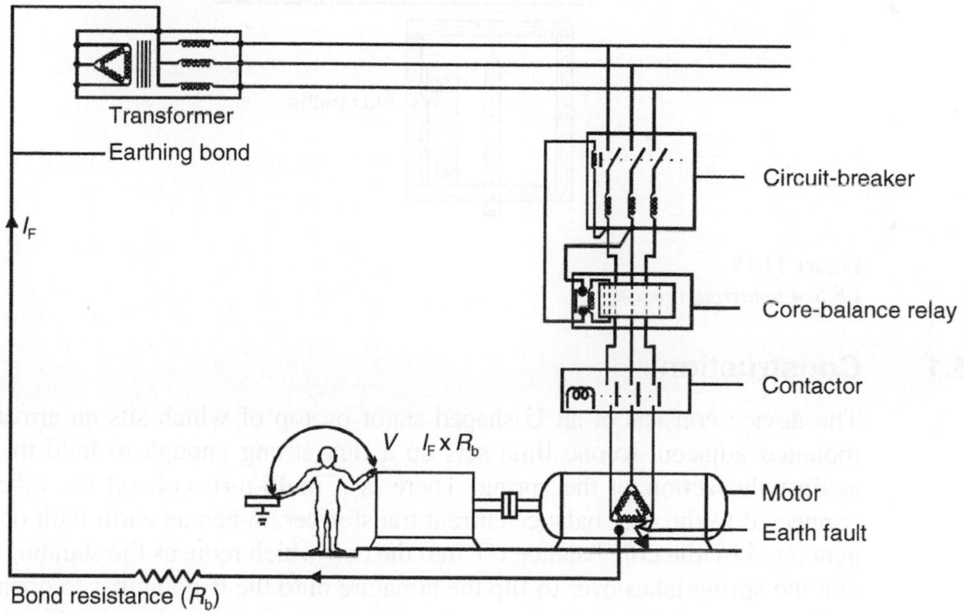

Figure 11.18
Protection against indirect contact

Safety codes specify that in mining and industrial installations any voltage above the range of 25–40 V is considered to be unsafe. These figures are derived from the current level that causes ventricular fibrillation – 80 mA times the minimum resistance of the human body which can be in the range of 300 Ω (3 × safety factor) 500 Ω (2 × safety factor). Please refer to Chapter 4.

It would not be possible to utilize the sensitive domestic earth leakage devices (30 mA, 30 ms) in these applications because of the transient spill currents that occur during motor starting. Instantaneous tripping would occur and the machine would never get started.

Tests have been carried out in coalmines to determine the maximum resistance that could occur on an open earth bond. This was measured as 100 Ω. With 25 V specified as the safe voltage, a current of 250 mA can be regarded as the minimum sensitivity level (derived from dividing 25 V by 100 Ω). This level was found to be stable for motor starting. It is however above the 80 mA fibrillation level of the heart, so speed is now of the essence if we are to save human life.

The earth leakage relays used in industrial applications should therefore operate in 30 ms. Modern earth leakage relays can achieve this and one such method is to use a unique sensitive polarized release as illustrated in Figure 11.19.

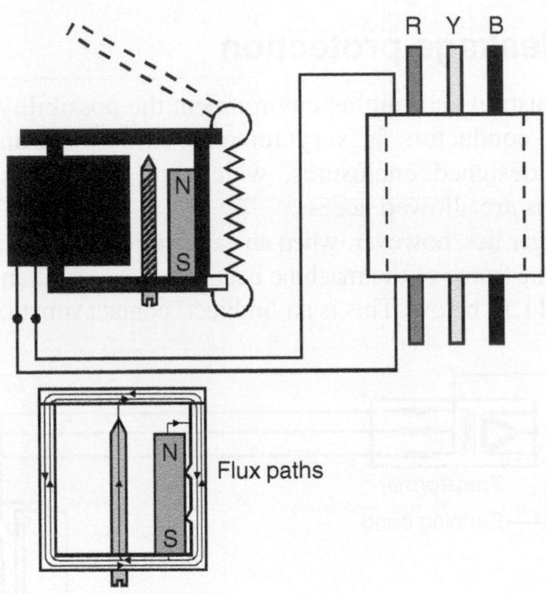

Figure 11.19
I.E.S. 4 polarized release

11.5.1 Construction

The device consists of an U shaped stator on top of which sits an armature. The magnet mounted adjacent to one limb sets up a flux strong enough to hold the armature closed against the action of the spring. There is a multi-turn coil on the other limb, which is connected to the core balance current transformer. When an earth fault occurs, an output is generated by the core balance CT into the coil which reduces the standing flux to the extent that the spring takes over to flip the armature onto the tripping bar to open the breaker. The calibration grub screw is a magnetic shunt. Screwing it in bleeds off magnetism from the main loop making the release more sensitive. Screwing it out allows more magnetism around the main loop, making the armature attraction stronger, hence less sensitive.

The burden of the release is only 400 microVA (10 mV, 40 mA), which allows extremely high sensitivities to be achieved. The release can be complimented by the addition of some electronics in order to produce a series of inverse time/current tripping curves (see Figure 11.20).

11.5.2 Description of operation

When an earth fault or earth leakage condition occurs on the system, the core balance CT mc generates an output. On the positive half cycle, the secondary current flows through diode D_1, resistor R and charges up capacitor C. On the negative half cycle, the current flows through diode D_2, resistor R and charges up capacitor C even further.

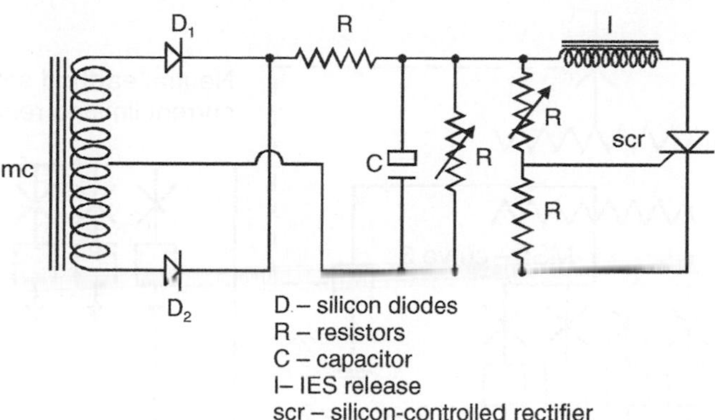

Figure 11.20
Internal circuitry

The voltage across the capacitor *C* is monitored by the resistor divider and once it reaches a pre-set voltage level the gate of the scr is triggered. All of the energy stored in the capacitor now flows through the release *I* to cause operation of the relay. The capacitor is now fully discharged enabling the relay to be reset immediately.
By varying the values of *R* and *C* the charge-up time can be varied.

11.5.3 Application and coordination of earth leakage relays

A family of relays has been designed to provide coordinated earth fault protection for low-voltage distribution systems. Using the above-mentioned technology, the following time/current inverse curves have been developed (see Figure 11.21). This allows coordinated sensitive earth fault protection to be applied to a typical distribution system. They afford 'back-up' protection to the end relay, which provides instantaneous protection to the apparatus where operators are most likely to be working (see Figure 11.22).

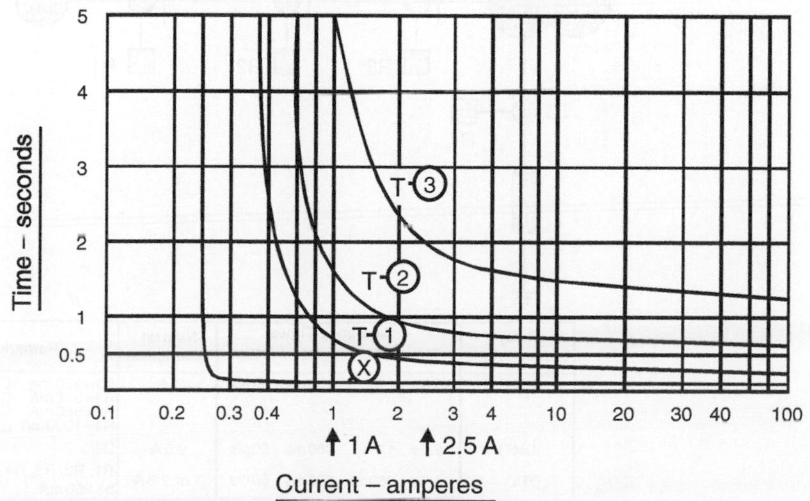

Figure 11.21
Time/current response curves

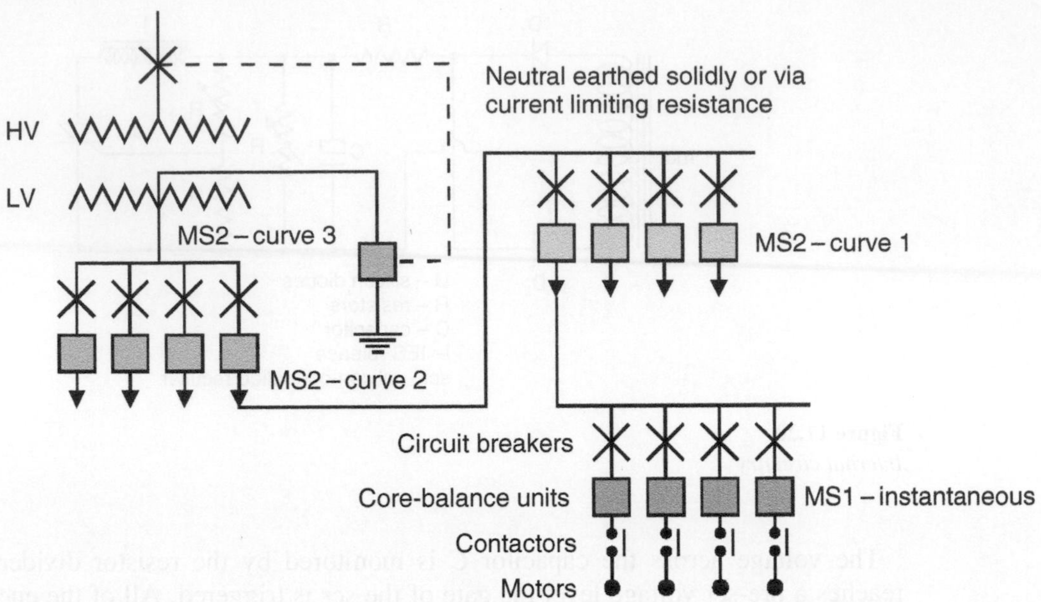

Figure 11.22
Typical LV distribution system

11.5.4 Optimum philosophy

It is important to note that the choice of relay settings cannot be considered in isolation. They are influenced by the manner of neutral earthing, current pick-up levels and time grading intervals, which in turn will be dictated by the system configuration.

All are interdependent and in the following example, it will be seen that optimum philosophy for the system would be a definite time lag philosophy (DTL) as opposed to an inverse definite time lag philosophy (IDMTL) as faster clearance times can be achieved (see Figure 11.23).

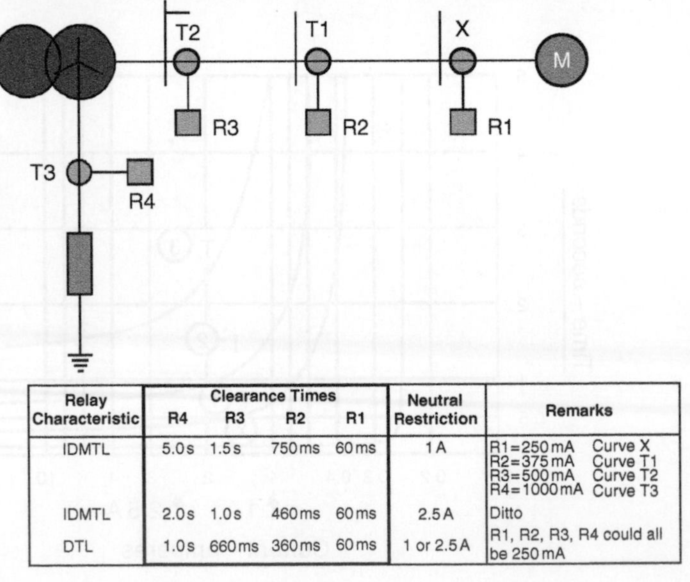

Relay Characteristic	Clearance Times R4	R3	R2	R1	Neutral Restriction	Remarks
IDMTL	5.0 s	1.5 s	750 ms	60 ms	1 A	R1 = 250 mA Curve X R2 = 375 mA Curve T1 R3 = 500 mA Curve T2 R4 = 1000 mA Curve T3
IDMTL	2.0 s	1.0 s	460 ms	60 ms	2.5 A	Ditto
DTL	1.0 s	660 ms	360 ms	60 ms	1 or 2.5 A	R1, R2, R3, R4 could all be 250 mA

Figure 11.23
Optimum philosophy

12

Mine underground distribution protection

12.1 General

A typical colliery underground network is shown in Figure 12.1. The protection required for the medium-voltage (11 kV) network from the surface substation to the mobile transformer will require considerations about the trailing cable. However, this could be the standard protection that is followed in a normal industrial substation, taking care of short-circuit and overcurrent conditions.

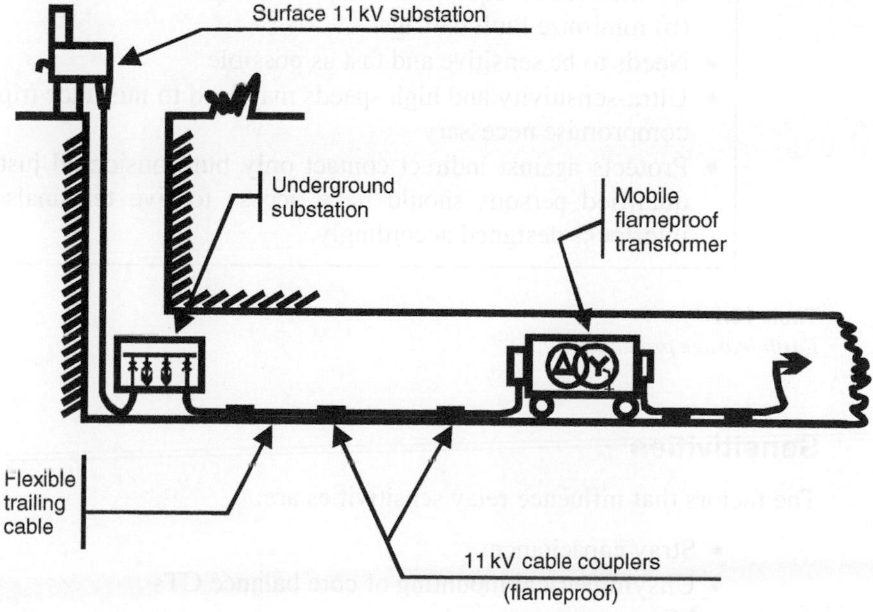

Figure 12.1
Typical colliery UG network

However, it is important to pay particular attention to the protection of the low-voltage 'front-end electrics' – especially at the coalface where most activity takes place, increasing the possibility of electrical faults occurring.

In coal mining, the normal protection found in the flameproof gate-end boxes comprises:

- Earth-leakage protection
- Pilot wire monitors (ground continuity monitors)
- Earth fault lockout
- NERM (neutral earthing resistor monitors).

These features will now be discussed in detail point by point.

12.2 Earth-leakage protection

Earth-leakage protection is primarily employed to protect life. It must therefore detect and isolate faulty equipment as soon as possible to protect the rest of the system and to minimize fault damage.

Consequently, it needs to be as sensitive and as fast as possible. However, ultra-sensitivity and high speeds can lead to nuisance tripping so a compromise is necessary. Generally, one only needs to consider protecting against indirect contact. This is considered justified, as only qualified persons should have access to live terminals, equipment and interlocks being designed accordingly (see Table 12.1).

Earth-Leakage Protection
• Primarily employed to protect human life • Isolates faulty equipment asap: to (a) protect rest of the system; and (b) minimize fault damage • Needs to be sensitive and fast as possible • Ultra-sensitivity and high speeds may lead to nuisance tripping – hence compromise necessary • Protects against indirect contact only but considered justified as only qualified persons should have access to live terminals. Equipment/interlocks designed accordingly.

Table 12.1
Earth-leakage protection

12.2.1 Sensitivities

The factors that influence relay sensitivities are:

- Stray capacitance
- Unsymmetrical mounting of core balance CTs
- Motor starting in-rush currents
- Transients:
  - Switching surges/point-on-wave switching
  - Lightning
  - Voltage dips
  - Harmonics (especially 3rd and 9th, etc.).

Unbalances can be caused by one or more combinations of the above. Relay sensitivities of 250 mA were found to be immune from the above whereas levels of 100 mA were susceptible so they had to be time delayed by 100 ms to ride through the transient disturbances.

Using 250 mA instantaneous sensitivity, typical relay coordination for earth-leakage protection is shown in Figure 12.2 and Table 12.2.

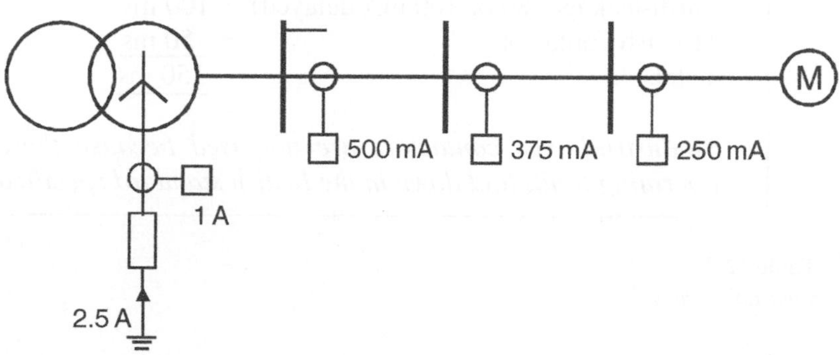

Figure 12.2
Typical earth-leakage current sensitivities

Earth-leakage Sensitivities
• Factors influencing relay sensitivities: – Stray capacitance – Unsymmetrical mounting of core balance CTs – Motor starting in-rush currents – Transients (a) Switching surges/point-on-wave switching (b) Lightning (c) Voltage dips – *Harmonics (especially 3rd, 9th, etc.)*: Unbalances could be caused by one or more combinations of the above. – *Relay coordination*: Grading is achieved on a current/time basis as shown in the diagram.

Table 12.2
Earth-leakage sensitivities

12.2.2 Clearance times

Typical clearance times for South African equipment compared to UK equipment are shown in Table 12.3.

The faster speeds are desirable as they are much less than the 'T' phase resting period of the heart.

Earth Fault Clearance Times	
• South African equipment:	
Earth leakage (250 mA instantaneous)	= 30 ms
Industrial type contactor	= 30 ms
Total	60 ms
• UK equipment (NCB):	
Earth-leakage (80 or 100 mA delayed)	= 100 ms
*British contactor	= 50 ms
Total	150 ms

**Industrial type contactors are not used because these are outside the operating limits laid down in the British standard specification.*

Table 12.3
Clearance times

12.3 Pilot wire monitor

This is a very important and sophisticated relay as it carries out the following functions (see Figure 12.3):

- Prevents on-load uncoupling of cable couplers
- Ensures continuity and measures earth bond resistance
- Detects pilot-to-earth short-circuit
- Permits remote start/stop of contactor using pilot
- Unit has to be fail safe.

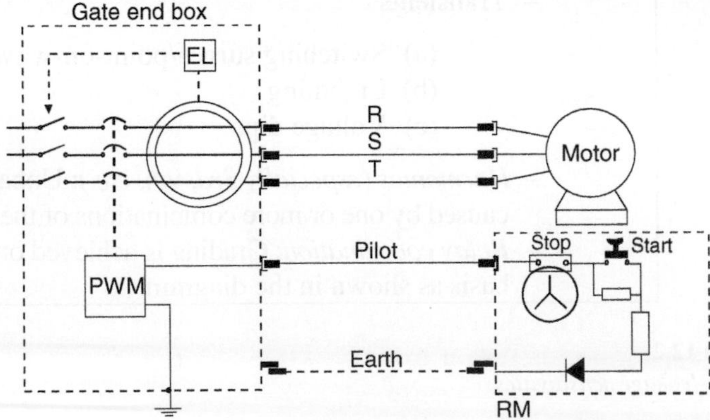

Figure 12.3
Pilot wire monitor (earth continuity monitor)

It is designed to meet very fine tolerances as shown in Figure 12.4. As it is continuously monitoring the resistance of the earth bond to keep equipment within safe touch – potential limits, it can be regarded as important, if not more so, than the earth-leakage relay.

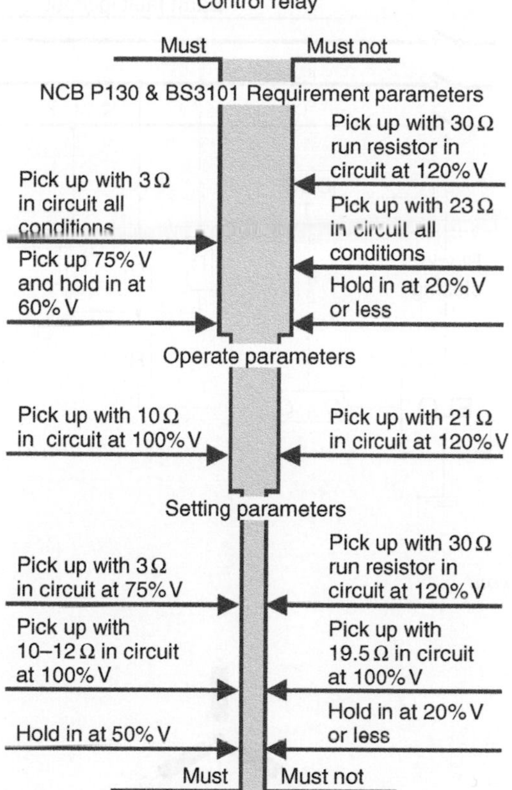

Note: The contactor must be capable of following the relay under all of the above conditions.

Figure 12.4
Pilot wire monitor operating characteristics

12.4 Earth fault lockout

As an additional safety measure, an earth fault lockout feature is installed, typically as shown in Figure 12.5. After the contactor has been tripped, a DC signal is injected onto the power conductor via the resistor bank to monitor the insulation. Closing is prevented if this drops below the pre-set value. As an example, this would ensure safety on start-up should a rock fall have occurred during the off-shift period.

12.5 Neutral earthing resistor monitor (NERM)

The final element in the protection system is this relay, which ensures the integrity of the neutral earthing resistor. If this latter device should open or short-circuit, the NERM will operate to either alarm on trip (see Figure 12.6). The problems experienced with solid earthing are as shown in Figure 12.7, namely:

- High fault currents – only limited by inherent impedance of power system
- Solid earthing means high earth fault currents
- This damages equipment extensively
- Leading to long outage times – lost production, lost revenue
- Heavy currents in earth bonding gives rise to high touch potentials – dangerous to human life
- Large fault currents are more hazardous in igniting gases (explosion hazard).

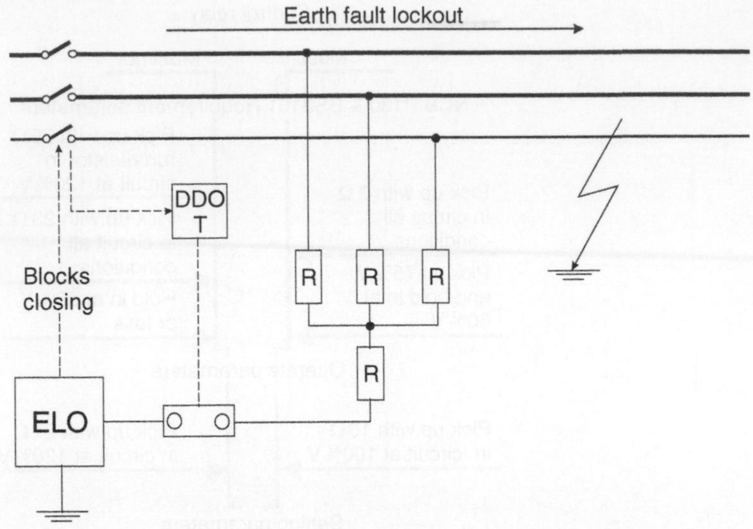

Figure 12.5
Earth fault lockout

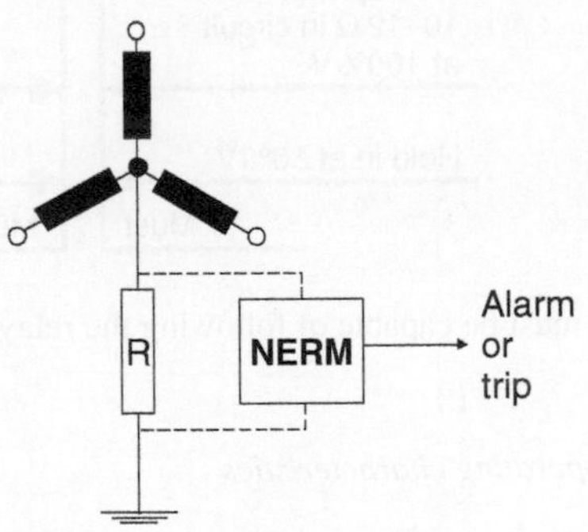

Figure 12.6
NER monitor

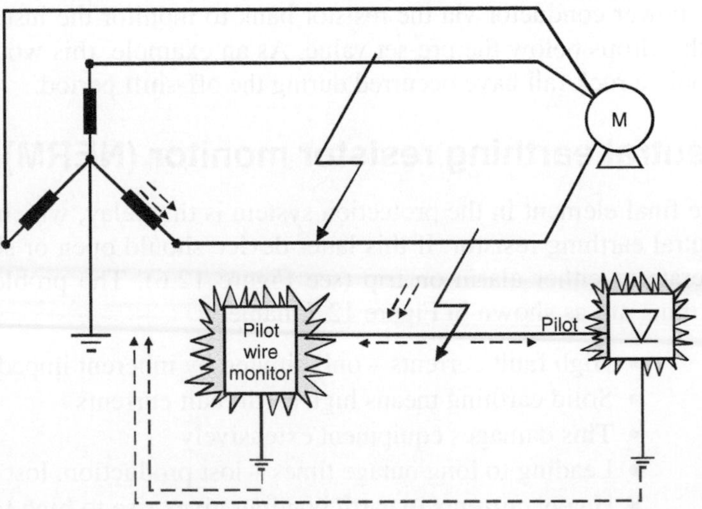

Figure 12.7
Problems

These can be overcome by introducing an earth barrier between the phases so that all faults become earth faults then controlling the earth fault current levels by the neutral earthing resistor (see Figures 12.8–12.10 and Table 12.4).

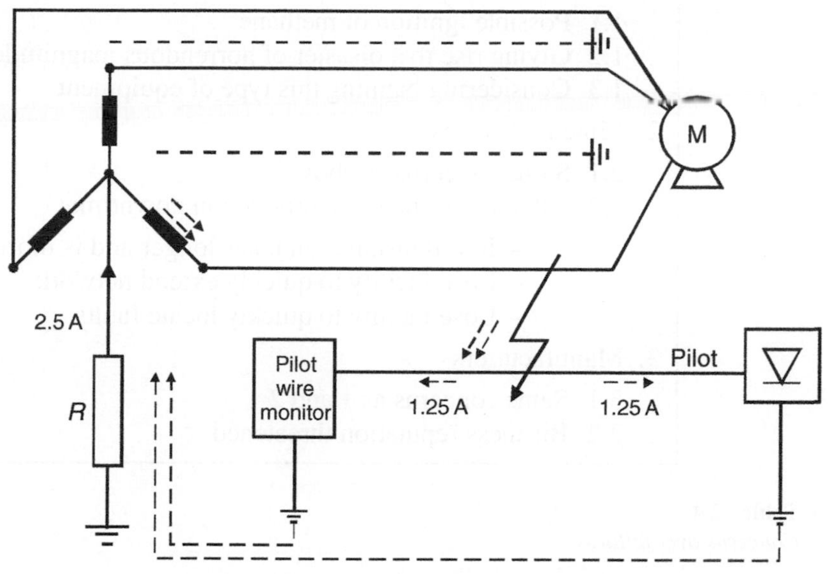

Figure 12.8
Solutions

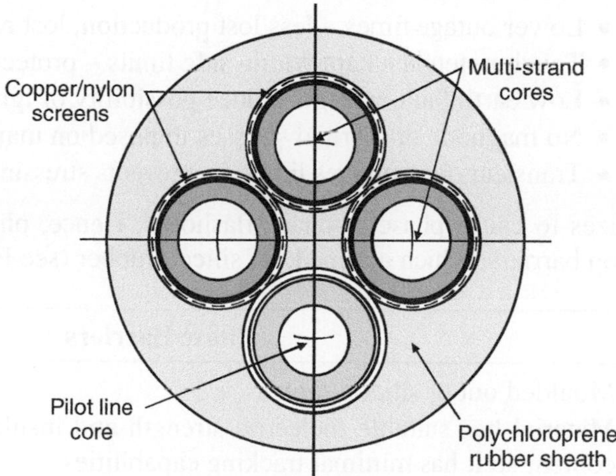

Figure 12.9
Screened trailing cable

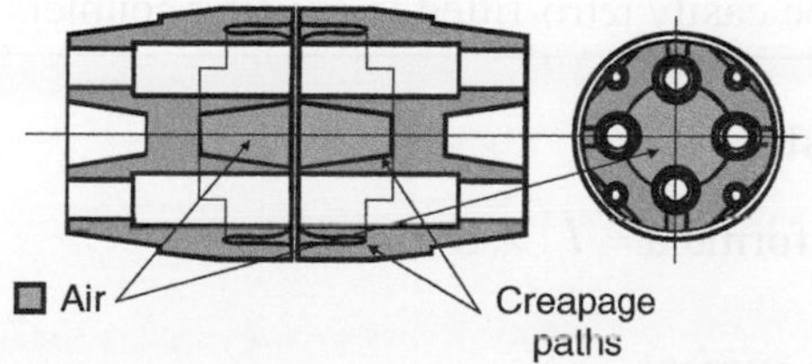

Figure 12.10
Investigation

Grave Concern
Over failure of 11 kV cable couplers

1. Government mining engineer
   - 1.1 Possible ignition of methane
   - 1.2 Giving rise to a disaster of horrendous magnitude
   - 1.3 Considering banning this type of equipment
2. Mine authorities
   - 2.1 Same concerns as above
   - 2.2 Alternatives increase production downtime:
     - Initial installation takes longer and is more difficult
     - Lose facility to quickly extend network
     - Lose facility to quickly locate faults
3. Manufacturers
   - 3.1 Same concerns as 1 and 2
   - 3.2 Business reputation threatened

Table 12.4
Concerns over failures

- Phase segregation eliminates phase-to-phase faults
- Resistance earthing means low earth fault currents
- Fault damage minimal – reduces fire hazard
- Lower outage times – less lost production, lost revenue
- Touch potentials kept within safe limits – protects human life
- Low earth fault currents reduce possibility of igniting gases (explosion hazard)
- No magnetic or thermal stresses imposed on major plan during fault
- Transient overvoltages limited – prevents stressing of insulation, MCB restrikes.

Air ionizes to cause phase-to-phase flashover. Hence, phase segregation is achieved by insulation barriers, which are made of silicon rubber (see Figure 12.11).

Phase Barriers
• Moulded out of silicon rubber • Material has suitable dielectric strength and insulating properties; no carbon content so it has minimal tracking capabilities • Flexible – allows push-on fit over insulator ports and gives a cleaning action • Is compressed as coupler halves bolt together • Acts as a 'heat sink' for any arc that occurs, thereby assisting quenching • Can be easily retro-fitted to existing couplers in service.

Proving tests

Fault energy formula = $I^2 \times R \times t$

Where

I = fault current
R = resistance of fault arc
t = time in seconds fault is on.

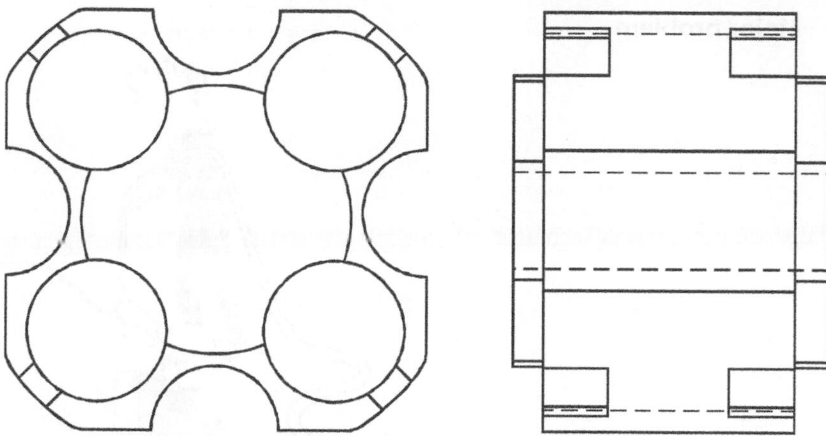

Figure 12.11
Solution–rubber phase barrier

Khutala fault throwing tests

Fault current = 4000 A
Clearance time = 350 ms
Assume arc resistance of 1 Ω
Fault energy $= 4000 \times 4000 \times 1 \times 0.35$
$= 5.6$ MJ
If clearance time reduced to 100 m/s
Fault energy $= 4000 \times 4000 \times 1 \times 0.1$
$= 1.6$ MJ

70% reduction

If steps could be taken to also reduce level of fault current then major strides would be made.

Fault location

In mining industries, identification of a fault location is a critical requirement. The following sketches/ pictures show how couplers are useful in identifying the fault location (see Figures 12.12–12.14 and Tables 12.5–12.8).

Flags to Identify Fault Location
• Flags indicate path to earth fault • Flags automatically reset on restoration of normal load current • Just plugs in between new or existing couplers • Obviates uncoupling and recoupling cable couplers in wet polluted environment when fault-finding.

Major problem

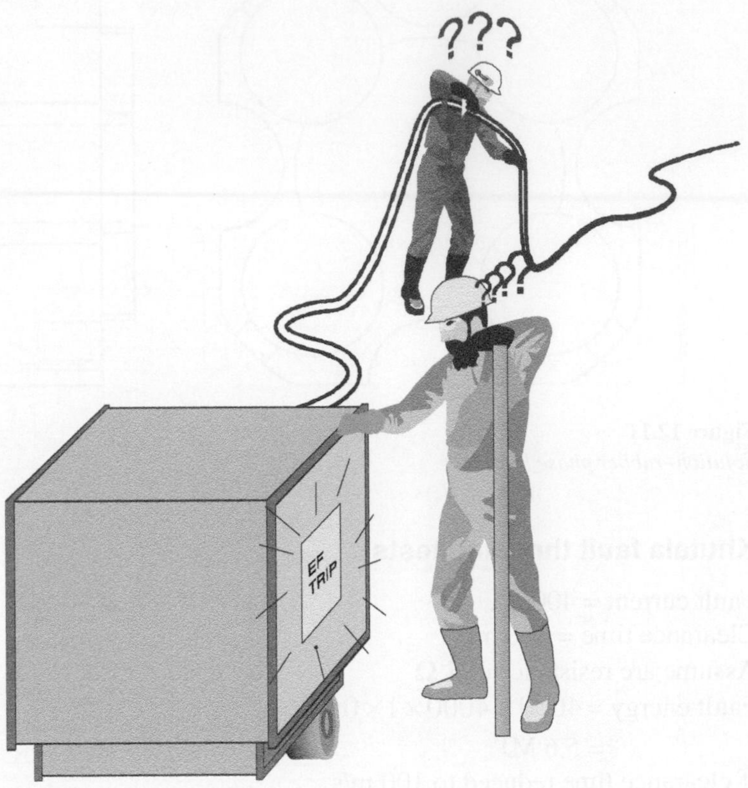

Figure 12.12
Great difficulty in locating earth faults

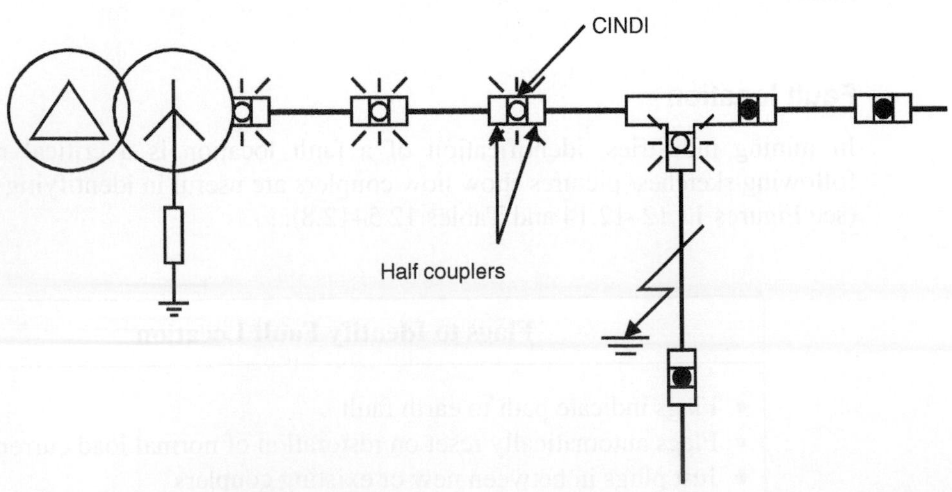

Figure 12.13
Fast fault location

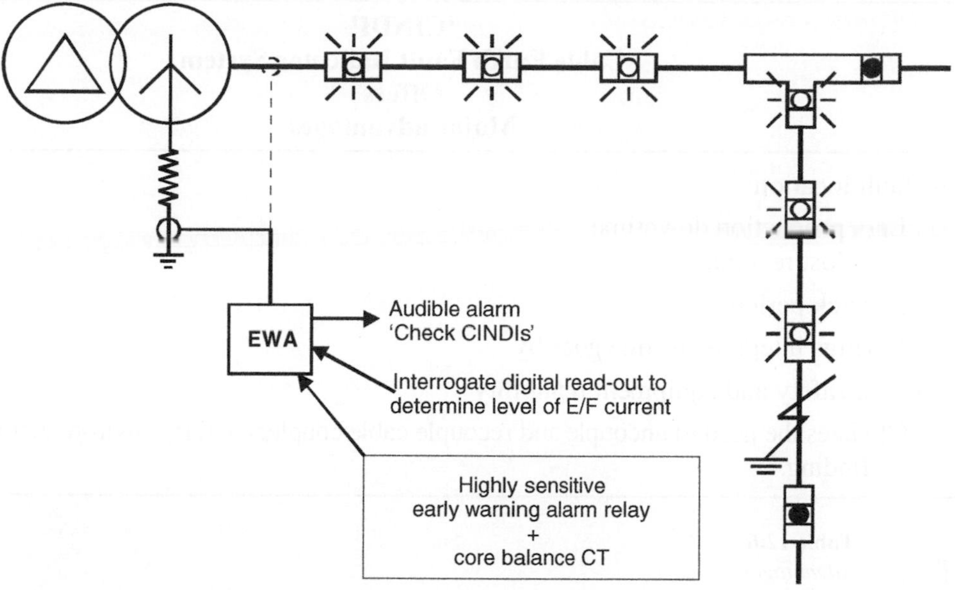

Figure 12.14
Early warning alarm system

Features	
• Universal	Voltage independent
	Any voltage up to 3.3 kV (500 V, 1000 V, 3300 V)
• High sensitivity	250 mA earth fault, 40 ms
	Coordinates with upstream relays
	Operates when closing onto a fault
• Self-reset	Automatically resets of restoration of normal load current
• Self-powered	No separate power supply required
• Stable flag	Magnetic disk pulse operated
	Can be mounted in any position
	Retains status through shock, vibration and power failures, etc.
• Installation	Quick and easy
	Just plugs in between two half couplers
	Ideal for retro-fitting
	No special tools required
	No breakdown of prepared joints
• Tamperproof	Electronics potted in epoxy-resin
	Intrinsically safe circuitry
• Flameproof	AS/SABS/BS tested
	GME certificate obtained

Table 12.5
Features

'CINDI' Cable Earth Fault Indicator System Offers Major advantages
• Fast fault location – Less production downtime – Less lost revenue • Fast payback period – Savings integrate as time goes by • Increased safety and equipment reliability – Obviates the need to uncouple and recouple cable couplers in dirty environment when fault finding.

Table 12.6
Advantages

CINDI-Major Advantages Other factors	
Economics	
• Potential to save mining industry millions of dollars in lost revenue due to production down time	
• Cost-effective	Same price as a half coupler
• Fast payback period	Cost of one extended outage would more than pay for equipping mine with these units
• Return on investment	Excellent. Benefits integrate as time goes by
New developments	
• Field experience so far has been excellent and requests have been received to develop:	
11 kV (1 A) model	This is the network that grows. Prototypes now in the field
3.3 kV (80 mA) model	For transformer and continuous miner To coordinate with 100 mA E/L relay Prototypes also in field.

Table 12.7
Other factors

Summary
• Phase segregation a must • System earthing – restricts neutrals currents to a low value with a resistor • Install fast, highly sensitive, earth-leakage protection • Pilot wire monitor, an essential and equal partner • Fit an earth fault lockout system • Install phase barriers and cable earth fault indicators to reduce production downtime and improve safety levels.

Table 12.8
Summary of recommendations

13

Principles of unit protection

13.1 Protective relay systems

The basic function of protection is to detect faults and to clear them as soon as possible. It is also important that in the process the minimum amount of equipment should be disconnected. The ability of the protection (i.e. relays and circuit breakers) to accomplish the latter requirement is referred to as 'selectivity'.

Speed and selectivity may be considered technically as figures of merit for a protection scheme. In general; however greater the speed and/or selectivity, the greater is the cost. Hence, the degree of speed or selectivity in any scheme is not purely a technical matter, it is also an economic one.

13.2 Main or unit protection

The graded overcurrent systems described earlier do not meet the protection requirements of a power system. As seen in Chapter 10, the grading is not possible to be achieved in long and thin networks and also it can be noticed that grading of settings may lead to longer tripping times closer to the sources, which are not always desired. These problems have given way to the concept of 'unit protection' where the circuits are divided into discrete sections without reference to the other sections.

Ideally, to realize complete selectivity of protection, the power system is divided into discrete zones. Each zone is provided with relays and circuit breakers to allow for the detection and isolation of its own internal faults.

This ideal selective zoning is illustrated in Figure 13.1. The protection used in this manner – essentially for internal faults in a particular zone – is referred to as main or unit protection.

13.3 Back-up protection

It is necessary to provide additional protection to ensure isolation of the fault when the main protection fails to function correctly. This additional protection is referred to as 'back-up' protection. For example, referring to the above figure, assume that a fault has occurred on the feeder and that the breaker at A fails to open. To clear this fault, the circuits which are able to feed current to the fault through the stuck breaker A must be opened. The fault is outside the zones of the main protection and can only be cleared by the separate back-up protection. Back-up protection must be time delayed to allow for the selective isolation of the fault by the main or unit protection.

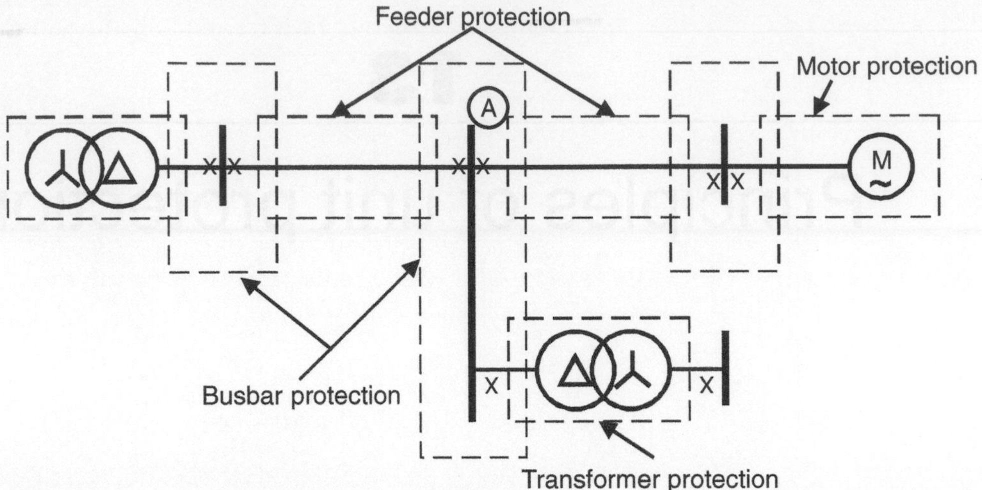

Figure 13.1
Overall schematic indicating busbar, feeder, transformer and motor protection

13.4 Methods of obtaining selectivity

The most positive and effective method of obtaining selectivity is the use of differential protection. For less important installations, selectivity may be obtained, at the expense of speed of operation, with time-graded protection.

The principle of unit protection was initially established by Merz and Price who were the creators of the fundamental differential protection scheme. These systems basically employ the direction of current rather than their actual values, protecting a particular zone by means of detecting the circulating currents through pilot wires and relays. The basic principles of these well-known forms of protection will now be considered.

13.5 Differential protection

Differential protection, as its name implies, compares the currents entering and leaving the protected zone and operates when the differential between these currents exceeds a pre-determined magnitude. This type of protection can be divided into two types, namely balanced current and balanced voltage.

13.5.1 Balanced circulating current system

The principle is shown in Figure 13.2. The CTs are connected in series and the secondary current circulates between them. The relay is connected across the midpoint thus the voltage across the relay is theoretically nil, therefore no current through the relay and hence no operation for any faults outside the protected zone. Similarly under normal conditions the currents, leaving zone A and B are equal, making the relay to be inactive by the current balance.

Under internal fault conditions (i.e. between the CTs at end A and B) relay operates. This is basically due to the direction of current reversing at end B making the fault current to flow from B to A instead of the normal A to B condition in the earlier figure (see Figure 13.3).

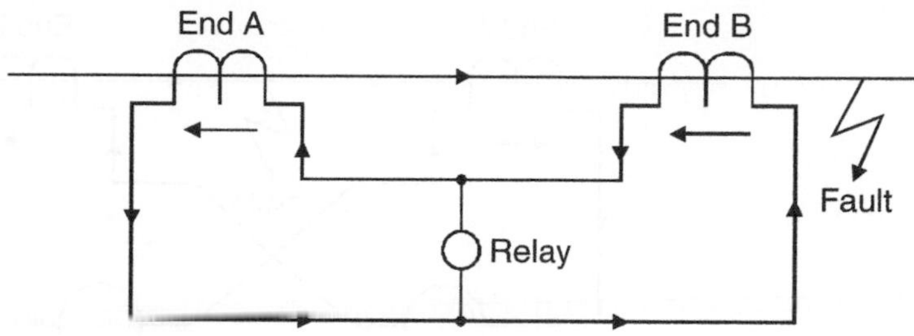

Figure 13.2
Balanced circulating current system, external fault (stable)

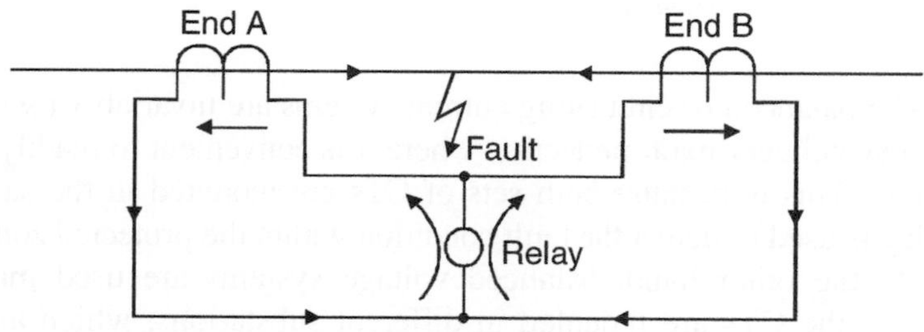

Figure 13.3
Balanced circulating current system, internal fault (operate)

The current transformers are assumed identical and are assumed to share the burden equally between the two ends. However, it is not always possible to have identical CTs and to have the relay at a location equidistant from the two end CTs. It is a normal practice to add a resistor in series with the relay to balance the unbalance created by the unequal nature of burden between the two end circuits. This resistor is named as 'stabilizing resistance'.

13.5.2 Balanced voltage system

As the name implies, it is necessary to create a balanced voltage across the relays in end A and end B under healthy and out-of-zone fault conditions. In this arrangement, the CTs are connected to oppose each other (see Figure 13.4). Voltages produced by the secondary currents are equal and opposite; thus no currents flow in the pilots or relays, hence stable on through-fault conditions. Under internal fault conditions relays will operate (see Figure 13.5).

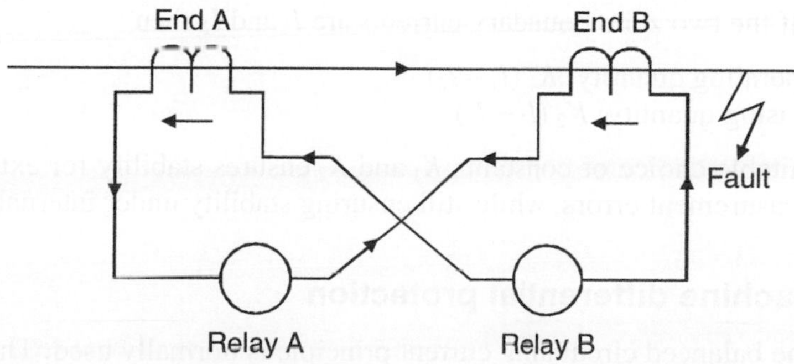

Figure 13.4
Balanced voltage system – external fault (stable)

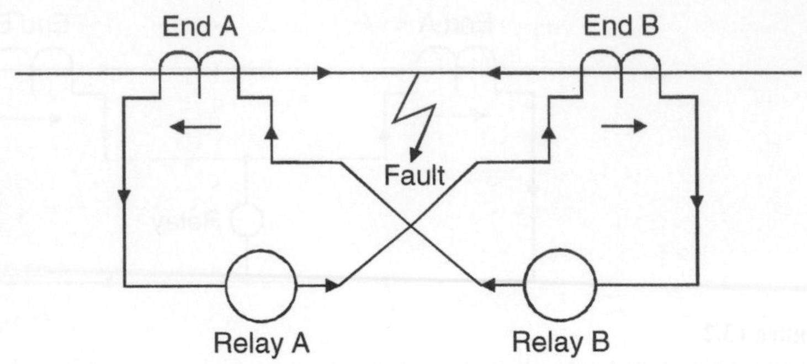

Figure 13.5
Balanced voltage system, internal fault (operate)

The balanced or circulating current systems are invariably used for generator, transformer and switchgear main protection where it is convenient to readily access the midpoint of the pilots. This is because both sets of CTs are mounted in the same substation and a single relay is used to detect the fault condition within the protected zone.

On the other hand, balanced voltage systems are used mainly on feeder protection where the CTs are mounted in different substations, which are some distance apart. As there are two relays involved, one at each end, they can each be mounted in their respective substation.

Although similar, the various forms of differential protection differ considerably in detail. The differences are concerned with the precautions taken to ensure stability – i.e. to ensure that the protection does not operate incorrectly for a through fault.

13.5.3 Bias

The spill current in the differential relay due to the various sources of errors is dependent on the magnitude of the through current. Hence it is necessary to consider the setting of the differential relay to be more than or proportional to the worst spill current likely to occur under through-fault conditions. Because of the wide range of fault current magnitudes, it is not always satisfactory to make the relay insensitive to lower-spill current values. This problem had been overcome by adjusting the operating level of the relay according to the total amount of fault current. This was done originally by providing a restraining winding or electromagnet which carries the total fault current while an operating electromagnet was allowed to carry only the differential current. This principle of bias is applied to circulating current protection to ensure proper operation under all fault conditions.

If the two zone boundary currents are I_1 and I_2, then

Operating quantity: $K_1 (I_1 - I_2)$
Biasing quantity: $K_2 (I_1 + I_2)$

Suitable choice of constants K_1 and K_2 ensures stability for external fault currents despite measurement errors, while still ensuring stability under internal fault conditions.

13.5.4 Machine differential protection

The balanced circulating current principle is normally used. The bias feature is introduced to ensure stability despite possible small differences in the performance of the two nominally balanced sets of current transformers.

The sensitivity of this protection is normally of the order of 10%, which means that the protection will operate when the differential current is greater than 10% of the normal full load. Without bias, for a through-fault current of ten times full load, the protection would operate if the 'spill' or differential current exceeded 10% of full load or 1% of the through-fault current. To avoid the necessity of matching current transformers to this degree of accuracy the protection is biased with through current.

13.6 Transformer differential protection

A typical transformer differential protection system also adopts the circulating current principle. The first point to notice is that the CTs on one side are connected in delta whilst they are connected in star on the other. This has been done for two reasons:

1. To correct for the phase-shift through the transformer in order to obtain cophasal currents at the relay.
2. To prevent the relay from operating incorrectly for an external earth fault on the side of the power transformer where the windings are connected in star with the neutral earthed.

Through-current bias is necessary on these relays not only for the inherent unbalances of the CTs but also to take care of any voltage tappings on the transformer provided by the tap-changer. For example, a transformer having a nominal ratio of 132/40 kV having a tap change range of +15 to –5% on the 40 kV side would have the CT ratios selected to be balanced at the midtap, namely 132/42 kV.

The above is discussed in more detail in Chapter 15.

13.7 Switchgear differential protection

In switchgear differential protection, all the currents entering and leaving the protected zone are added and if the resultant is zero then the busbars are healthy. However, if the current exceeds the chosen setting, the protection will operate and trip all associated circuit breakers.

The stability of this type of protection is obviously of vital importance since an incorrect operation could result for example in the shutdown of a power station. On account of the large number of circuits involved, all carrying different currents, stability is also a more difficult problem than with machine or transformer differential protection.

A number of different schemes are used for this protection, normally referred to as 'bus zone protection'. The schemes differ mainly in the principle adopted to obtain stability and these are discussed in detail in Chapter 16.

13.8 Feeder pilot-wire protection

Pilot-wire protection is similar to differential protection in that it normally compares the current entering the circuit at the one end with the current leaving at the other end. Its field of application is the protection of power cables and short transmission lines. For these circuits the distance between the current transformers at the two ends of the protected zone is too great for the circulating current differential protection of the type described previously for machines and transformers, etc. The pilot-wire provides the communicating channel for conveying the information relative to conditions at the one end of the feeder to protective relays at the other end of the feeder and vice versa. These relays or groups of relays, at the two ends are able to make a comparison between local and remote conditions and thus determine if there is an internal fault. Each relay normally trips only its associated circuit breaker. There are many different types of pilot wire protection schemes, but the most commonly used are of the opposing voltage type, an example of which is illustrated in Chapter 14.

13.9 Time taken to clear faults

With the inherently selective forms of protection, apart from ensuring that the relays do not operate incorrectly due to initial transients, no time delay is necessary. Operating times for the protection, excluding the breaker tripping/clearing time are generally of the following order:

Machine differential – few cycles
Transformer differential – 10 cycles
Switchgear (busbar) differential – 4 cycles
Feeder differential – few cycles

These operating times are practically independent of the magnitude of fault current.

13.10 Recommended unit protection systems

- *Cable feeders*: Pilot wire differential
- *Transformer*: HV-balanced (restricted) earth fault
  - HV high-set instantaneous overcurrent (low transient over-reach)
  - LV-restricted earth fault
  - Buchholz
- *Busbars*: Medium-/low-impedance schemes for strategic busbars (including busbars operating on closed rings)
  - Busbar blocking schemes for radial networks
- *Unit protection*: Should be used where possible throughout the network to remove the inverse time. Relays (IDMT) from the front line
- *IDMT*: Must be retained as back-up only to cover for a failure of the main protection.

13.11 Advantages of unit protection

13.11.1 Fast and selective

Unit protection is fast and selective. It will only trip the faulty item of plant, thereby ensuring the elimination of any network disruptions.

13.11.2 Easy to set

Unit protection is easy to set and once installed very rarely requires changing, as it is independent of whatever happens elsewhere on the system.

13.11.3 No time constraints

Time constraints imposed by the supply authorities do not become a major problem anymore. They only need consideration when setting up the back-up inverse time (IDMT).

13.11.4 Maximum operating flexibility

The system can be operated in any switching configuration without fear of a loss of discrimination.

13.11.5 Better continuity of supply

In many applications rings can be run closed, so that switching would not be necessary to restore loads resulting in better continuity of supply.

13.11.6 Future expansion relatively easy

Any future expansion that may require another in-feed point can be handled with relative ease without any change to the existing protection.

14

Feeder protection cable feeders and overhead lines

14.1 Introduction

Two commercially available systems have been chosen as typical examples to illustrate the concepts. However there are other similar products on the market using essentially the same technology.

14.2 Translay (see Figure 14.1)

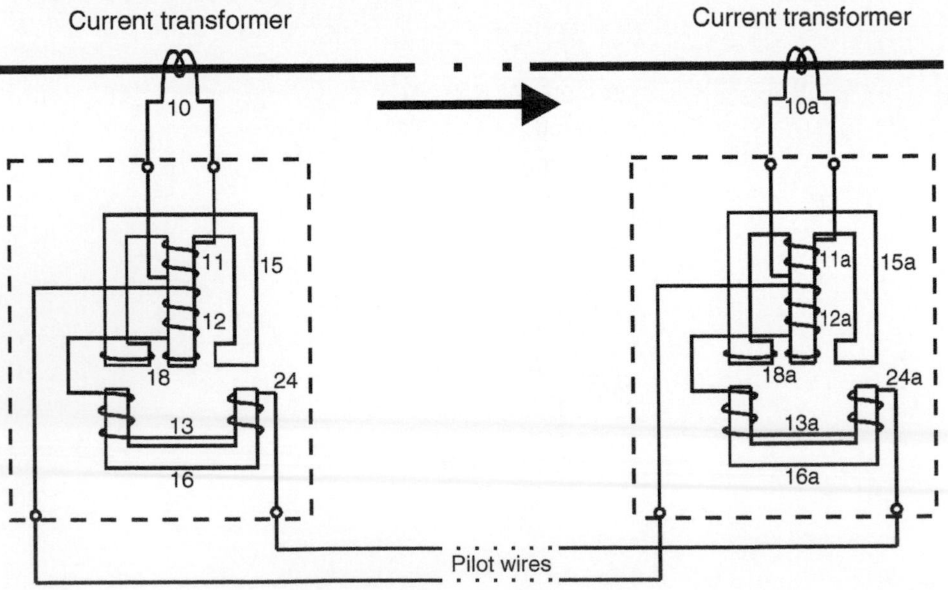

Figure 14.1
Simplified connections illustrating principles of operation

14.2.1 Translay is a voltage balance system

Whilst the feeder is healthy, the line CTs at each end carry equal currents. Equal and opposite voltages are induced in the secondary windings 12 and 12a and no current flows in the pilots. No magnetic flux is set up in the bottom magnets 16 and 16a so the relays do not operate. Under heavy through-fault conditions there may be a small circulating current due to line CT mismatch. A restraint torque is produced by bias loop 18, which also stabilizes the relay against pilot capacitance currents. A fault fed from one end causes current to circulate in the pilots and the relay at that end will operate to trip. A fault fed from both ends will cause a current reversal in the remote CTs, making the circulating current additive so that both ends operate to trip.

14.3 Solkor protection

Solkor unit protection is used where solid metallic pilot wires are available. The system is a differential protection system and is available as Solkor R/Rf. Optional equipment includes pilot wire supervision and injection intertripping systems.

Solkor protection can be configured in two modes. The R mode caters for systems where pilot/earth insulation levels are 5 kV or less. The Rf mode is used in newer systems with either 5 or 15 kV insulation. Solkor Rf gives faster clearance times for internal faults whilst its stability for through-faults is the same high value as Solkor R.

The standard relay has a pilot circuit to earth withstand of 5 kV, and interposing transformers are available to cater for circumstances when a 15 kV pilot insulation level is required. Solkor equipment will trip both ends of a faulty feeder even if current is fed from one end only. Solkor R/Rf relays are designed to use telephone type pilot cables having a loop resistance of up to 2000 Ω and a maximum capacitance between cores of 2.5 uF.

The relays incorporate diodes, which act as switches and permit the use of a single pilot loop for two-way signaling using a system of time sharing switched at the power frequency (see Figures 14.2–14.6). Normally, only two pilot wires are used for interconnecting the relays at the two ends of the feeder. For this reason, summation transformers are incorporated in the protection (see Figure 14.7). Because of the summation transformer the sensitivity of the protection is dependent upon phases that are involved in short circuit. Typical figures for sensitivity are as follows:

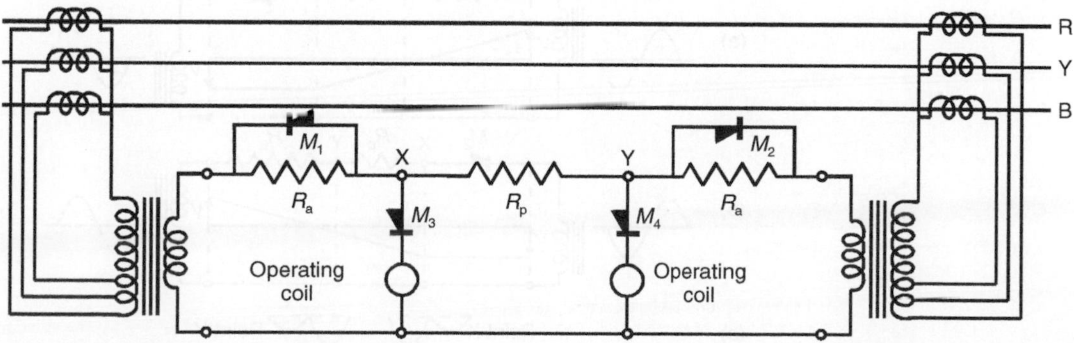

Figure 14.2
Basic circuit of Solkor R protection system

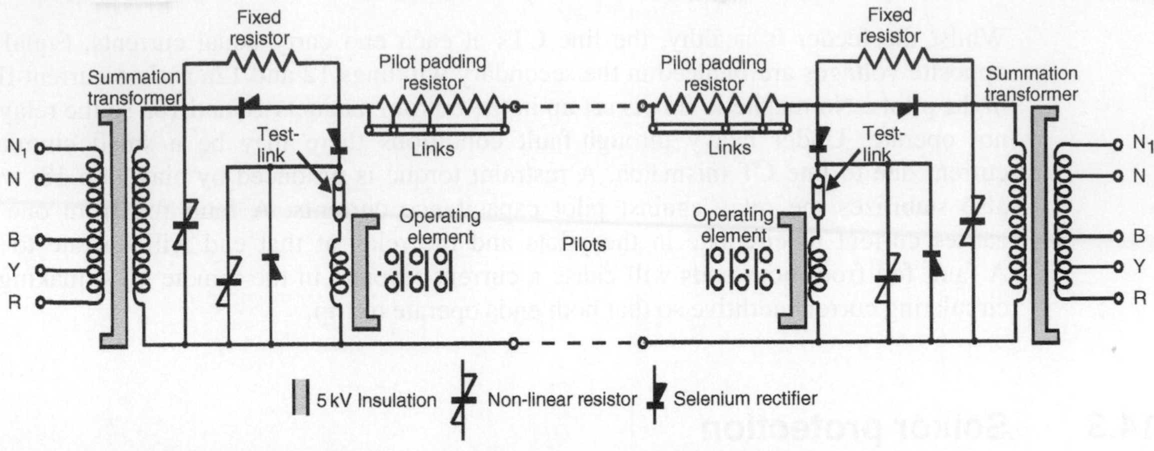

Figure 14.3
Schematic diagram of complete 5 kV Solkor R protective system

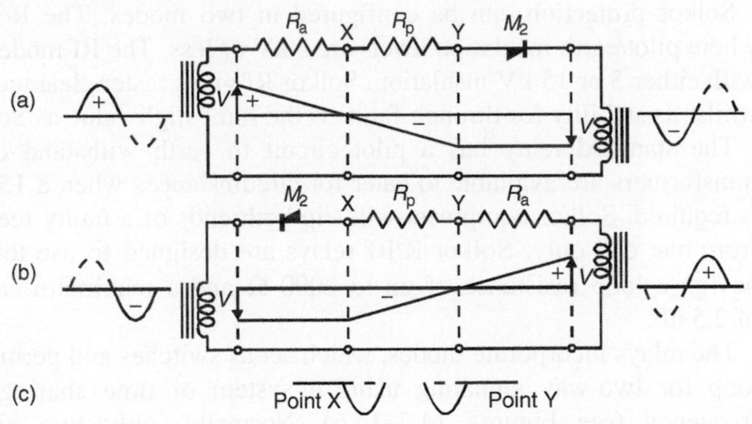

Figure 14.4
Behavior of basic circuit under external fault conditions when $R_a = R_p$ *(a) and (b) show the effective circuit during successive half-cycles; (c) indicates the voltages across relaying points X and Y during one cycle*

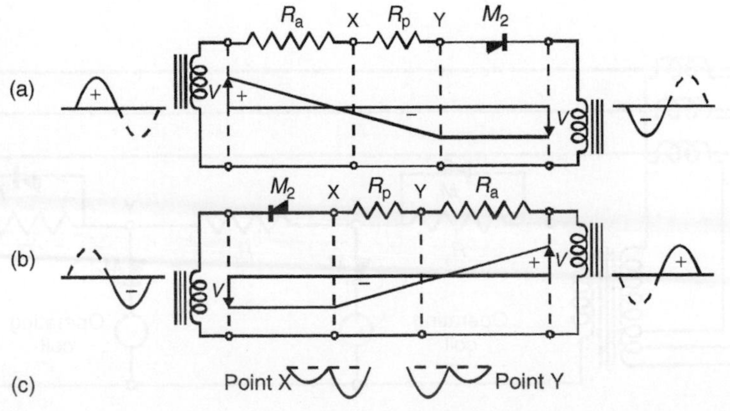

Figure 14.5
(a) Behavior of basic circuit under external fault conditions when $R_a = R_p$*; (b) show the effective circuit during successive half-cycles; (c) indicates the voltages across relaying points X and Y during one cycle*

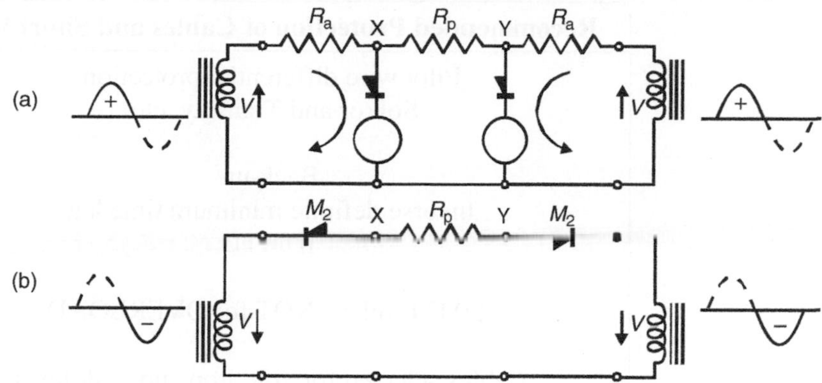

Figure 14.6
Behavior of basic circuit under internal-fault conditions (fault fed from both ends (a) and (b) show effective circuits during successive half cycle)

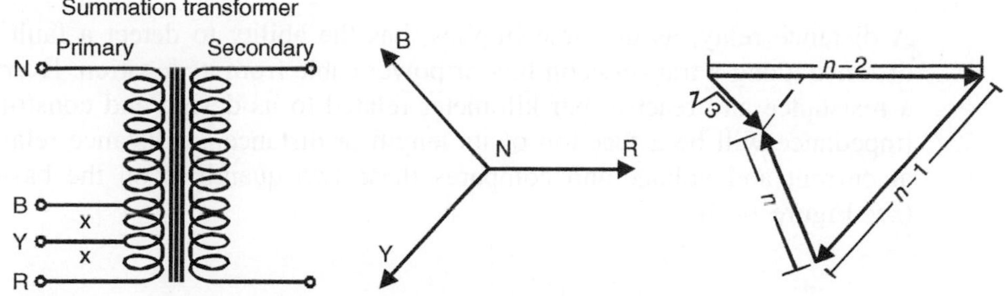

Figure 14.7
Schematic diagram, table of characteristics and vectoral demonstration of summation transformer

Type of Fault	Sensitivity (%)
Red to earth	25%
White to earth	32%
Blue to earth	42%
Red to white	125%
White to blue	125%
Blue to red	62%
Three phase	72%

Type of Fault	Fault Settings			
	5 kV Equipment		15 kV Equipment	
	N Tap	N1 Tap	N Tap	N1 Tap
Red-earth	25	18	33	23.5
Yellow-earth	32	21	41	27.5
Blue-earth	42	25	55	33
Red-yellow	125		165.0	
Yellow-blue	125		165.0	
Red-blue	62		82.5	
Three-phase	72		95.0	

Recommended Protection of Cables and Short Lines
Pilot wire differential protection Solkor and Translay, etc. Back-up Inverse definite minimum time lag IDMTL IDMTL relays NOT for OVERLOAD Cables selected for volt drop and fault level

14.4 Distance protection

14.4.1 Basic principle

A distance relay, as its name implies, has the ability to detect a fault within a pre-set distance along a transmission line or power cable from its location. Every power line has a resistance and reactive per kilometer related to its design and construction so its total impedance will be a function of its length or distance. A distance relay therefore looks at current and voltage and compares these two quantities on the basis of Ohm's law (see Figure 14.8).

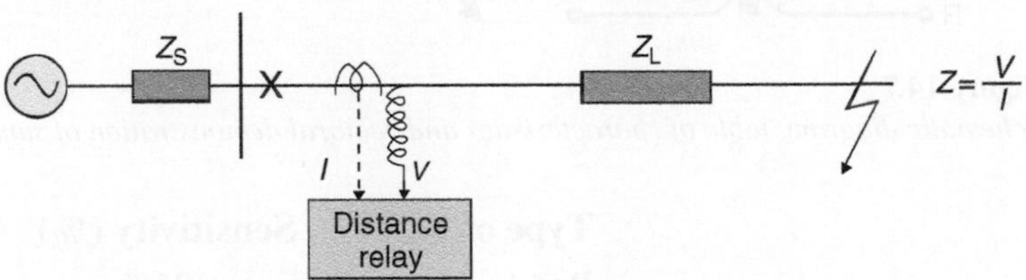

Figure 14.8
Basic principle of operation

The concept can best be appreciated by looking at the pioneer-type balanced beam relay (see Figure 14.9). The voltage is fed onto one coil to provide restraining torque, whilst the current is fed to the other coil to provide the operating torque. Under healthy conditions, the voltage will be high (i.e. at full-rated level), whilst the current will be low (at normal load value), thereby balancing the beam, and restraining it so that the contacts remain open. Under fault conditions, the voltage collapses and the current increase dramatically, causing the beam to unbalance and close the contacts.

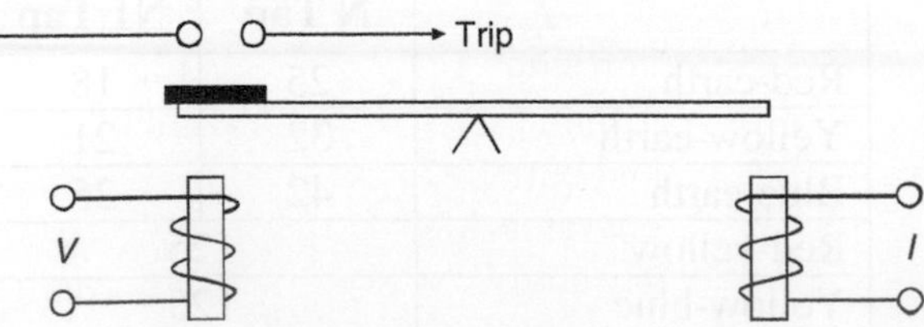

Figure 14.9
Balanced beam principle

By changing the ampere-turns relationship of the current coil to the voltage coil, the ohmic reach of the relay can be adjusted. A more modern technique for achieving the same result is to use a bridge comparator (see Figure 14.10).

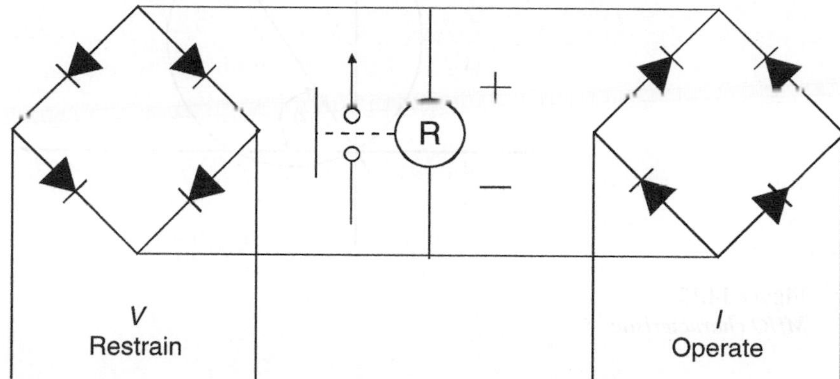

Figure 14.10
Bridge comparator

14.4.2 Tripping characteristics

If the relay's operating boundary is plotted, on an *R/X* diagram, its impedance characteristic is a circle with its center at the origin of the coordinates and its radius will be the setting (reach) in ohms (Figure 14.11).

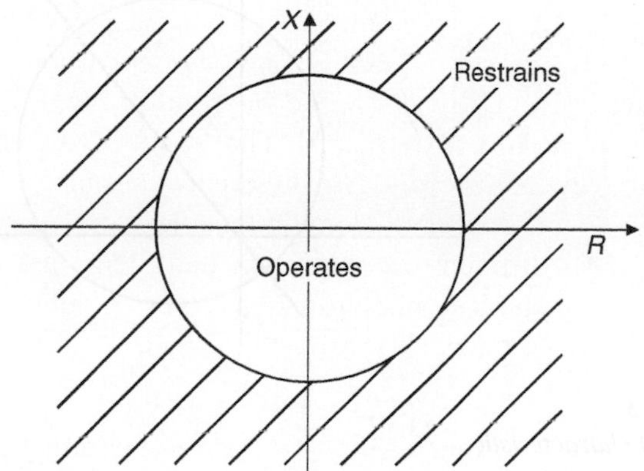

Figure 14.11
Plain impedance characteristic

The relay will operate for all values less than its setting i.e. is for all points within the circle.

This is known as a plain impedance relay and it will be noted that it is non-directional, in that it can operate for faults behind the relaying point. It takes no account of the phase angle between voltage and current.

This limitation can be overcome by a technique known as self-polarization. Additional voltages are fed into the comparator in order to compare the relative phase angles of voltage and current, so providing a directional feature. This has the effect of moving the circle such that the circumference of the circle now passes through the origin. Angle θ is known as the relay's characteristic angle (see Figure 14.12).

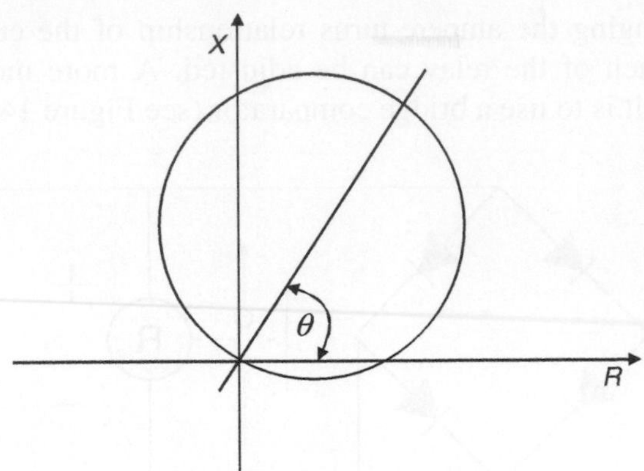

Figure 14.12
MHO characteristic

This is known as the MHO relay, so called as it appears as a straight line on an admittance diagram. By the use of a further technique of feeding in voltages from the healthy phases into the comparator (known as cross polarization) a reverse movement or offset of the characteristic can be obtained (see Figure 14.13). This is called the offset MHO characteristic.

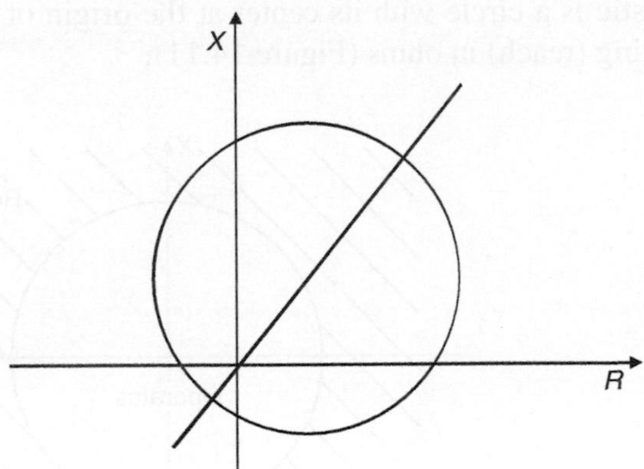

Figure 14.13
Offset MHO characteristic

14.4.3 Application onto a power line

Correct coordination of the distance relays is achieved by having an instantaneous directional zone 1 protection and one or two more time-delayed zones. A transmission line has a resistance and reactance proportional to its length, which also defines its own characteristic angle. It can therefore be represented on an *R*/*X* diagram as shown below.

Zone 1

The relay characteristic has also been added, from which it will be noted that the reach of the measuring element has been set at approximately 80% of the line length (see Figure 14.14). This 'under-reach' setting has been purposely chosen to avoid over-reaching into the next line section to ensure sound selectivity, for the following reasons:

- It is not practical to accurately measure the impedance of a transmission line, which could be very long (say 100 km). Survey lengths are normally used and these could have errors up to 10%.
- Errors are also present in the current and voltage transformers, not to mention the possible transient performance of these items.
- Manufacturing tolerances on the relay's ability to measure accurately, etc.

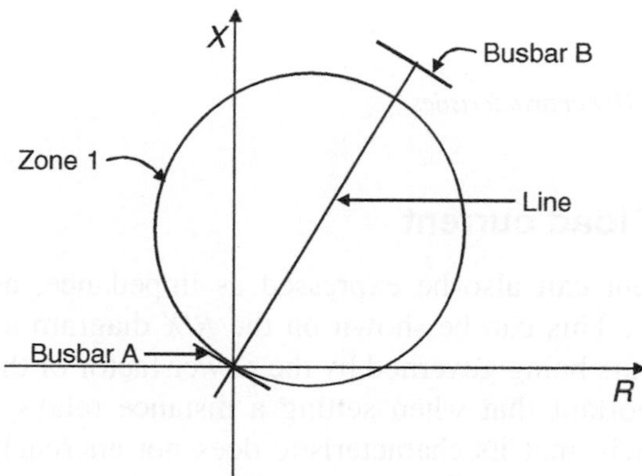

Figure 14.14
Zone 1 MHO characteristic

This measuring element in known as zone 1 of the distance relay and is instantaneous in operation.

Zone 2

To cover the remaining 20% of the line length, a second measuring element can be fitted, set to over-reach the line, but it must be time delayed by 0.5 s to provide the necessary coordination with the downstream relay. This measuring element is known as zone 2. It not only covers the remaining 20% of the line, but also provides backup for the next line section should this fail to trip for whatever reason.

Zone 3

A third zone is invariably added as a starter element and this takes the form of an offset mho characteristic. This offset provides a closing-onto-fault feature, as the mho elements may not operate for this condition due to the complete collapse of voltage for the nearby fault. The short backward reach also provides local backup for a busbar fault. This element can also be used for starting a carrier signal to the other end of the line – see later section.

The zone 3 element also has another very useful function. As a starter it can be used to switch the zone 1 element to zone 2, reach after say 0.5 s, thereby saving the installation of a second independent zone 2 measuring element so reducing its cost (see Figure 14.15).

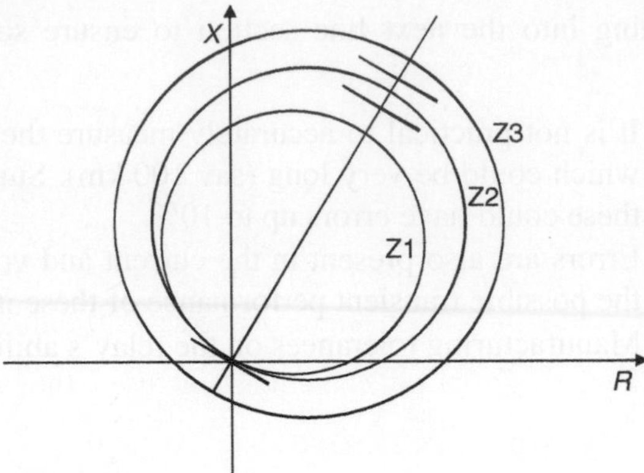

Figure 14.15
Three zone MHO characteristics

14.4.4 Effect of load current

Load current can also be expressed as impedance, again by the simple application of Ohm's law. This can be shown on the *R*/*X* diagram as depicted by the shaded area, the angular limits being governed by the power factor of the load.

It is important that when setting a distance relays, especially zone 3, which has the longest reach, that its characteristic does not encroach on the load area, as unnecessary tripping will undoubtedly occur (see Figure 14.16).

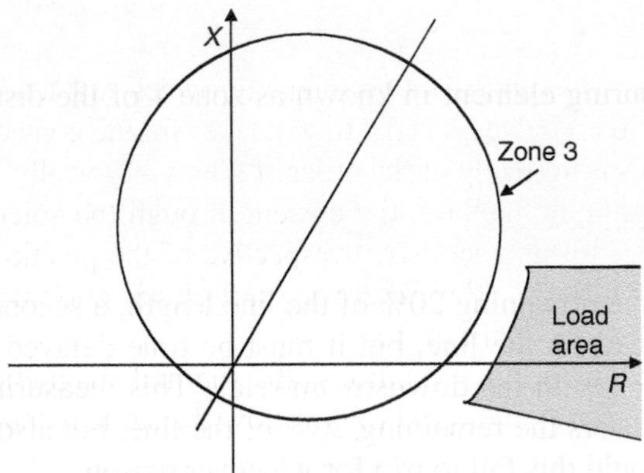

Figure 14.16
Load encroachment

14.4.5 Effect of arc resistance

Resistance of the fault arc can also have an impact on the performance of a distance relay, as can be seen on the following *R*/*X* diagram (see Figure 14.17).

It will be noted that the resistance of the fault arc takes the fault impedance outside the relay's tripping characteristic, so that it does not detect this condition. Alternatively, it is only picked up by either zone 2 or zone 3 in which case tripping will be unacceptably delayed. The effect of arc resistance is most significant on short lines where the reach of the relay setting is small. It can be a problem if the fault occurs near the end of the reach.

High fault-arc resistances tend to occur during midspan flashovers to ground during a veldt fire or on transmission lines carried on wood poles without earth wires. These problems can usually be overcome by using relays having different shaped characteristics such as described below.

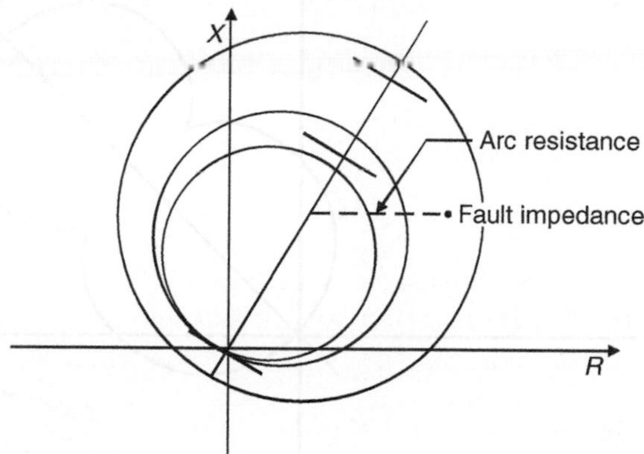

Figure 14.17
Effect of arc resistance

14.4.6 Different shaped characteristics

To overcome the problems of load encroachment and arc resistance, distance relays have been developed having different-shaped tripping characteristics, some examples of which are as follows:

- Circular (as illustrated above)
- Lenticular
- Figure of eight
- Trapezoidal.

With the advent of modern digital technology, many shapes are now possible to suit a variety of applications (see Figures 14.18(a)–(c)).

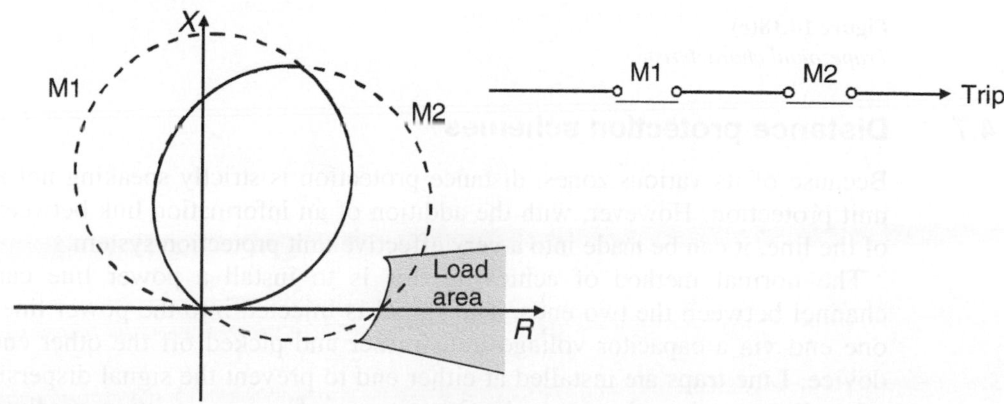

Figure 14.18(a)
Lenticular characteristic

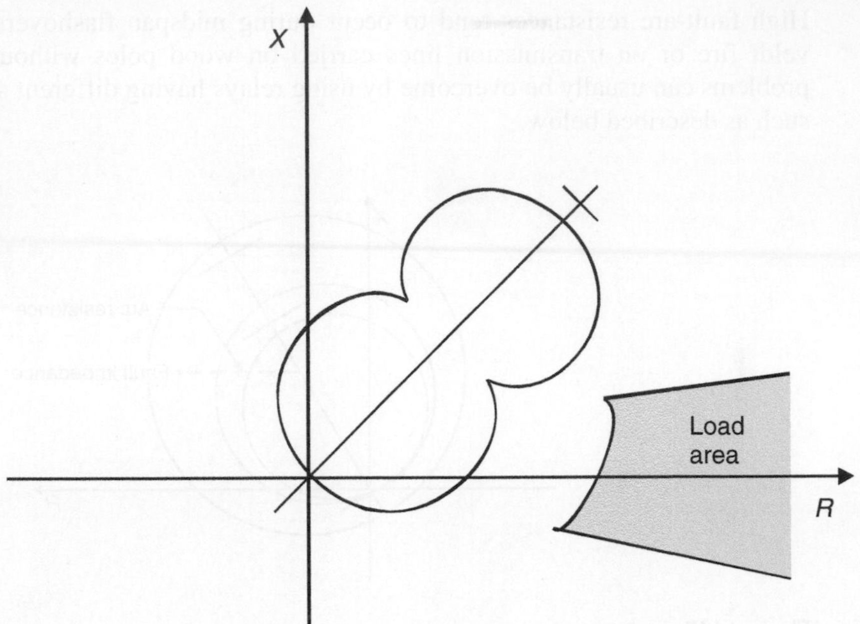

Figure 14.18(b)
Figure-of-eight characteristic

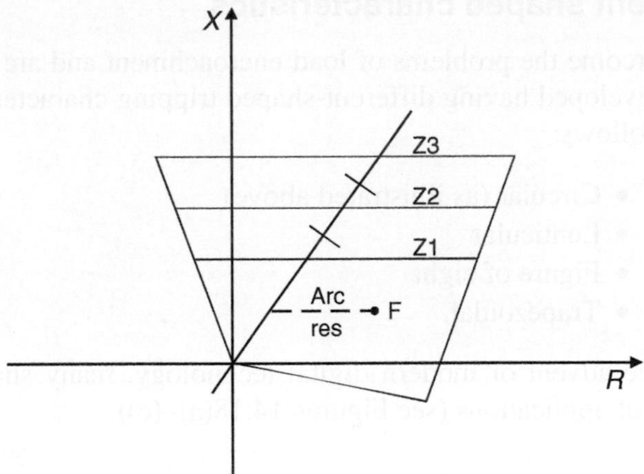

Figure 14.18(c)
Trapezoidal characteristic

14.4.7 Distance protection schemes

Because of its various zones, distance protection is strictly speaking not a pure form of unit protection. However, with the addition of an information link between the two ends of the line, it can be made into a very effective unit protection system.

The normal method of achieving this is to install a power line carrier signaling channel between the two ends. The signal is injected into the power line conductors at one end via a capacitor voltage transformer and picked off the other end by a similar device. Line traps are installed at either end to prevent the signal dispersing through all other lines, etc. in the network. Other types of communication medium can be used such as copper or fiber-optic pilots or microwave radio could be considered if line-of-site is available.

Conventional distance scheme

When carrier or signaling equipment is not available, the conventional distance scheme illustrated in Figures 14.19(a)–(c) is used; however faults in the end 20% of the line will only be cleared in zone 2 time, namely 0.5 s.

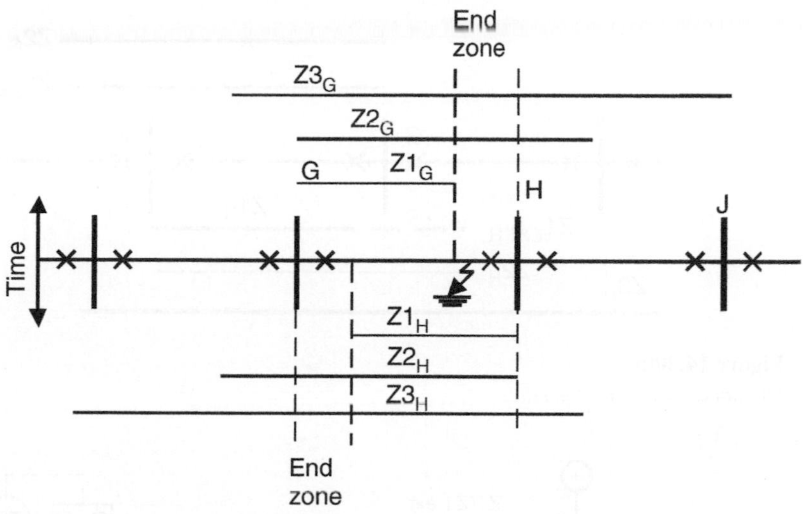

Figure 14.19(a)
Stepping time/distance characteristics

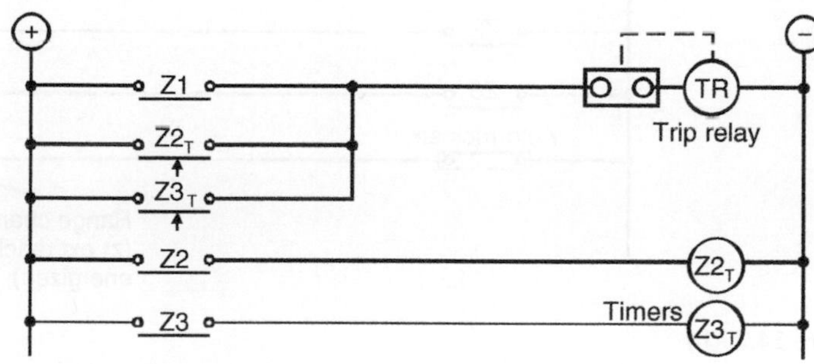

Figure 14.19(b)
Trip circuit (contact logic)

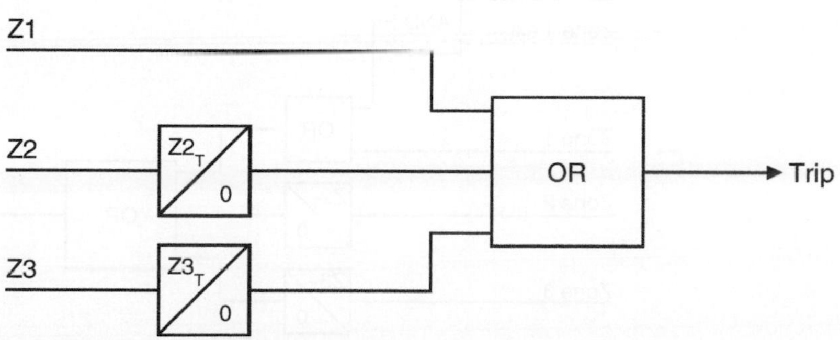

Figure 14.19(c)
Trip circuit (solid-state logic)

Zone 1 extension or overlap

Fast tripping for these portions can be achieved by extending the reach of zone 1 to 120% of the line and cutting it back to 80% reach after tripping before auto-closing the breaker. The logic is shown in Figures 14.20(a)–(c).

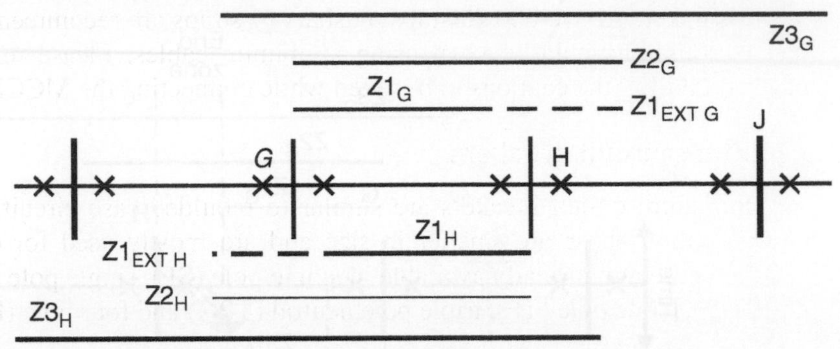

Figure 14.20(a)
Distance/time characteristics

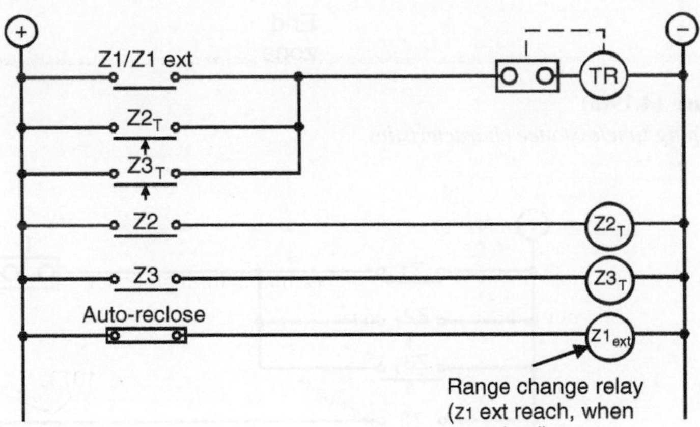

Figure 14.20(b)
Trip circuit (contact logic)

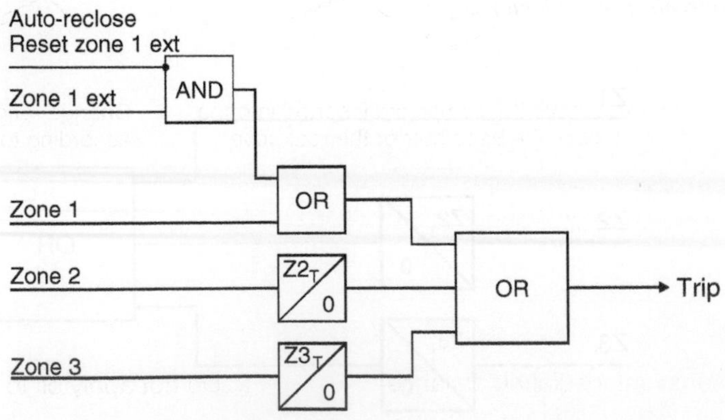

Figure 14.20(c)
Simplified solid-state logic

Direct transfer trip (under-reaching scheme)

When carrier/signaling is available, the simplest way to speed up fault clearance at the terminal which clears an end-zone fault in zone 2 time is to adopt a direct transfer trip or intertrip technique as shown in Figures 14.21(a)–(c). A zone 1 contact is used to send a carrier signal to the remote end to directly trip that breaker via a receive relay.

The disadvantage of this scheme is the possibility of undesired tripping by accidental or mal-operation of the signaling equipment.

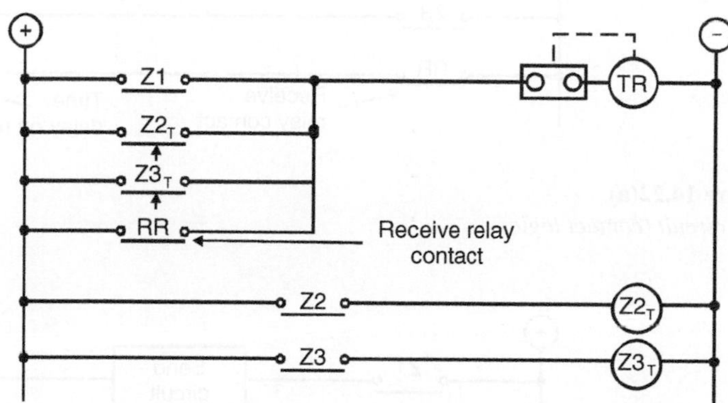

Figure 14.21(a)
Trip circuit contact logic

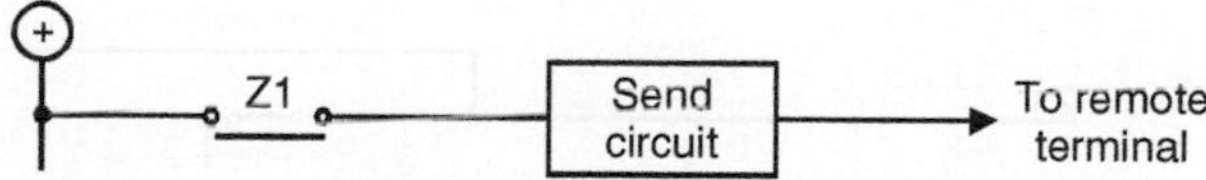

Figure 14.21(b)
Signaling channel send arrangement (contact logic)

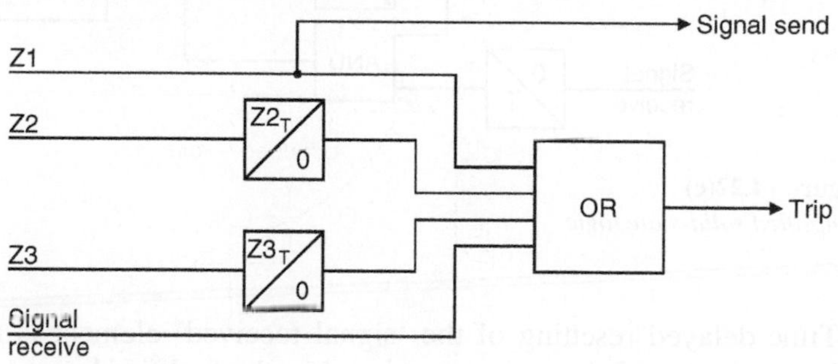

Figure 14.21(c)
Simplified solid-state logic

Permissive under-reach scheme

The direct transfer trip scheme is made more secure by monitoring the received signal with an instantaneous zone 2 operation before allowing tripping, as shown in Figures 14.22(a)–(c).

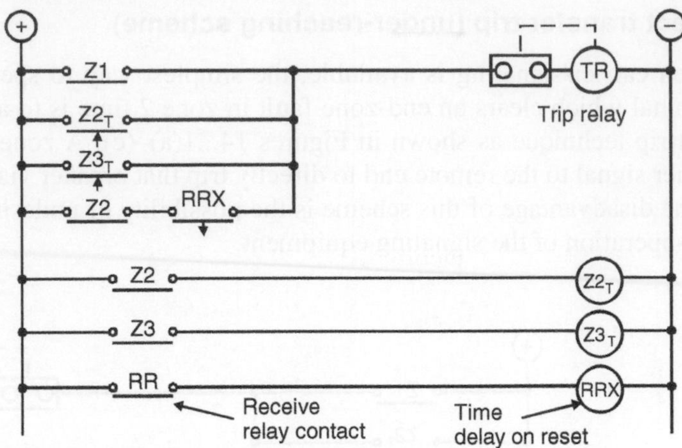

Figure 14.22(a)
Trip circuit (contact logic)

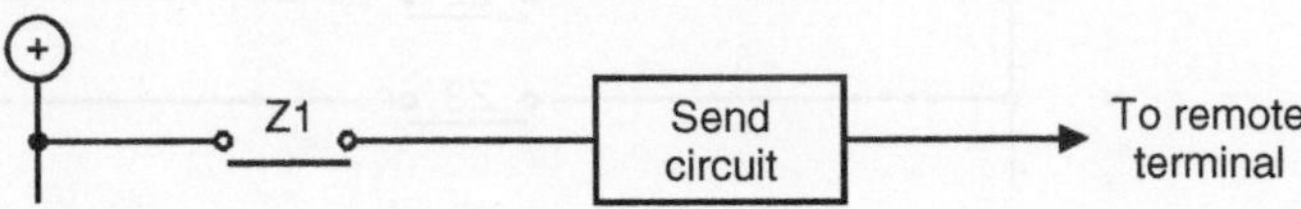

Figure 14.22(b)
Signaling send arrangement (contact logic)

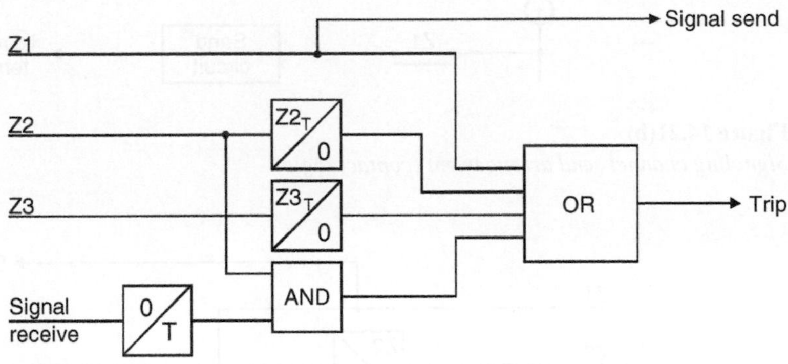

Figure 14.22(c)
Simplified solid-state logic

Time-delayed resetting of the 'signal received' element is required to ensure that the relays at both ends have time to trip when the fault is close to one end. When the breaker at one end is open, instantaneous clearance cannot be achieved for end-zone faults near the breaker open terminal.

Acceleration scheme

This scheme is similar to the permissive under-reach scheme in its principle of operation, but is applicable only to zone-switched distance relays, which share the same measuring elements for zones 1 and 2.

In this scheme, the incoming carrier signal switches the zone 1 reach to zone 2 immediately without waiting for the zone 2 timer (0.5 s) to switch the reach. This accelerates the fault clearance at the remote end. The scheme is shown in Figures 14.23(a)–(d).

The longer reach of the measuring elements gives better arc resistance coverage so it is better suited to short lines. This scheme is sometimes referred to as a 'directional comparison' scheme.

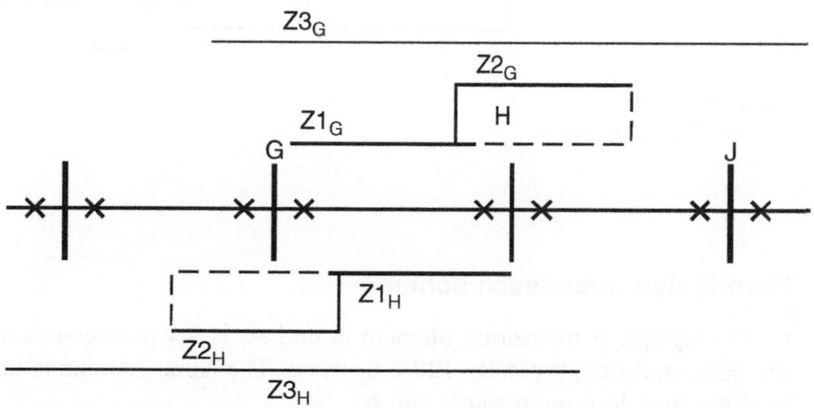

Figure 14.23(a)
Distance/time characteristics

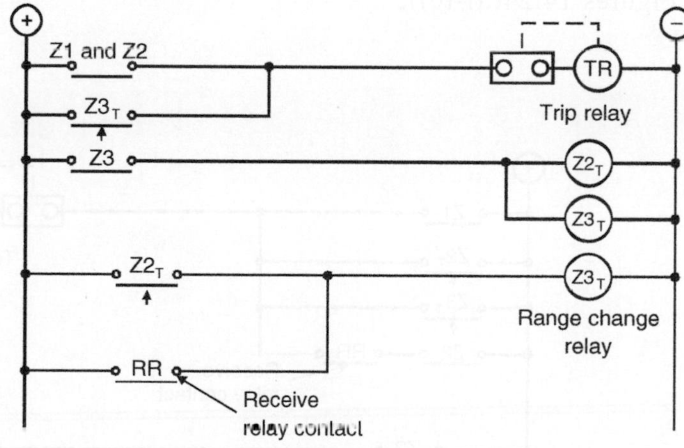

Figure 14.23(b)
Trip circuit (contact logic)

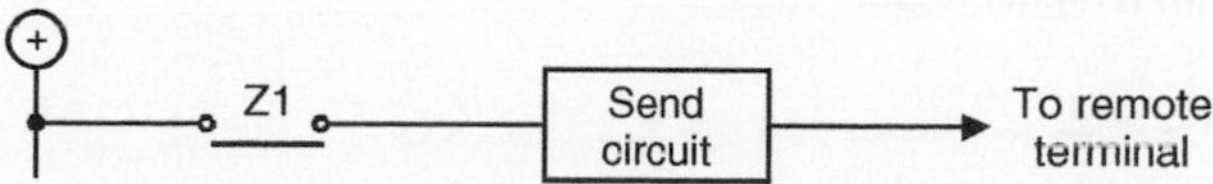

Figure 14.23(c)
Signaling channel send arrangement (contact logic)

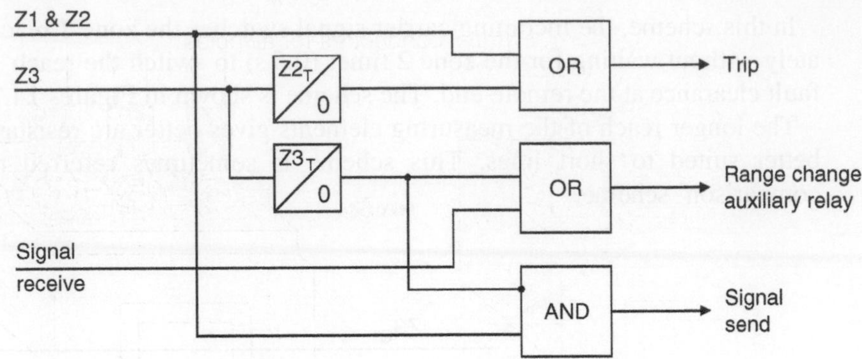

Figure 14.23(d)
Simplified solid-state logic

Permissive over-reach scheme

In this scheme a measuring element at end A, is set to over-reach beyond the far end of the protected line, typically 120% or more. The same scheme is installed at the other end B of the line, looking towards end A.

This element sends an intertrip signal to the remote end. It is arranged to trip its own breaker immediately after a signal is received. If no signal is received, it will trip its own breaker after zone 2 time (0.5 s). The longer reach of the measuring elements gives better arc resistance coverage so it is better suited to short lines. This scheme is sometimes referred to as a 'directional comparison' scheme (see Figures 14.24(a)–(c)).

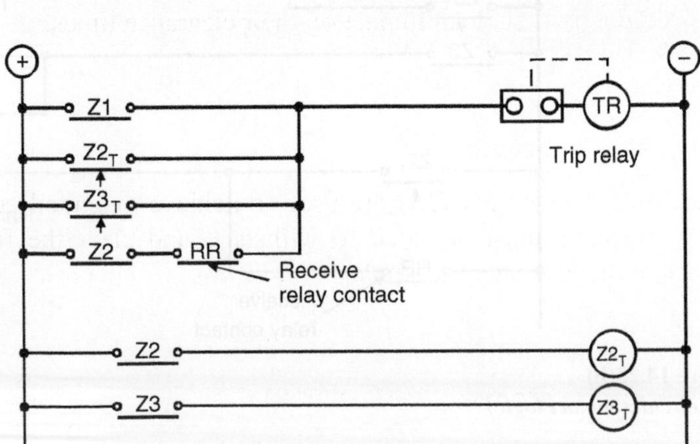

Figure 14.24(a)
Trip circuit (contact logic)

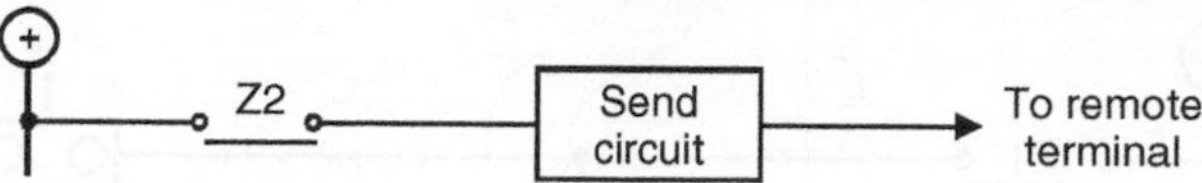

Figure 14.24(b)
Signaling send arrangement (contact logic)

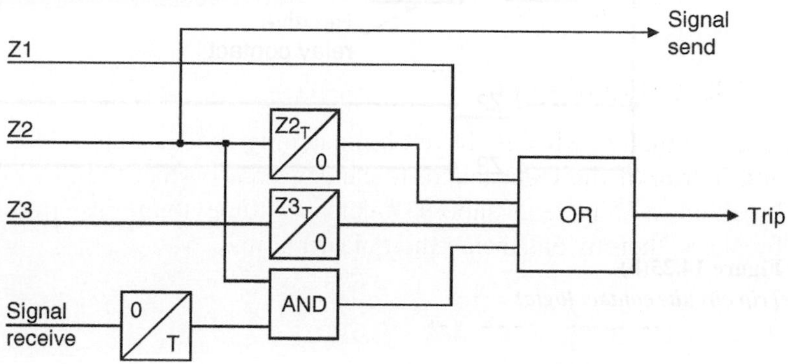

Figure 14.24(c)
Simplified solid-state logic

Blocking scheme

The schemes described above have used a signaling channel to transmit a tripping instruction, normally through the fault.

A blocking scheme uses inverse logic, the signal preventing tripping. Signaling is initiated only for external faults and takes place over healthy line sections. Fast fault clearance occurs when no signal is received and the over-reaching zone 2 measuring elements looking into the line operate.

The signaling channel is keyed, by a reverse-looking distance element (zone 3). An ideal blocking scheme is shown in Figures 14.25(a)–(d).

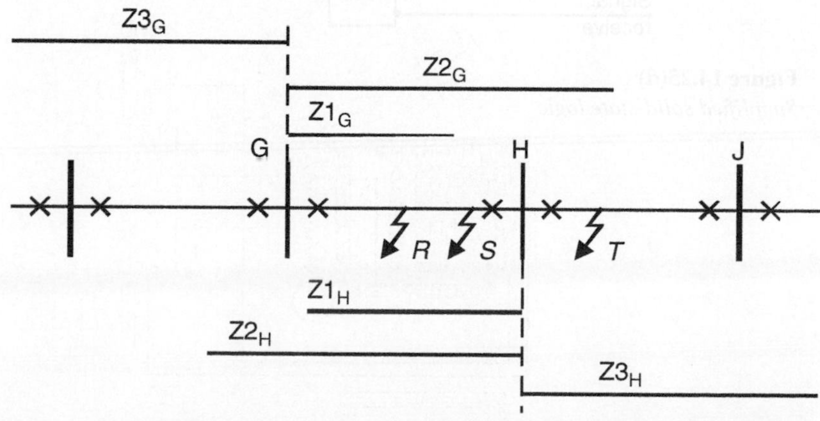

Figure 14.25(a)
Distance/time characteristics

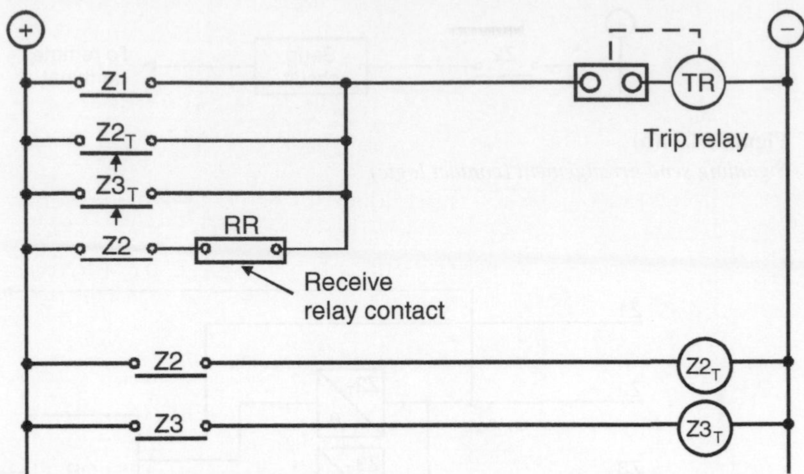

Figure 14.25(b)
Trip circuit (contact logic)

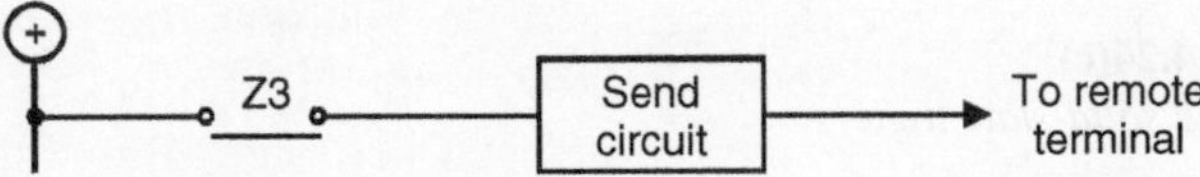

Figure 14.25(c)
Signaling channel send arrangement (contact logic)

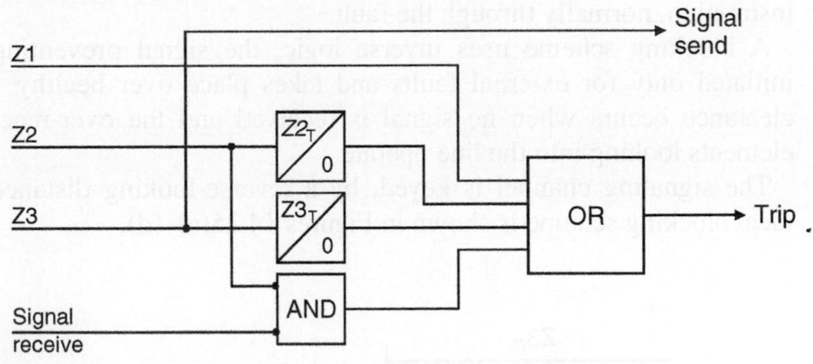

Figure 14.25(d)
Simplified solid-state logic

15

Transformer protection

15.1 Winding polarity

A transformer consists of two windings viz., primary and secondary coupled to a common magnetic core. International standards define the polarity of the primary and secondary windings sharing the same magnetic circuit as follows.

If the core flux induces an instantaneous emf from a low-number terminal to a high-number terminal in one winding, then the direction of induced emf in all other windings linked by that flux will also be from a low-number terminal to a high-number terminal. In the following sketch the induced emf on primary winding E_p is from A1 to A2 in the A phase when a primary voltage V is applied across A as shown.

The secondary emf E_s is also from a1 to a2 in secondary a phase (see Figure 15.1). From the laws of induction, it will be seen that the current flow in the windings is in the opposite direction.

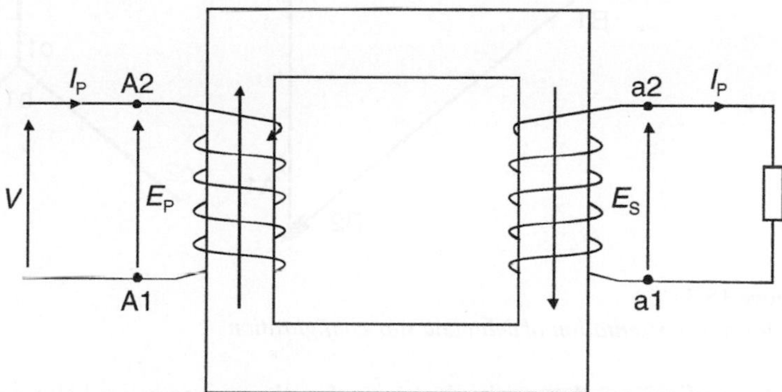

Figure 15.1
Principle of operation of a transformer

15.2 Transformer connections

Transformer windings can be connected either in a star (Y) or delta (D) configuration, bearing in mind that each phase will be displaced 120° from the other.

Figures 15.2 and 15.3 show the three windings of a three-phase core type transformer. This shows the primary connected in delta while the secondary windings are connected in star. The vectorial representation of primary and secondary voltages are also indicated.

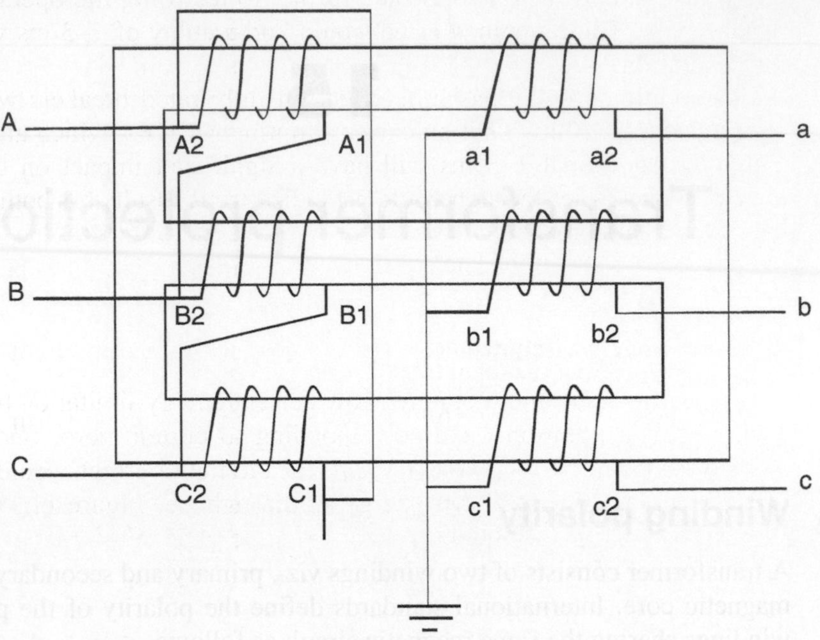

Figure 15.2
Physical connection of delta (D) or star (Y) configuration

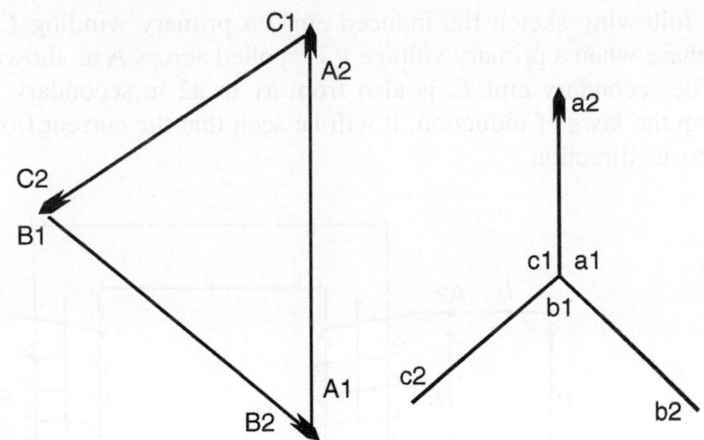

Figure 15.3
Vectorial representation of delta and star configuration

Depending on the method chosen for the primary and the secondary, a phase-shift can take place between the corresponding phases in the primary and secondary voltages of a transformer (see Figure 15.4).

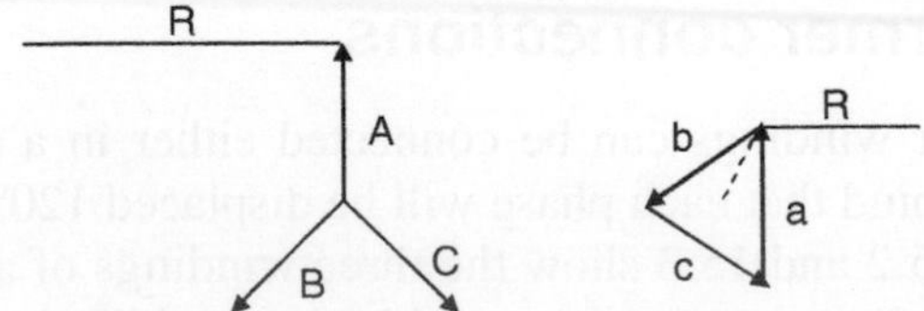

Figure 15.4
Phase shift of transformer

Clockface numbers are used to represent phase shifts, the highest voltage winding being used as the reference. A 360° shift corresponds to a full 12 h of a clock with each 30° shift being represented by 1 h. For example, 30° corresponds to 1 o'clock position, 150° shift corresponds to 5 o'clock position and 330 (or –30)° shift corresponds to 11 o'clock position.

The vector grouping and phase shift can then be expressed using a simple code. The primary winding connection is represented by capital letter while small letter represents the secondary connection. The 'N' means the primary neutral has been brought out.

For example:
YNd1= Primary winding connected in star with neutral brought out.
Secondary winding connected in delta.
Phase shift of secondary 30° from 12 to 1 o'clock compared to primary phase angle.

A knowledge of the primary connections and polarities of the windings enables connections of the CT secondary leads to be correctly determined, which are very important for sensing the fault currents, the basic need for correct protection.

15.3 Transformer magnetizing characteristics

When a transformer is energized, it follows the classic magnetization curve given in Figure 15.5. For efficiency reasons transformers are generally operated near to the 'knee-point' of the magnetic characteristic. Any increase above the rated terminal voltage tends to cause core saturation and therefore demands an excessive increase in magnetization current.

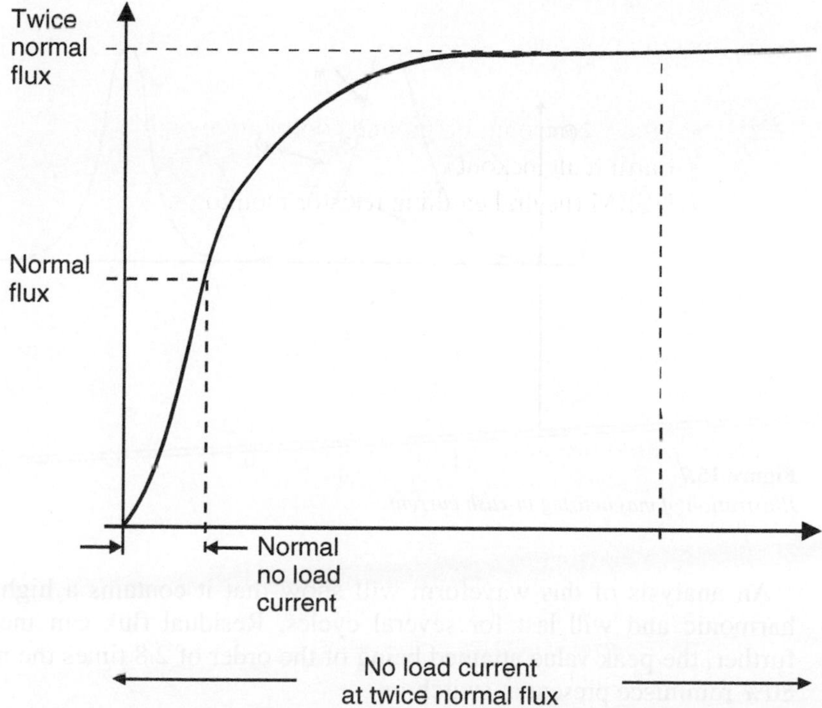

Figure 15.5
Transformer magnetizing characteristics

15.4 In-rush current

Under normal steady-state conditions, the magnetizing current required to produce the necessary flux is relatively small, usually less than 1% of full load current (see Figure 15.6).

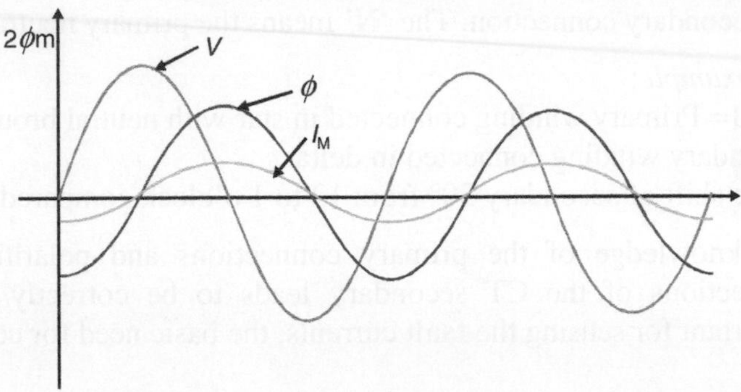

Figure 15.6
Steady-state conditions

However, if the transformer is energized at a voltage zero then the flux demand during the first half voltage cycle can be as high as twice the normal maximum flux. This causes an excessive unidirectional current to flow, referred to as the magnetizing in-rush current as shown in Figure 15.7.

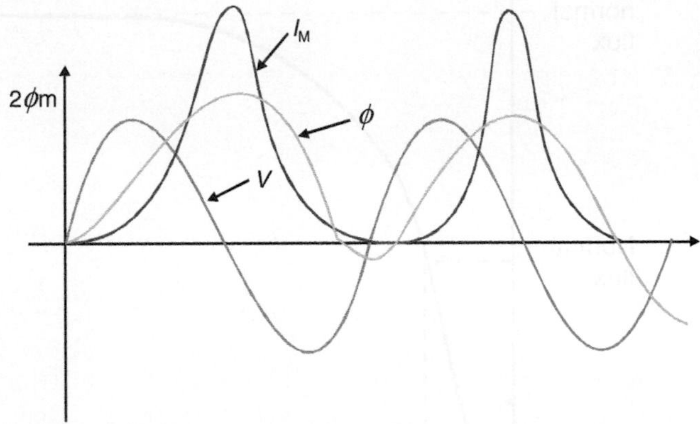

Figure 15.7
Illustration of magnetizing in-rush current

An analysis of this waveform will show that it contains a high proportion of second harmonic and will last for several cycles. Residual flux can increase the current still further, the peak value attained being of the order of 2.8 times the normal value if there is 80% reminisce present at switch-on.

As the magnetizing characteristic is non-linear, the envelope of this transient in-rush current is not strictly exponential. In some cases, it has been observed to be still changing up to 30 min after switching on (see Figure 15.8). It is therefore important to be aware of this transient phenomenon when considering differential protection of transformers, which will be discussed later.

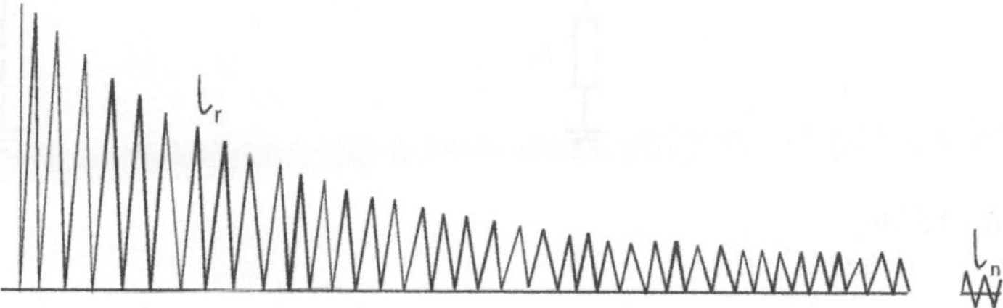

Figure 15.8
Typical transient current-rush when switching in a transformer at instant when E=O

15.5 Neutral earthing

It is important that the neutral of a power system be earthed otherwise this could ‘float’ all over with respect to true ground, thereby stressing the insulation above its design capability. This is normally done at the power transformer as it provides a convenient access to the neutral point.

On HV systems (i.e. 6 kV and above), it is a common practice to effectively earth the primary neutral by means of a solid copper, in which case the system is referred to as an effectively earthed system (see Figure 15.9).

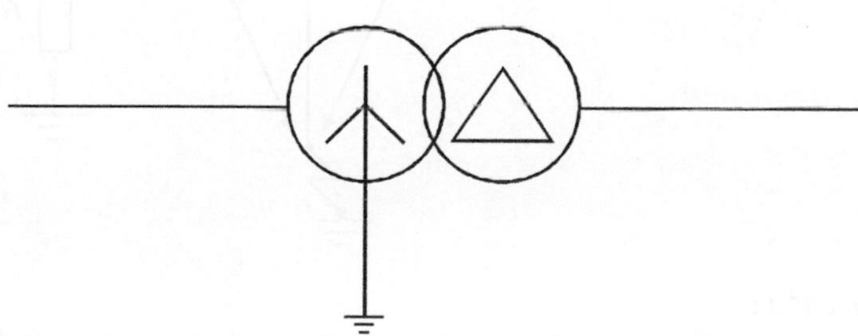

Figure 15.9
Earthing of the neutral

This has the advantage that when an earth fault occurs on one phase, the two healthy phases remain at phase-to-neutral voltage above earth. This allows insulation of the transformer windings to be graded towards the neutral point, resulting in a significant saving in cost. All other primary plant need only phase-to-neutral insulation and surge arrestors in particular need only be rated for 80% line-to-line voltage. This provides an enormous saving in capital expenditure and explains why Eskom’s HV system are invariably solidly earthed.

The disadvantage is that when an earth fault occurs an extremely high current flows (approximately equal to three-phase fault current), stressing the HV windings both electromagnetically and thermally. The forces and heat being proportional to the current squared. Earthing of the LV system neutral can be achieved as shown in Figure 15.10.

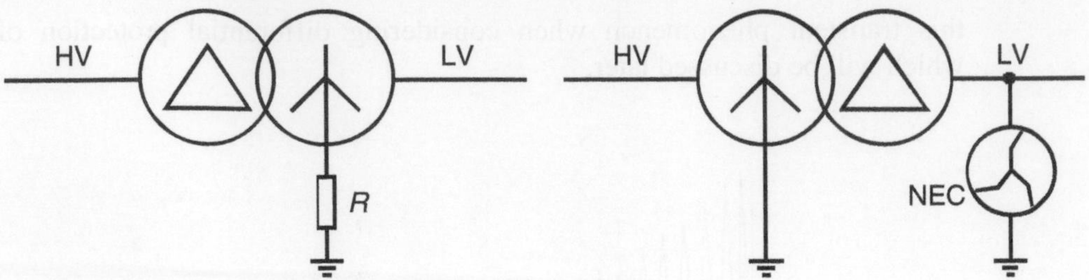

Figure 15.10
Earthing of the LV system

It will be noted that the LV system is impedance and/or resistance earthed. This allows the earth fault current to be controlled to manageable levels, normally of the order of the transformer full load current, typically 300 A.

Here, the transformer does not get a shock on the occurrence of each earth fault; however, the phase conductors now rise to line potential above earth during the period of the earth fault (see Figure 15.11).

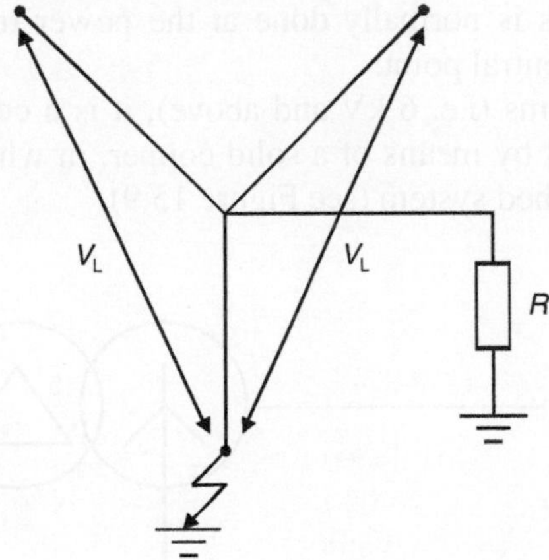

Figure 15.11
Phase diagram illustrating phase conductors rising to phase voltage on fault

Phase-to-earth insulation of all items of primary plant must therefore withstand line-to-line voltage.

15.6 On-load tap changers

On-load tap changers are very necessary to maintain a constant voltage on the LV terminals of the transformer for varying load conditions.

This is achieved by providing taps, generally on the HV winding because of the lower current levels. The tap changer changes the turns ratio between primary and secondary, thereby maintaining a nominal LV voltage within a specific tolerance (see Figure 15.12). A typical range of taps would be +15 to –5% giving an overall range of 20% (see Figure 15.13). The tap changer is usually mounted in a separate compartment to the main tank with a barrier board in between. This sometimes has a vent between the two to equalize the pressures.

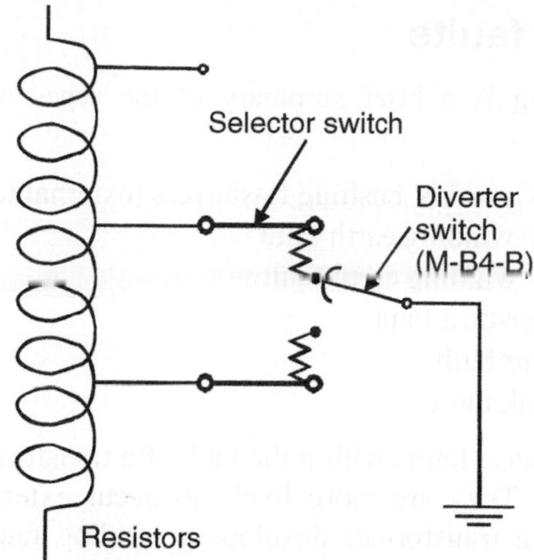

Figure 15.12
On-load tap changer

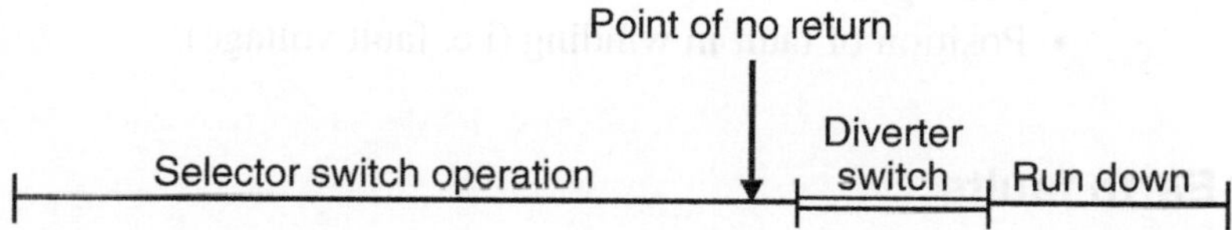

Figure 15.13
Top changer range of operations

15.7 Mismatch of current transformers

Current transformers are provided on the HV and LV sides of a power transformer for protection purposes. If we consider a nominal 132/11 kV 10 MVA transformer, the HV and LV full load currents would be as shown in Figure 15.14.

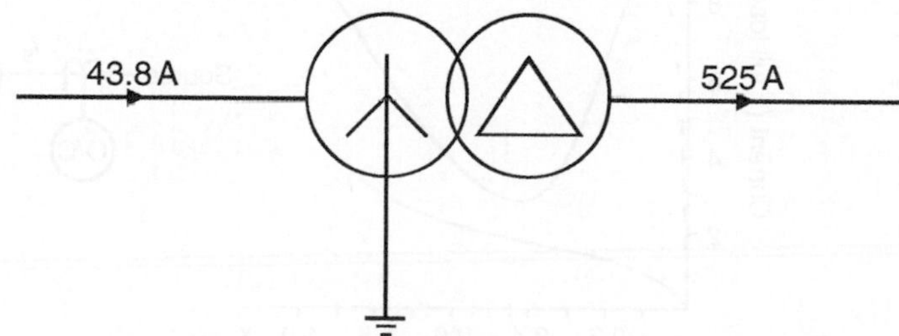

Figure 15.14
Nominal 132/11 kV 10 MVA transformer

A ratio of 50/1 A would most likely be chosen for the HV current transformers, as it is not possible to obtain fractions of a turn. 525/1 could be achieved comfortably for the LV current transformers. We therefore have a mismatch of current transformer ratios. Furthermore, it is more than likely that the HV CTs will be supplied by a different manufacturer than the LV CTs. There is therefore no guarantee that the magnetization curves will be the same, so adding to the mismatch.

15.8 Types of faults

The following is a brief summary of the types of faults that can occur in a power transformer:

- HV and LV bushing flashovers (external to the tank)
- HV winding earth fault
- LV winding earth fault
- Inter-turn fault
- Core fault
- Tank fault.

Phase-to-phase faults within the tank of a transformer are relatively rare by virtue of its construction. They are more likely to occur external to the tank on the HV and LV bushings. If a transformer develops a winding fault, the level of fault current will be dictated by:

- Source impedance
- Method of neutral earthing
- Leakage reactance
- Position of fault in winding (i.e. fault voltage).

15.8.1 Earth faults

Effectively earthed neutral

The fault current in this case is controlled mainly by the leakage reactance, which varies in a complex manner depending on the position of the fault in the winding. The reactance decreases towards the neutral so that the current actually rises for faults towards the neutral end (see Figure 15.15).

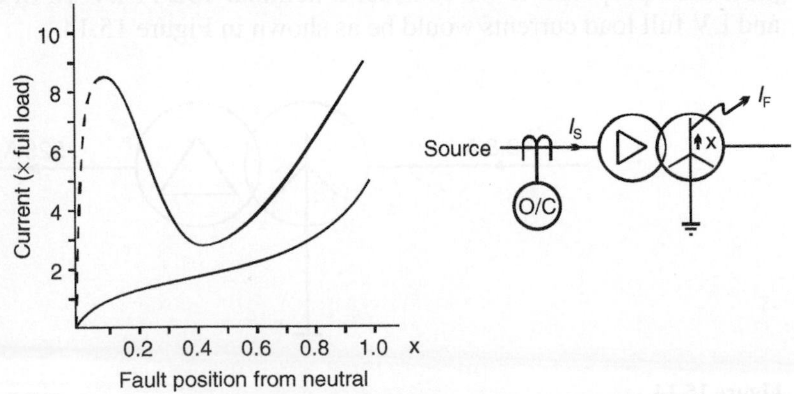

Figure 15.15
Relationship of fault current to position from neutral (earthed)

The input primary current is modified by the transformation ratio and is limited to 2–3 times the full load current of the transformer for fault positions over a major part of the star winding. An overcurrent relay on the HV side will therefore not provide adequate protection for earth faults on the LV side.

Resistance earthed neutral

For this application, the fault current varies linearly with the fault position, as the resistor is the dominant impedance, limiting the maximum fault current to approximately full load current (see Figure 15.16).

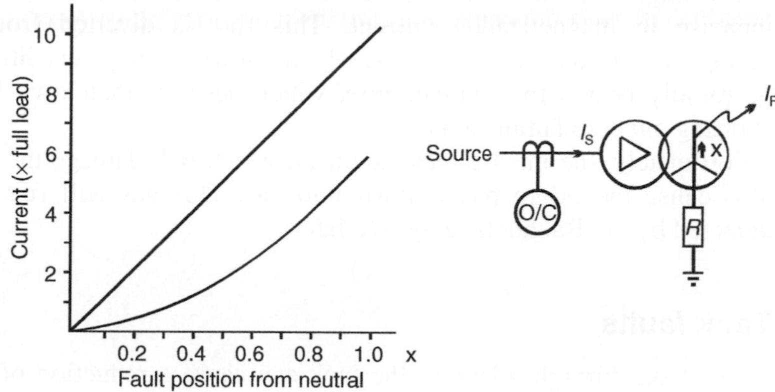

Figure 15.16
Relationship of fault current to positions from neutral (resistance earthed)

The input primary current is approximately 57% of the rated current making it impossible for the HV overcurrent relay to provide any protection for LV earth faults. Restricted earth fault protection is therefore strongly recommended to cover winding earth faults and this will be covered in more detail in a later section.

15.8.2 Inter-turn faults

Insulation between turns can break down due to electromagnetic/mechanical forces on the winding causing chafing or cracking. Ingress of moisture into the oil can also be a contributing factor.

Also an HV power transformer connected to an overhead line transmission system will be subjected to lightning surges sometimes several times rated system voltage. These steep-fronted surges will hit the end windings and may possibly puncture the insulation leading to a short-circuited turn. Very high currents flow in the shorted turn for a relatively small current flowing in the line (see Figure 15.17).

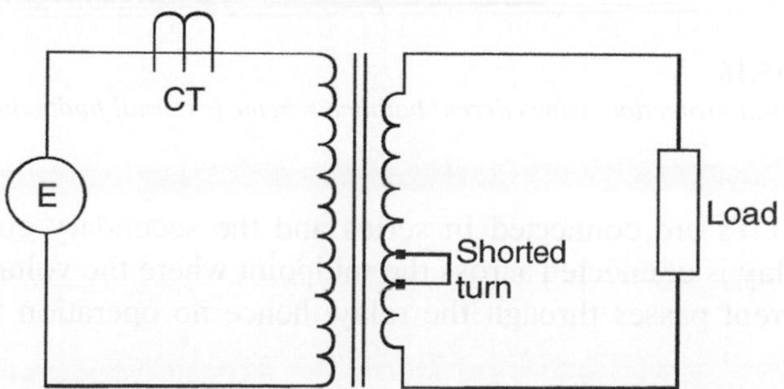

Figure 15.17
Inter-turn faults

15.8.3 Core faults

Heavy fault currents can cause the core laminations to move, chafe and possibly bridge causing eddy currents to flow, which can then generate serious overheating.

The additional core loss will not be able to produce any noticeable change in the line currents and thus cannot be detected by any electrical protection system. Power frequency overvoltage not only increases stress on the insulation but also gives an excessive increase in magnetization current. This flux is diverted from the highly saturated laminated core into the core bolts, which normally carry very little flux. These bolts may be rapidly heated to a temperature, which destroys their own insulation, consequently shorting out core laminations.

Fortunately, the intense localized heat, which will damage the winding insulation, will also cause the oil to break down into gas. This gas will rise to the conservator and detected by the Buchholz relay (see later).

15.8.4 Tank faults

Loss of oil through a leak in the tank can cause a reduction of insulation and possibly overheating on normal load due to the loss of effective cooling. Oil sludge can also block cooling ducts and pipes, contributing to overheating, as can the loss of forced cooling pumps and fans generally fitted to the larger transformer.

15.9 Differential protection

Differential protection, as its name implies, compares currents entering and leaving the protected zone and operates when the differential current between these currents exceed a pre-determined level.

The type of differential scheme normally applied to a transformer is called the current balance or circulating current scheme as shown in Figure 15.18.

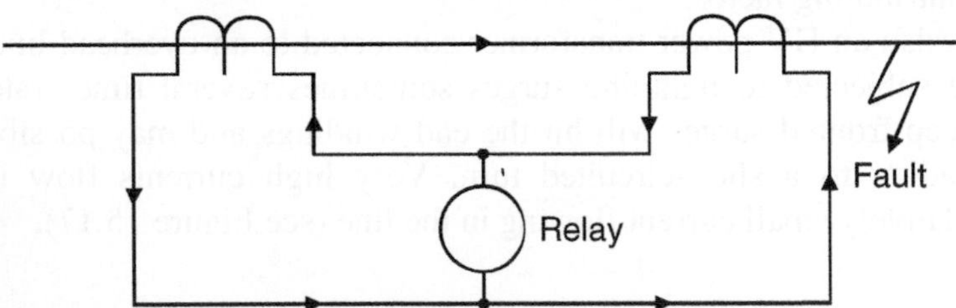

Figure 15.18
Differential protection using current balance scheme (external fault conditions)

The CTs are connected in series and the secondary current circulates between them. The relay is connected across the midpoint where the voltage is theoretically nil, therefore no current passes through the relay, hence no operation for faults outside the protected zone.

Under internal fault conditions (i.e. faults between the CTs) the relay operates, since both the CT secondary currents add up and pass through the relay as seen in Figure 15.19.

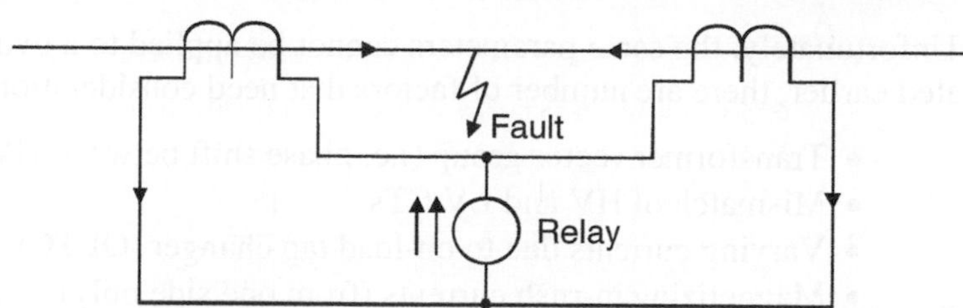

Figure 15.19
Differential protection and internal fault conditions

This protection is also called unit protection, as it only operates for faults on the unit it is protecting, which is situated between the CTs. The relay therefore can be instantaneous in operation, as it does not have to coordinate with any other relay on the network. This type of protection system can be readily applied to auto-transformers as shown in Figure 15.20. All current transformer ratios remain the same and the relays are of the high-impedance (voltage-operated) type, instantaneous in operation (see Figure 15.21).

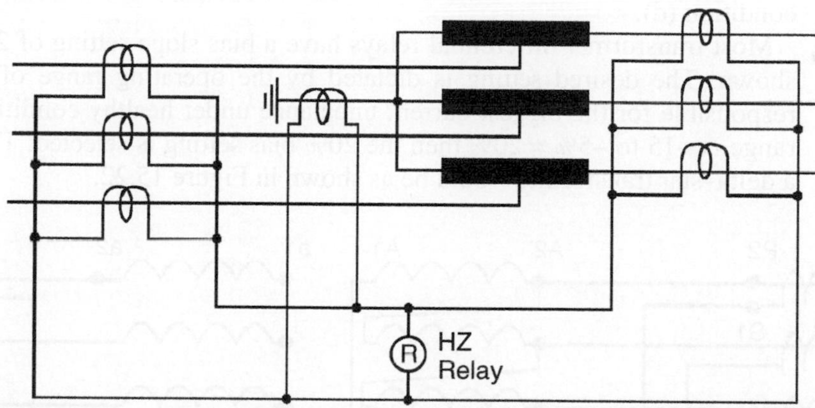

Figure 15.20
Differential protection applied to auto-tranformers

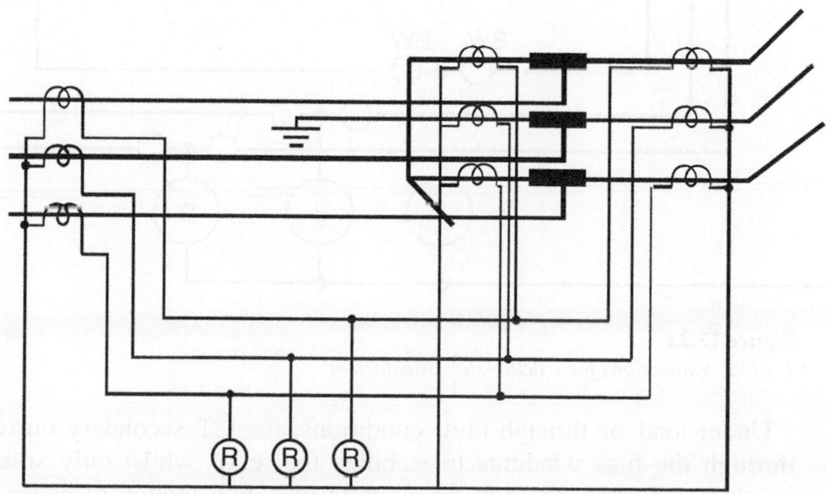

Figure 15.21
Auto-transformer – phase and earth fault scheme

Unfortunately, the same parameters cannot be applied to a two-winding transformer. As stated earlier, there are number of factors that need consideration:

- Transformer vector group (i.e. phase shift between HV and LV)
- Mismatch of HV and LV CTs
- Varying currents due to on-load tap changer (OLTC)
- Magnetizing in-rush currents (from one side only)
- The possibility of zero sequence current destabilizing the differential for an external earth fault.

Factor (a) can be overcome by connecting the HV and LV CTs in star/delta respectively (or vice versa) opposite to the vector group connections of the primary windings, so counteracting the effect of the phase shift through the transformer.

The delta connection of CTs provides a path for circulating zero sequence current, thereby stabilizing the protection for an external earth fault as required by factor (e). It is then necessary to bias the differential relay to overcome the current unbalances caused by factor (b) mismatch of CTs and (c) OLTC. Finally, as the magnetizing current in-rush is predominantly 2nd, harmonic filters are utilized to stabilize the protection for this condition (d).

Most transformer differential relays have a bias slope setting of 20%, 30% and 40% as shown. The desired setting is dictated by the operating range of the OLTC, which is responsible for the biggest current unbalance under healthy conditions; e.g. if the OLTC range is +15 to –5% = 20% then the 20% bias setting is selected. Typical connections for a delta–star transformer would be as shown in Figure 15.22.

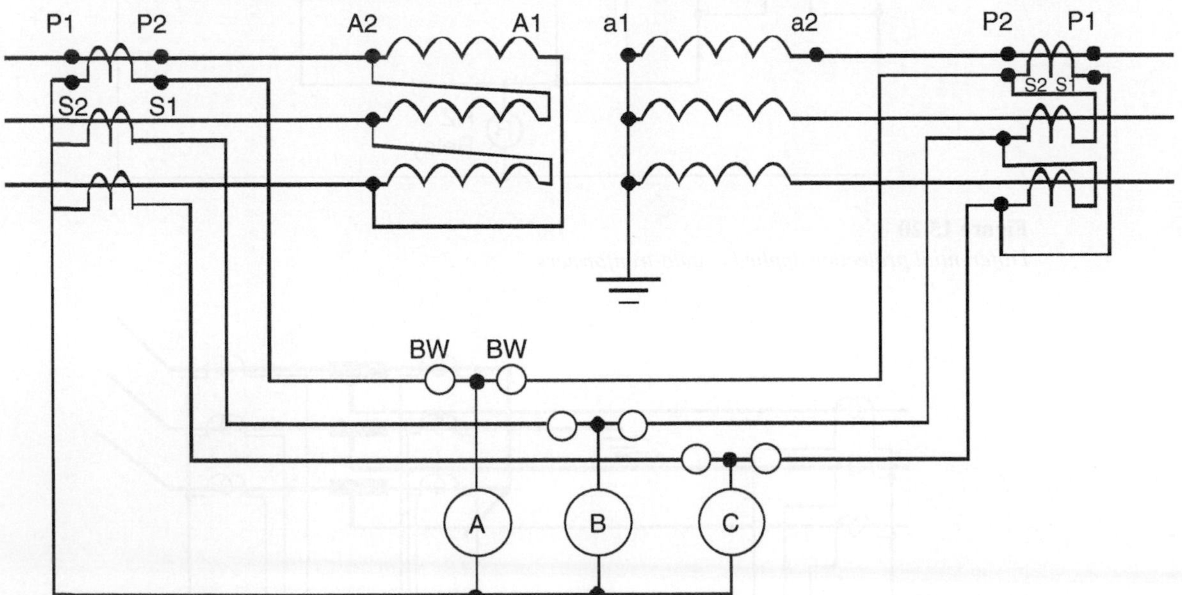

Figure 15.22
Typical connections for a delta–star transformer

Under-load or through-fault conditions, the CT secondary currents circulate, passing through the bias windings to stabilize the relay, whilst only small out-of-balance spill currents will flow through the operate coil, not enough to cause operation. In fact the higher the circulating current the higher will be the spill current required to trip the relay, as can be seen from Figures 15.23 and 15.24.

Operating current in multiples of relay rating
Bias current in multiples of relay rating
40% Bias setting
30% Bias setting
20% Bias setting
0
1
2
3
4
5
6
7
8
9
10
11
12
13
14
15
16
17
18
19
20

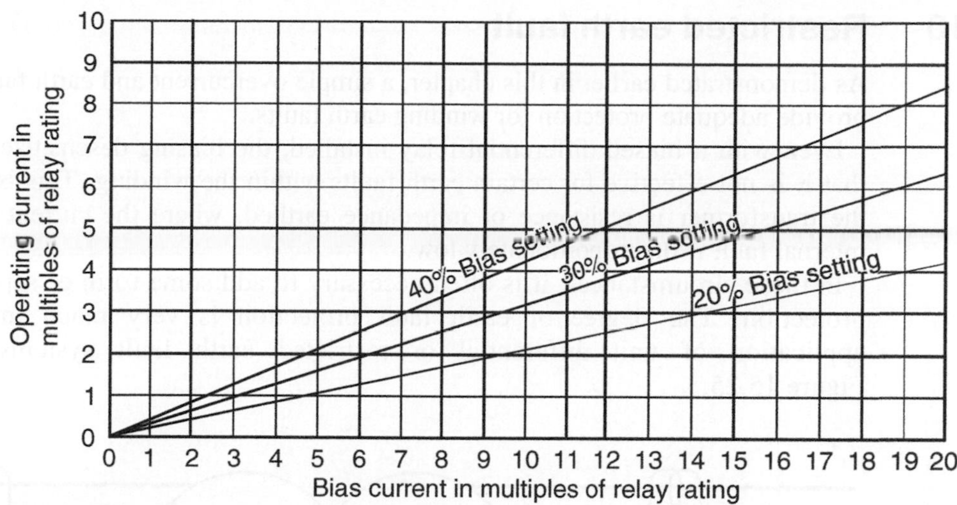

Figure 15.23
Operating current vs bias current

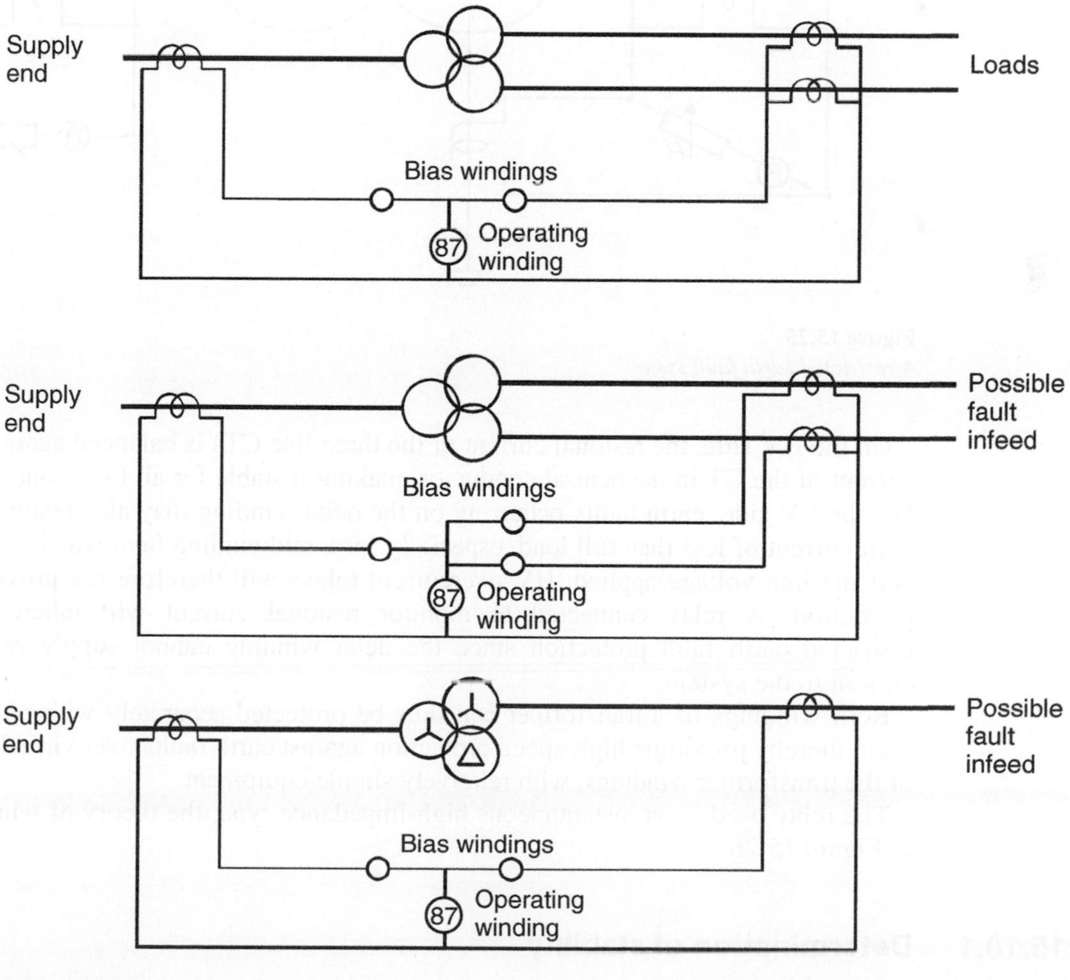

Supply end
Loads
Bias windings
87
Operating winding
Supply end
Possible fault infeed
Bias windings
87
Operating winding
Supply end
Possible fault infeed
Bias windings
87
Operating winding

Figure 15.24
Biased differential configurations

15.10 Restricted earth fault

As demonstrated earlier in this chapter, a simple overcurrent and earth fault relay will not provide adequate protection for winding earth faults.

Even with a biased differential relay installed, the biasing desensitizes the relay such that it is not effective for certain earth faults within the winding. This is especially so if the transformer is resistance or impedance earthed, where the current available on an internal fault is disproportionately low.

In these circumstances, it is often necessary to add some form of separate earth fault protection. The degree of earth fault protection is very much improved by the application of unit differential or restricted earth fault systems as shown in Figure 15.25.

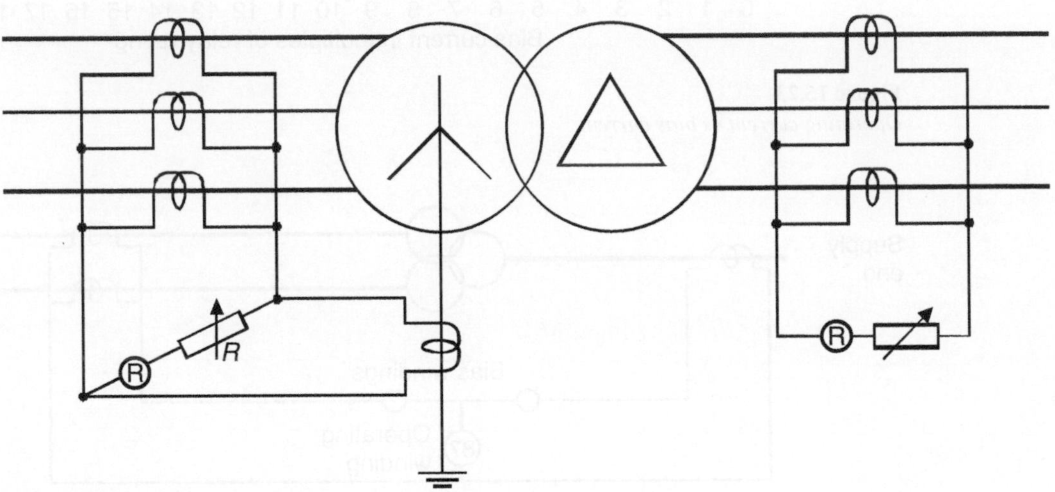

Figure 15.25
A restricted earth fault system

On the HV side, the residual current of the three line CTs is balanced against the output current of the CT in the neutral conductor, making it stable for all faults outside the zone. For the LV side, earth faults occurring on the delta winding may also result in a level of fault current of less than full load, especially for a midwinding fault which will only have half the line voltage applied. HV overcurrent relays will therefore not provide adequate protection. A relay connected to monitor residual current will inherently provide restricted earth fault protection since the delta winding cannot supply zero sequence current to the system.

Both windings of a transformer can thus be protected separately with restricted earth fault, thereby providing high-speed protection against earth faults over virtually the whole of the transformer windings, with relatively simple equipment.

The relay used is an instantaneous high-impedance type, the theory of which is shown in Figure 15.26.

15.10.1 Determination of stability

The stability of a current balance scheme using high-impedance relay depends upon the relay voltage setting being greater than the maximum voltage which can appear across the relay for a given through-fault condition. This maximum voltage can be determined by means of a simple calculation, which makes the following assumptions:

- One current transformer is fully saturated, making its excitation impedance negligible.
- The resistance of the secondary winding of the saturated CT together with lead resistance constitute the only burden in parallel with the relay.
- The remaining CTs maintain their ratio.

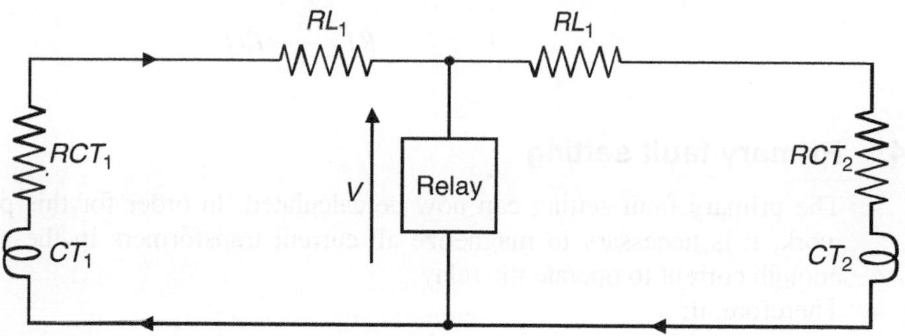

Figure 15.26
Basic circuit of high-impedance current balance scheme

Hence, the maximum voltage is given by Equation (15.1):

$$V = I(Rct + R1) \qquad (15.1)$$

Where

I = CT secondary current corresponding to the maximum steady state through fault current

Rct = Secondary winding resistance of CT

$R1$ = Largest value of lead resistance between relay and current transformer.

For stability, the voltage setting of the relay must be made equal to or exceed the highest value of V calculated above.

Experience has shown that if this method of setting is adopted the stability of the protection will be very much better than calculated. This is because a CT is normally not continuously saturated and consequently any voltage generated will reduce the voltage appearing across the relay circuit.

15.10.2 Method of establishing the value of stabilizing resistor

To give the required voltage setting the high-impedance relay operating level is adjusted by means of an external series resistor as follows:

Let v = operating voltage of relay element
Let i = operating current of relay equipment and
V = maximum voltage as defined under 'determination of stability' above.

Then the required series resistor setting;

$$R = \frac{V - v}{i}$$

It is sometimes the practice to limit the value of series resistor to say 1000 Ω, and to increase the operating current of the relay by means of a shunt-connected resistor, in order to obtain larger values of relay operating voltage.

15.10.3 Method of estimating maximum pilot loop resistance for a given relay setting

From Equation 15.1 above $V = I(Rct + Rl)$. Therefore

$$Rl = \frac{V}{I} - Rct$$

15.10.4 Primary fault setting

The primary fault setting can now be calculated. In order for this protection scheme to work, it is necessary to magnetize all current transformers in the scheme plus provide enough current to operate the relay.
Therefore, if;

I_r = relay operating current
I_1, I_2, I_3, I_4 = excitation currents of the CTs at the relay setting voltage
N = CT ratio.

Then the primary fault setting = $N \times (I_r + I_1 + I_2 + I_3 + I_4)$.

In some cases, it may be necessary to increase the basic primary fault setting as calculated above. If the required increase is small, the relay setting voltage may be increased (if variable settings are available on the relay), which will have the effect of demanding higher magnetization currents from the CTs I_1, I_2, etc. Alternatively, or when the required increase is large, connecting a resistor in parallel with the relay will increase the value of I_r.

15.10.5 Current transformer requirements

Class X CTs are preferably required for this type of protection, however experience has shown that most protection type CTs are suitable for use with high-impedance relays, providing the following basic requirements are met:

- The CTs should have identical turns ratio. Where turns error is unavoidable, it may be necessary to increase the fault setting to cater for this.
- To ensure positive operation, the relay should receive a voltage of twice its setting. The knee-point voltage of the CTs should be at least twice the relay setting voltage (Knee-point = 50% increase in mag. Current gives 10% increase in output voltage.)
- CTs should be of the low-reactance type.

15.10.6 Protection against excessively high voltages

As the relay presents very high impedance to the CTs the latter are required to develop an extremely high voltage. In order to contain this within acceptable limits, a voltage dependant resistor (VDR), or metrosil, is normally mounted across the relay to prevent external flashovers, especially in polluted environments (see Figure 15.27).

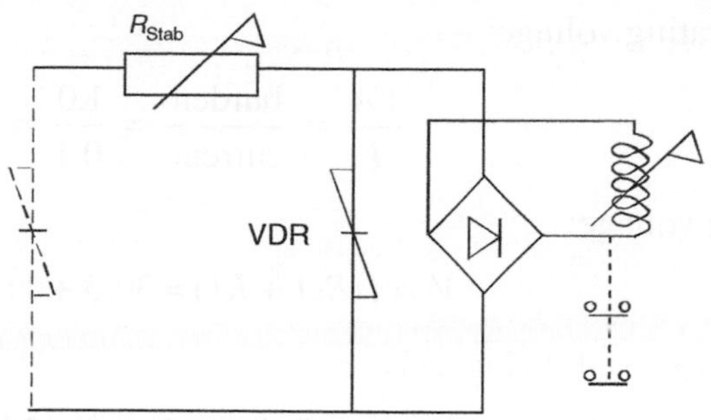

Figure 15.27
Protection against excessively high voltages

15.10.7 Example

Calculate the setting of the stabilizing resistor for the following REF protection. The relay is a type CAG14, 1 A, 10–40%, burden 1.0 VA (see Figure 15.28).

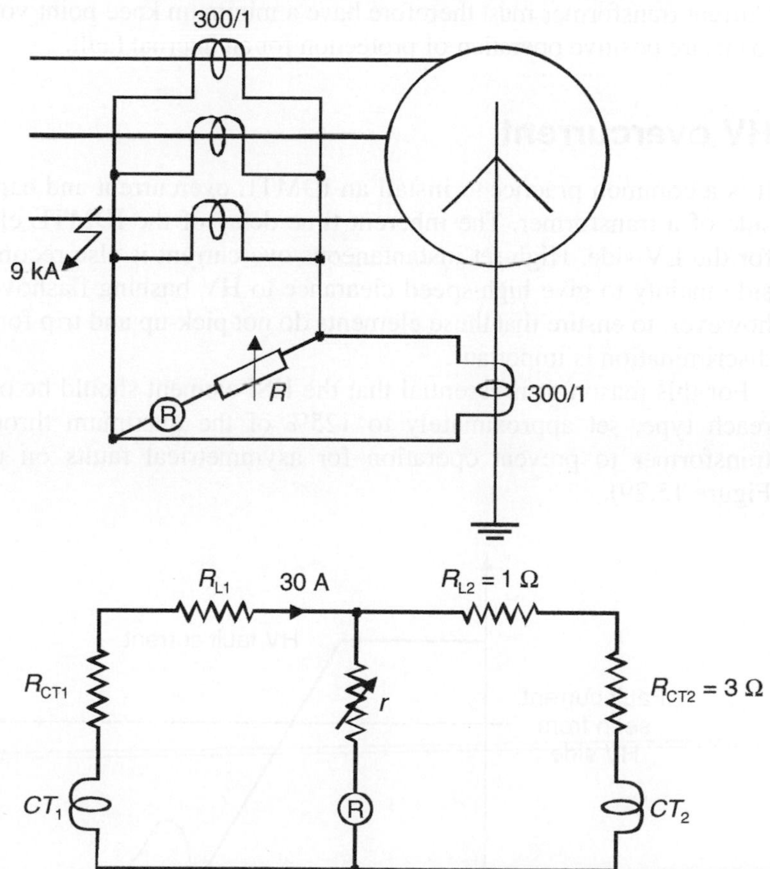

Figure 15.28
Example for calculation of setting of stabilizing resistor

$$\text{Secondary fault current} = 9000 \times \frac{1}{300} = 30\,\text{A}$$

Relay operating current:
Choose 10% tap on CAG14 = 1 A; 10% = 0.1 A

Relay operating voltage:

$$\frac{VA}{I} = \frac{\text{burden}}{\text{current}} = \frac{1.0}{0.1} = 10\,\text{V}$$

Stabilizing voltage:

$$V = I(Rct + R1) = 30(3 + 1) = 120\ \text{V}$$

Voltage across stabilizing resistor:

$$\text{Stabilizing voltage} - \text{Relay voltage} = 120 - 10 = 110\ \text{V}$$

$$\text{Stabilizing resistance} = \frac{\text{Voltage across resistor}}{\text{Current through resistor}}$$

$$= \frac{110}{0.1}$$

$$= 1100\ \Omega$$

Current transformer must therefore have a minimum knee-point voltage of 2 × 120 = 240 V to ensure positive operation of protection for an internal fault.

15.11 HV overcurrent

It is a common practice to install an IDMTL overcurrent and earth fault relay on the HV side of a transformer. The inherent time delay of the IDMTL element provides back-up for the LV side. High-set instantaneous overcurrent is also recommended on the primary side mainly to give high-speed clearance to HV bushing flashovers. Care must be taken, however, to ensure that these elements do not pick-up and trip for faults on the LV side as discrimination is important.

For this reason, it is essential that the HSI element should be of the low-transient overreach type, set approximately to 125% of the maximum through-fault current of the transformer to prevent operation for asymmetrical faults on the secondary side (see Figure 15.29).

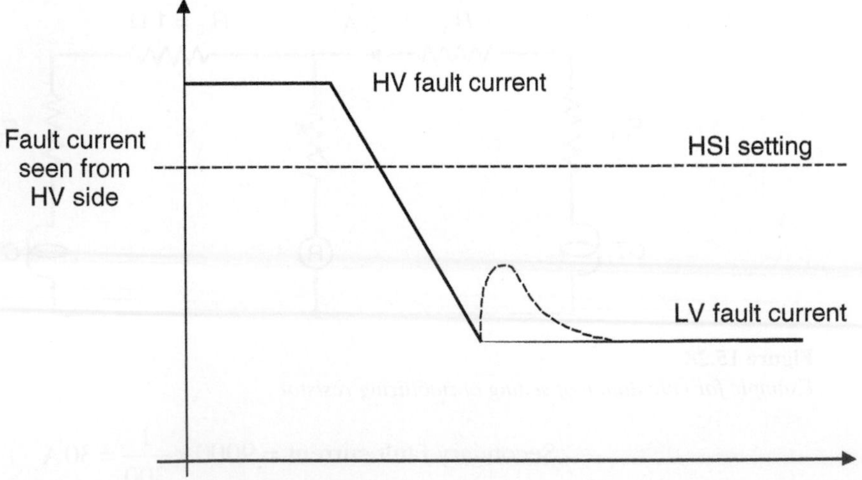

Figure 15.29
Fault current as seen from the HV side

This relay therefore looks into, but not through the transformer, protecting parts of the winding, so behaving like unit protection by virtue of its setting.

15.11.1 Current distribution

When grading IDMTL overcurrent relays across a delta–star transformer, it is necessary to establish the grading margin between the operating time of the star side relay at the phase-to-phase fault level and the operating time of the delta side relay at the three-phase fault level.

This is because, under a star side phase-to-phase fault condition, which represents a fault level of 86% of the three-phase fault level, one-phase of the delta side transformer will carry a current equivalent to the three-phase fault level (see Figures 15.30 and 15.31).

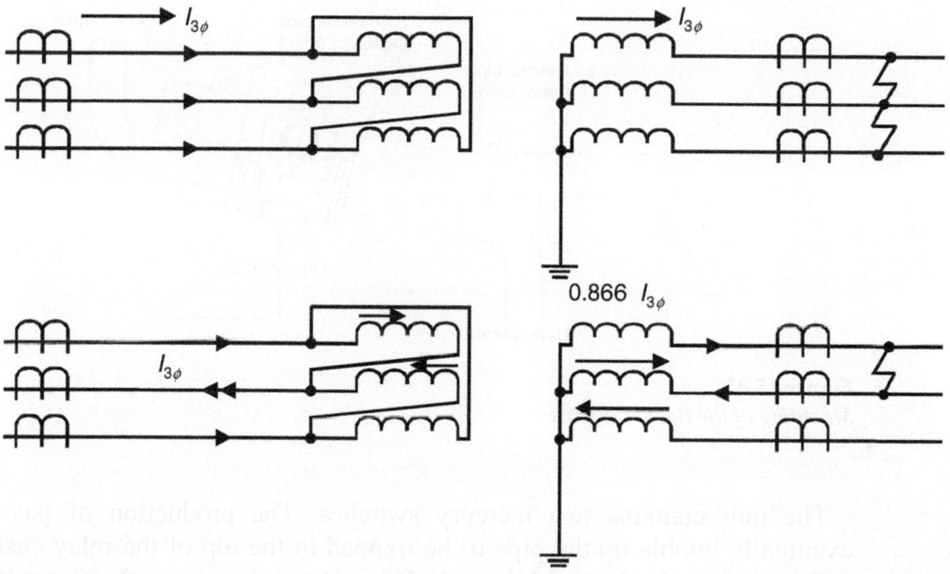

Figure 15.30
Delta star transformer configuration

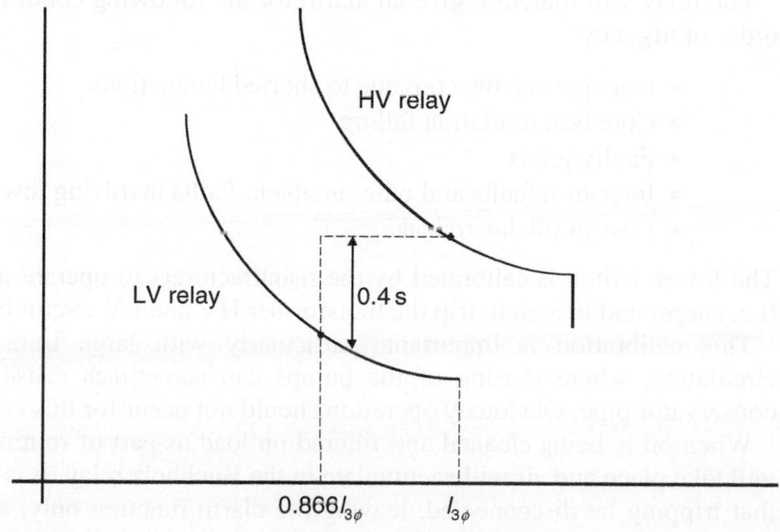

Figure 15.31
Graphical representation of the fault

15.12 Buchholz protection

Failure of the winding insulation will result in some form of arcing, which can decompose the oil into hydrogen, acetylene, methane, etc. Localized heating can also precipitate a breakdown of oil into gas.

Severe arcing will cause a rapid release of a large volume of gas as well as oil vapor. The action can be so violent that the build-up of pressure can cause an oil surge from the tank to the conservator.

The Buchholz relay can detect both gas and oil surges as it is mounted in the pipe to the conservator (see Figures 15.32 and 15.33).

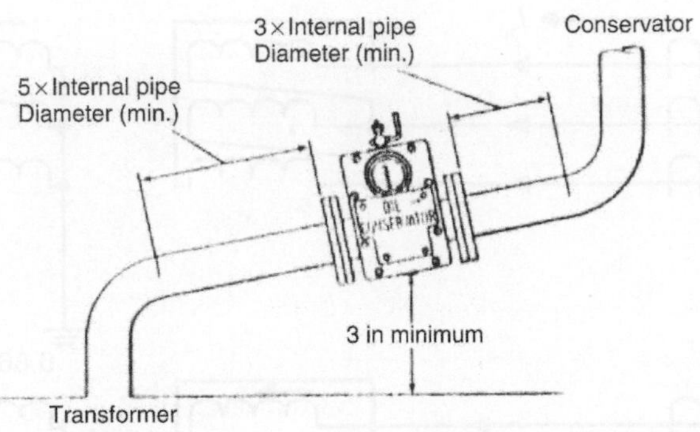

Figure 15.32
Mounting of the Buchholz relay

The unit contains two mercury switches. The production of gas in the tank will eventually bubble up the pipe to be trapped in the top of the relay casing, so displacing and lowering the level of the oil. This causes the upper float to tilt and operate the mercury switch to initiate the alarm circuit. A similar operation occurs if a tank leak causes a drop in oil level.

The relay will therefore give an alarm for the following conditions, which are of a low order of urgency:

- Hot spots on the core due to shorted laminations
- Core bolt insulation failure
- Faulty joints
- Inter-turn faults and other incipient faults involving low power
- Loss of oil due to leakage.

The lower switch is calibrated by the manufacturers to operate at a certain oil flow rate (i.e. surge) and is used to trip the transformer HV and LV circuit breakers.

This calibration is important, particularly with large transformers having forced circulation, where starting of the pumps can sometimes cause a rush of oil into the conservator pipe. Obviously operation should not occur for this condition.

When oil is being cleaned and filtered on load as part of routine maintenance, aeration will take place and air will accumulate in the Buchholz relay. It is therefore recommended that tripping be disconnected, leaving the alarm function only, during this oil treatment process and for about 48 h afterwards. Discretion must then be used when dealing with the alarm signals during this period.

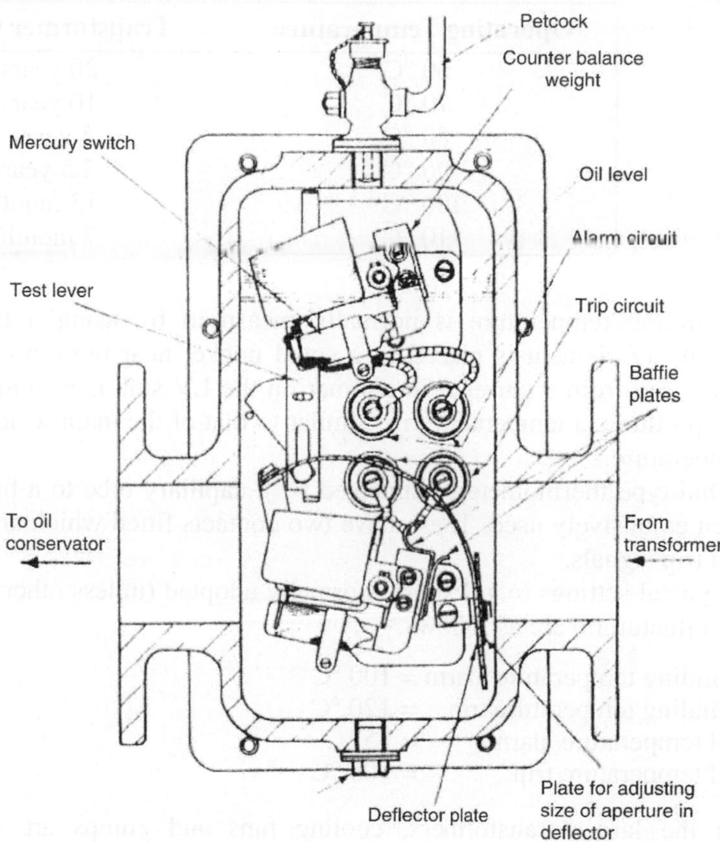

Figure 15.33
Details of the Bucholtz relay construction

Because of the universal response to faults within the transformer, some of which are difficult to protect by other means, the Buchholz relay is invaluable. Experience has shown that it can be very fast in operation. Speed as fast as 50 ms have been recorded, beating all other protection systems on the transformer in process. Gas sampling facilities are also provided to enable gas to be easily collected for analysis.

15.13 Overloading

A transformer is normally rated to operate continuously at a maximum temperature based on an assumed ambient. No sustained overload is usually permissible for this condition. At lower ambient it is often possible to allow short periods of overload but no hard and fast rules apply, regarding the magnitude and duration of the overload.

The only certain factor is that the winding must not overheat to the extent that the insulation is cooked, thereby accelerating ageing. A winding temperature of 98 °C is considered to be the normal maximum working value, beyond which a further rise of 8–10 °C, if sustained, is considered to half the life of the transformer. Oil also deteriorates from the effect of heat. It is for these reasons that winding and oil temperature alarm and trip devices are fitted to transformers.

Operating Temperature	Transformer Oil Life
60 °C	20 years
70 °C	10 years
80 °C	5 years
90 °C	2.5 years
100 °C	13 months
110 °C	7 months

Winding temperature is normally measured by using a thermal image technique. A sensing element is placed in a small pocket near the top of the main tank. A small heater fed from a current transformer on the LV side is also mounted in this pocket and this produces a temperature rise similar to that of the main winding, above the general oil temperature.

Dial-type thermometers connected by a capillary tube to a bulb in the oil pocket have been extensively used. These have two contacts fitted which are adjustable to give alarm and trip signals.

Typical settings (e.g. Eskom) normally adopted (unless otherwise recommended by the manufacturers) are as follows:

Winding temperature alarm = 100 °C
Winding temperature trip = 120 °C
Oil temperature alarm = 95 °C
Oil temperature trip = 105 °C

On the larger transformers, cooling fans and pumps are employed to control the temperature. In many cases, normal practice seems to be to use IDMTL overcurrent relays for overload protection, CT ratios being chosen on the basis of the transformer full-load current.

To use IDMTL overcurrent relays for overload is not a good practice for three reasons:

1. The BS142 tripping characteristic is not compatible with the thermal characteristic of a power transformer. Incorrect and often unnecessary tripping can occur for light overloads, whilst failure to trip for heavy overloads could shorten the life of the transformer dramatically.
2. If set too fine, there is also the danger of tripping on magnetizing in-rush current of its own or adjacent transformer.
3. It may not coordinate with LV circuit breaker protection for an LV fault, beating it in the process.

Overload protection should be done by oil and winding temperature devices, or relays that have similar tripping characteristics to the thermal time constant of the transformer.

15.13.1 Oil testing and maintenance

There are three important purposes of the oil in a transformer:

1. Good dielectric strength
2. Efficient heat transfer and cooling
3. To preserve the core and assembly
   - By filling voids (to eliminate partial discharge)
   - By preventing chemical attack of core, copper and insulation by having low gas content and natural resistance to ageing.

It is therefore vital that the oil is kept in tip-top condition and that regular testing and maintenance be carried out.

Samples should be drawn annually and tested to see if they comply with the following limits:

Dielectric strength = 50 kV/minimum
Moisture content = 30 ppm maximum
Acidity = 0.2 mg KOH/g maximum
Interfacial tension = 20 mN/m minimum
M.I.N. = 160 minimum

A typical test report is shown in Table 15.1 from which it would be immediately apparent if major problems are imminent and urgent action needs to be taken.

Serial No.: 7324/3	**Desig.: TXR3**
Customer:	**Rating: 10 MVA**
Site:	**Main Sub Voltage: 33/11 kV**

Sampling Date: 93-08-31	**Sample: BMT**
Sample Temp.: 25 °C	
Tests Done	**Results**
Moisture (ppm)	7
Acidity (mg KOH/g)	0.02
Dielectric strength (kV)	73
Interfacial tension (mN/m)	
MIN (–)	
Tan Delta at 90 °C (–)	
Recommended values according to BBT Document TPI 1171/1: Dielectric strength : 50 kV minimum Moisture content : 30 ppm maximum Acidity : 0.20 mg KOH/g maximum Interfacial tension : 20 mN/m minimum MIN : 160 minimum	

Table 15.1
Typical transformer oil test report

However, it is very important to also conduct a gas analysis on the samples. This analysis of the various constituents can provide some valuable information as to the rate of deterioration (or otherwise) of the transformer insulation.

Table 15.2 illustrates typical readings on an anonymous sample and it is important to interpret trends rather than absolute levels. There are no hard and fast rules that can be applied and even the oil filtration/purification companies fight shy of interpreting results. The production rates in Table 15.3 do, however, assist in drawing conclusions in order to pre-empt major problems occurring in future (see also Tables 15.4 and 15.5).

Serial No.: 7324/3	**Desig.: TXR3**
Customer:	**Rating: 10 MVA**
Site:	**Main Sub Voltage: 33/11 kV**

Sample: BMT						
Date	92-09-02	92-11-23	93-05-12	93-07-09	93-08-30	93-08-31
Next Date	93-09-02	93-11-23	94-05-12	04-07-09	94-02-26	94-08-31
Sample no.	1	2	3	4	5	6
Report no.	1346	1441	16009	1693	1718	1714
H_2	0	0	5	16	55	41
0_2	35119	22433	20421	24759	21705	21480
N_2	62326	58357	49992	59046	53134	63174
CO	19	16	15	22	2	0
CO_2	459	323	469	281	133	124
CH_4	5	1	1	19	0	0
C_2H_4	4	3	3	23	41	1
C_2H_6	1	0	0	8	109	5
C_2H_2	22	33	31	16	4	0
TCG	51	53	55	103	210	46
TGC%	9.7	8.1	7.6	8.5	7.7	8.7

Table 15.2
Typical gas analysis summary

A typical interpretation would be as follows:

Interpretation of historical results/trends
The high level of Ethane (C_2H_6) detected in sample no. 5 is a cause for concern. This is consistent with localized overheating having taken place in the transformer. The level of Ethylene (41 ppm) is also consistent with this conclusion.

Conclusion
The transformer appears to have had a localized hot spot between samples 4 and 5 but now appears to be fine. If the oil was purified between samples 5 and 6, then the results of sample 6 may not be significant and further samples should be drawn in 6 months time.

Tight control of these procedures and testing can prevent transformer faults occurring, whereas protection relays only operate after the event when the damage has been done. Gas analysis of samples taken from the Buchholz relay can also prove very illuminating and reveal potential major problems.

Serial No.: 7324/3	**Desig.: TXR3**
Customer:	**Rating: 10 MVA**
Site:	**Main Sub Voltage: 33/11 kV**

Sampling No.: 6		**Sample: BMT**
Sampling Date: 93-08-31		**Next Date: 94-08-31**
Gas Detected in Samples	**Sample Values**	**Production Rates**
Hydrogen (H_2)	41 ppm	–14.00 ppm/day
Oxygen (O_2)	21480 ppm	–
Nitrogen (N_2)	63174 ppm	–
Carbon Monoxide (CO)	0 ppm	–2.00 ppm/day
Carbon Dioxide (CO_2)	124 ppm	–9.00 ppm/day
Methane (CH_4)	0 ppm	0.0 ppm/day
Ethylene (C_2H_4)	1 ppm	– 40.00 ppm/day
Ethane (C_2H_6)	5 ppm	–104.0 ppm/day
Acetylene (C_2H_2)	0 ppm	– 4.00 ppm/day
Total Combustible Gas (TCG)	46 ppm	–164.0 ppm/day
Total Gas Content (TGC)	8.7%	–

Table 15.3
Typical gas analysis on transformer oil

Case 1: Transformer Rating: 250 MVA **Voltage: 400/30 kV** **Circumstances: Buchholz Trip but no Obvious Faults**		
Gas	**Main Tank**	**Buchholz Oil**
H_2	13	1458
CO	4	12
CH_4	3	376
CO_2	51	56
C_2H_4	3	204
C_2H_6	1	7
C_2H_2	6	576
Diagnosis:	Discharges of high energy, arcing, sparking and overheating	
Findings:	Flash over from dislocated connection in bushing turret	

Table 15.4
Case 1: Transformer rating: 250 MVA; voltage: 400/30 kV

Case 2: Transformer Rating: 11 MVA Voltage: 20/6.6 kV Circumstances: Old Unit in Service for +15 Years		
Gas	**Main Tank**	**Conservator**
H_2	219	51
CO	1791	2300
CH_4	1197	731
CO_2	14896	11152
C_2H_4	2273	1880
C_2H_6	663	526
C_2H_2	11	9
Diagnosis:	Thermal faults of high temperature. Overheated oil and cellulose	
Findings:	Interturn flash over between winding layers	

Table 15.5
Case 1: Transformer rating: 11 MVA; voltage: 20/6/6 kV

16

Switchgear (busbar) protection

16.1 Importance of busbars

Busbars are the most important component in a distribution network. They can be open busbars in an outdoor switch yard, up to several hundred volts, or inside a metal clad cubicle restricted within a limited enclosure with minimum phase-to-phase and phase-to-ground clearances. We come across busbars, which are insulated as well as those, which are open and are normally in small length sections interconnected by hardware.

They form an electrical 'node' where many circuits come together, feeding in and sending out power (see Figure 16.1).

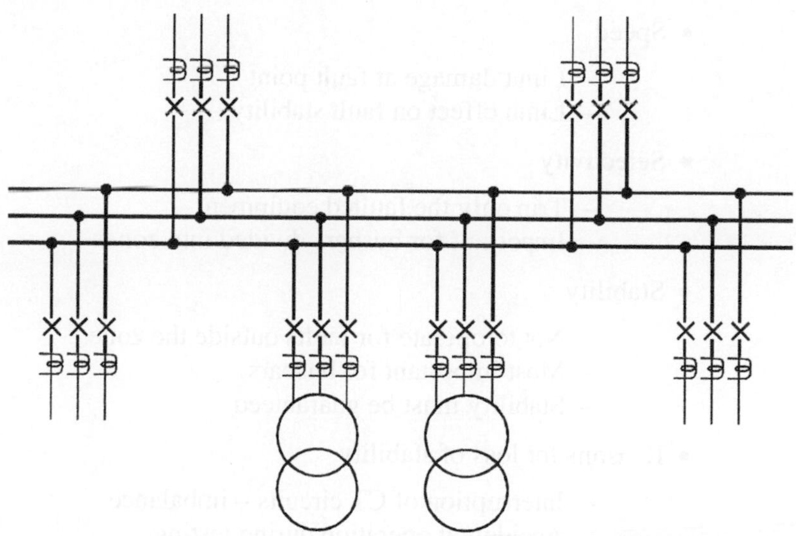

Figure 16.1
Schematic illustrating area of busbar zone

From the above diagram, it is very clear that for any reason the busbars fails, it could lead to shutdown of all distribution loads connected through them, even if the power generation is normal and the feeders are normal.

The important issues of switchgear protection can be summarized as:

- Loss very serious and sometimes catastrophic
- Switchgear damaged beyond repair
- Multi-panel boards not available 'off-the-shelf'

- Numerous joints
- Air enclosure
- Dust build-up
- Insect nesting
- Ageing of insulation
- Frequency of stress impulses
- Long earth fault protection tripping times.

16.2 Busbar protection

Busbars are frequently left without protection because:

- Low susceptibility to faults – especially metal clad switchgear
- Rely on system back-up protection
- Too expensive and expensive CT's
- Problems with accidental operation – greater than infrequent busbar faults
- Majority of faults are earth faults – limited earth fault current – fast protection not required.

However, busbar faults do occur.

16.3 The requirements for good protection

The successful protection can be achieved subject to compliance with the following:

- Speed
  - Limit damage at fault point
  - Limit effect on fault stability
- Selectivity
  - Trip only the faulted equipment
  - Important for busbars divided into zones
- Stability
  - Not to operate for faults outside the zone
  - Most important for busbars
  - Stability must be guaranteed
- Reasons for loss of stability
  - Interruption of CT circuits – imbalance
  - Accidental operation during testing
- Tripping can be arranged 'two-out-of-two'
  - Zone and check relays.

16.4 Busbar protection types

- Frame leakage
- High-impedance differential
- Medium-impedance biased differential
- Low-impedance biased differential
- Busbar blocking.

16.4.1 Frame leakage protection

This involves measurement of fault current from switchgear frame to the earth. It consists of a current transformer connected between frames to earth points and energizes an instantaneous ground fault relay to trip the switchgear. It generally trips all the breakers connected to the busbars.

Care must be taken to insulate all the metal parts of the switchgear from the earth to avoid spurious currents being circulated. A nominal insulation of 10 Ω to earth shall be sufficient. The recommended minimum setting for this protection is about 30% of the minimum earth fault current of the system (see Figures 16.2 and 16.3).

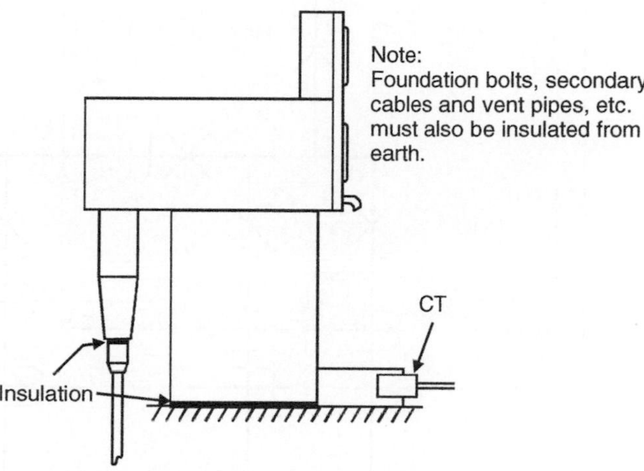

Figure 16.2
Requirements of frame leakage BB protection

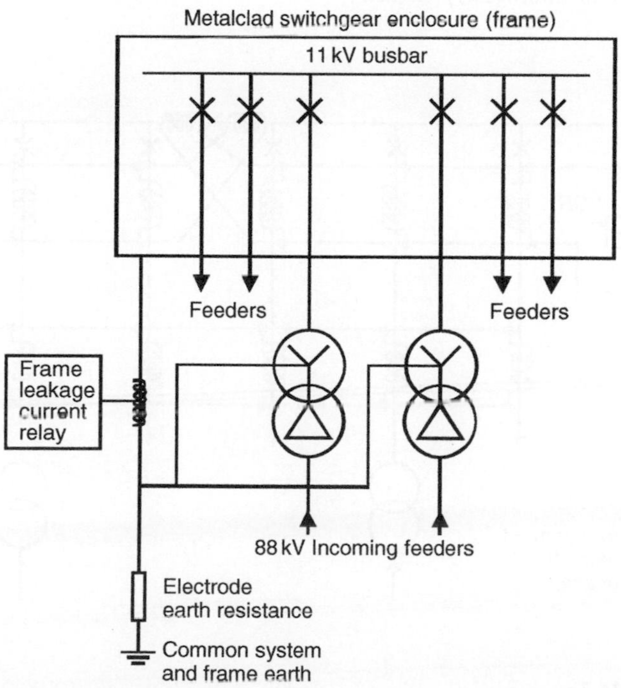

Figure 16.3
Schematic connections for frame leakage protection

16.4.2 Differential protection

This requires sectionalizing the busbars into different zones (see Figures 16.4–16.9).

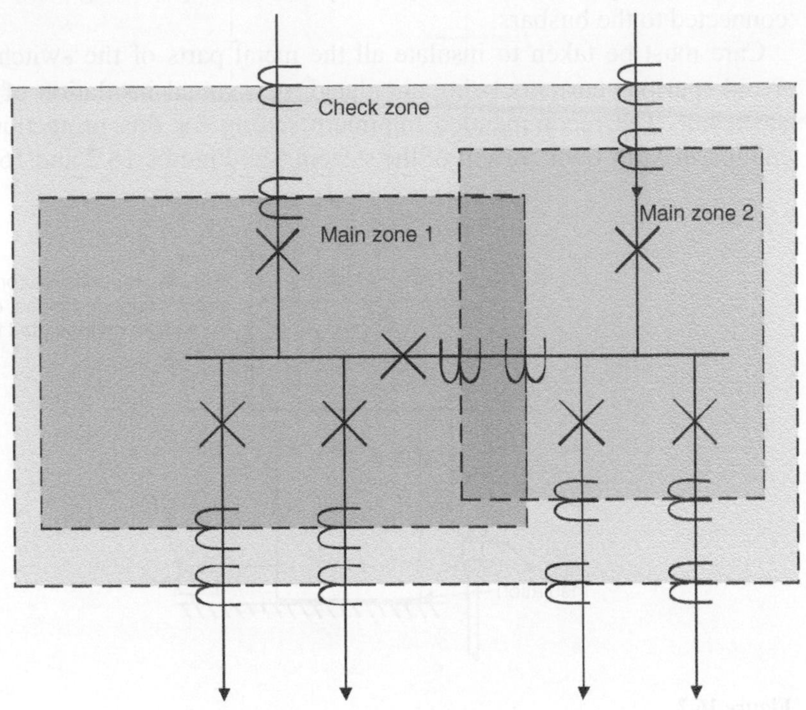

Figure 16.4
Zoned busbar (switchgear) protection

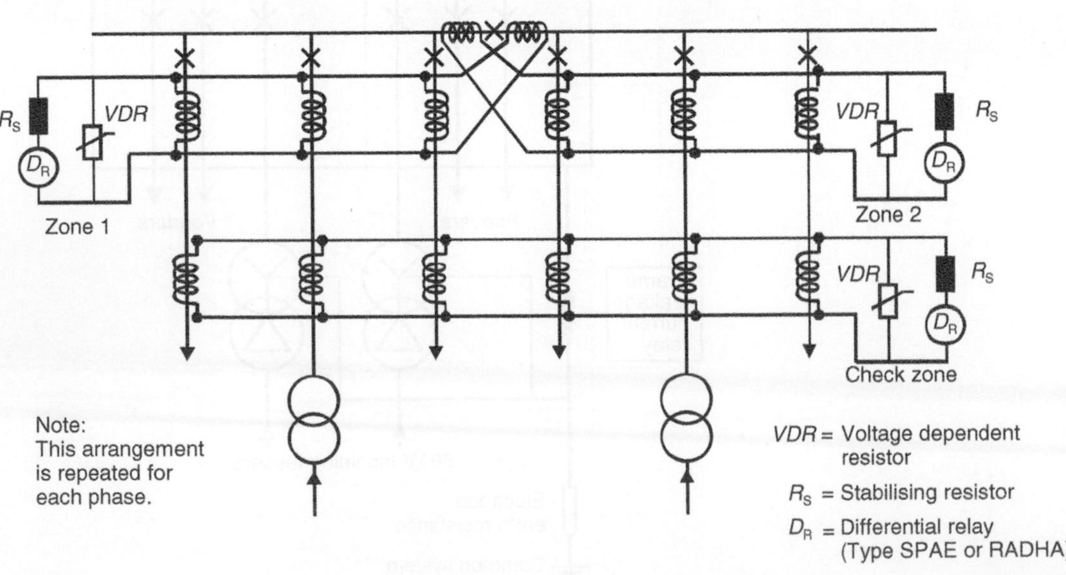

Figure 16.5
Single line diagram high-impedance busbar protection

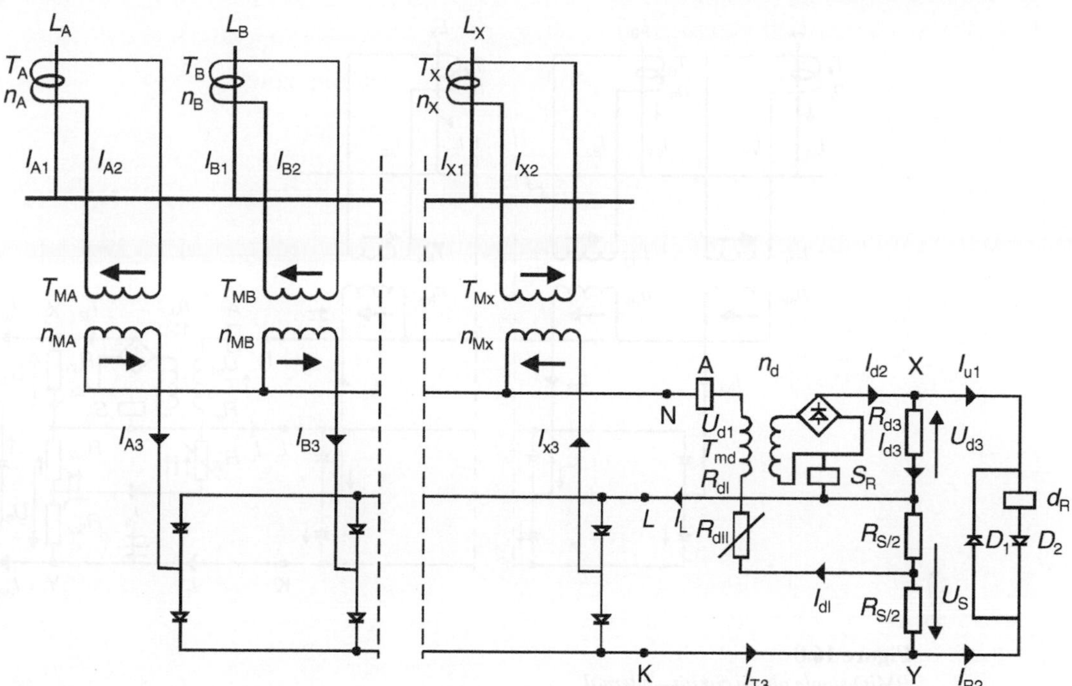

Figure 16.6
BMID single-phase circuit

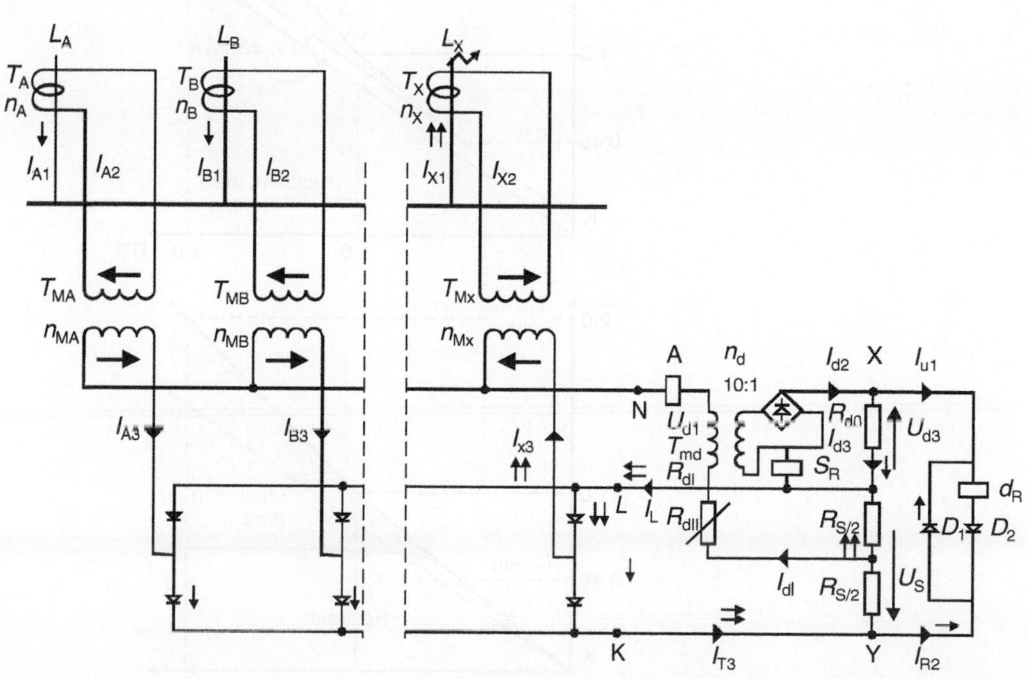

Figure 16.7
BMID single-phase circuit – external fault

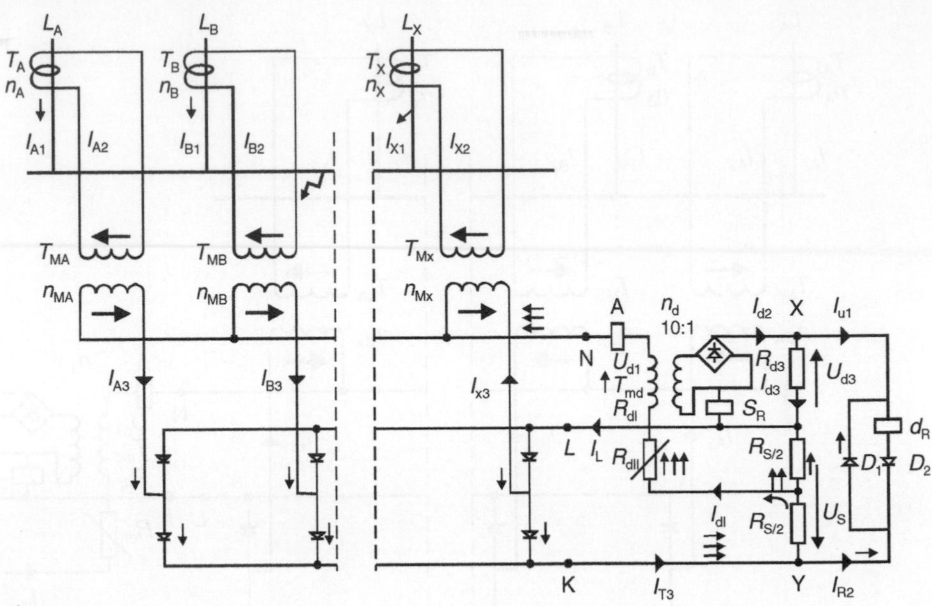

Figure 16.8
BMID single phase circuit – internal

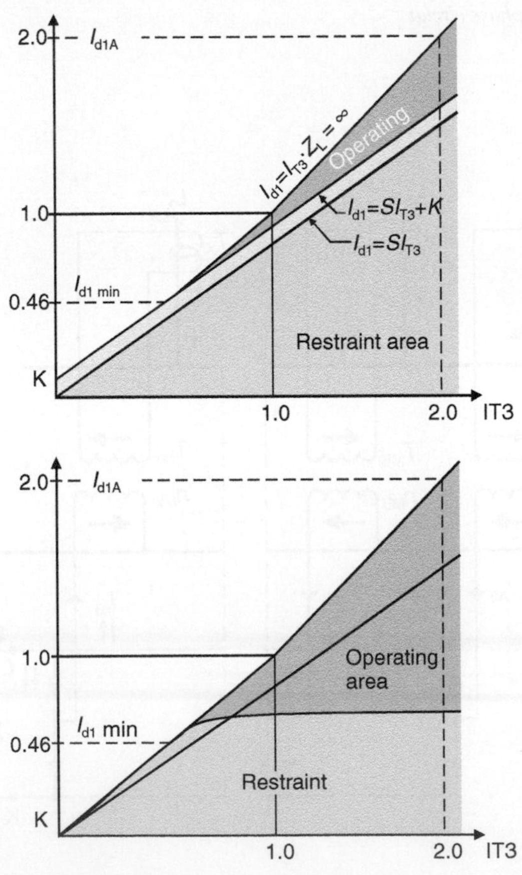

Figure 16.9
Stability characteristic busbar protection

High-impedance bus zone

Advantages

- Relays relatively cheap – offset by expensive CTs
- Simple and well proven
- Fast – 15–45 ms
- Stability and sensitivity calculations – easy providing data is available
- Stability can be guaranteed.

Disadvantages

- Very dependent on CT performance
- CT saturation could give false tripping on through faults
- Sensitivity must be decreased
- DC offset of CTs unequal – use filters
- Expensive class X CTs – same ratio – V_{knp} = 2 times relay setting
- Primary effective setting (30–50%)
- Limited by number of circuits
- Z-earthed system difficult for earth fault
- Duplicate systems – decreased reliability
- Require exact CT data
- V_{knp}, R_{sec}, i_{mag}, $V_{setting}$
- High voltages in CT circuits (±2.8 kV) limited by volt-dependent resistors.

Retrofitting

- Additional CTs six per circuit
- Space problems on metal clad switchgear
- Long shutdowns
- CT performance important
- Class X
- V_{knp} = 2 times setting
- Rsec must be low
- Limit on number of circuits
- CT polarity checks required
- Primary injection tests required
- Compete switchboard
- Separate relay cubicle
- Differential relays
- Auxiliary relays
- CT cabling
- Busbar tripping cabling.

Biased medium-impedance differential

Advantages

- High speed 8–13 ms
- Fault sensitivity ±20%
- Excellent stability for external faults
- Normal CTs can be used with minimal requirements

- Other protection can be connected to same CTs
- No limit to number of circuits
- Secondary voltages low (medium impedance)
- Well proven 10 000 systems worldwide
- Any busbar configuration
- No need for duplicate systems
- Retrofitting easy
- No work on primary CTs
- Biasing may prevent possibility of achieving a sensitive enough earth fault setting of Z-earthed systems.

Disadvantages

- Relays relatively expensive
- Offset by minimal CT requirements
- Relays with auxiliary CTs require a separate panel.

Low-impedance busbar protection

Principle: Merz-price circulating current biased differential CT saturation detector circuits (inhibit pulses) (see Figure 16.10). On through-fault, one CT may saturate – does not provide balancing current for other CT. Spill current ($i_1 - i_2$), then flows through the operating coil. Electronics detect CT saturation – shorts out differential path. Inhibit circuit only allows narrow spikes in differential coil. Relay stable.

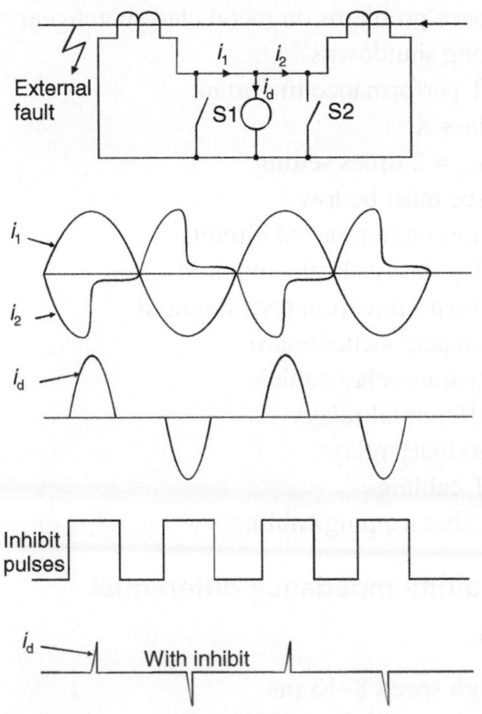

Figure 16.10
Low-impedance busbar protection

For an internal fault, differential current in phase with saturated CT current,

- Inhibit pulses remove insignificant portion of differential current. Relay operates
- *Setting range*: 20–200%
- *Operating time*: Less than 20 ms
- *CT supervision*: Alarms and blocks or trips after 3 s for CT open cct (see Figure 16.11).

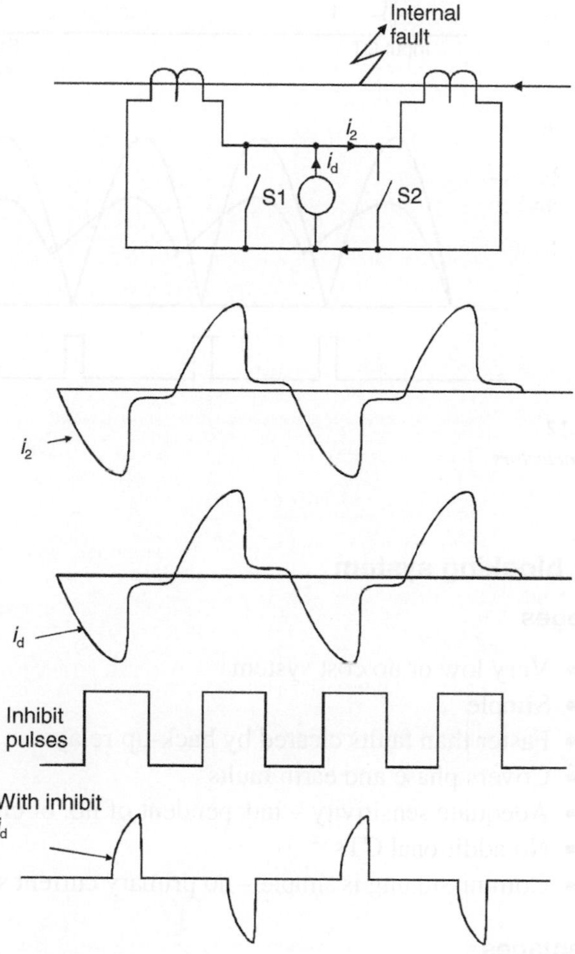

Figure 16.11
Low-impedance busbar protection

Saturation detectors

The CTs feed the differential circuit via auxiliary transformers, which in turn feed a typical saturation detector circuit shown in Figure 16.12:

- A voltage V_c is developed across the resistor R
- Capacitor C is charged to the peak value of that voltage
- A comparator compares voltage with 0.5 V stored in capacitor
- On saturation, V drops below 0.5 V capacitor voltage
- Comparator then turns on electronic switch across buswires
- Pulse width increases with optimum philosophy.

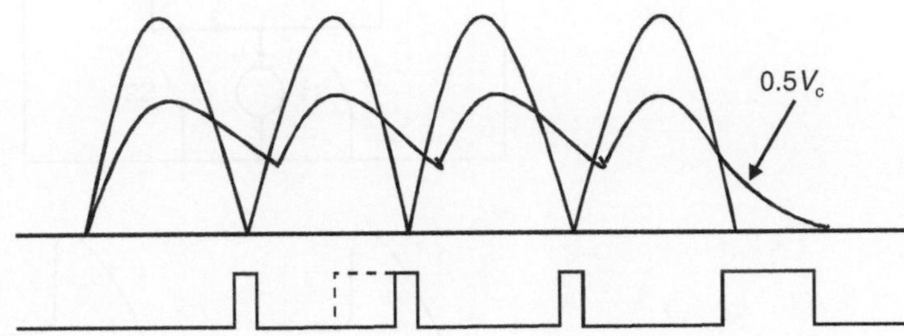

Figure 16.12
Saturation detectors

Busbar blocking system

Advantages

- Very low or no cost system
- Simple
- Faster than faults cleared by back-up relays
- Covers phase and earth faults
- Adequate sensitivity – independent of no. of circuits
- No additional CTs
- Commissioning is simple – no primary current stability tests (see Figure 16.13).

Disadvantages

- Only suitable for simple busbars
- Additional relays and control wiring for complex busbars
- Beware motor in-feeds to busbar faults
- Sensitivity limited by load current.

Retrofitting

- Easy – if starting contacts available, if not they need to be added
- Modern microprocessor relays have starters
- No need to work on CTs
- Most work is done with system operational
- Final commissioning requires very short shutdown. Injection to prove stability between up- and downstream relays.

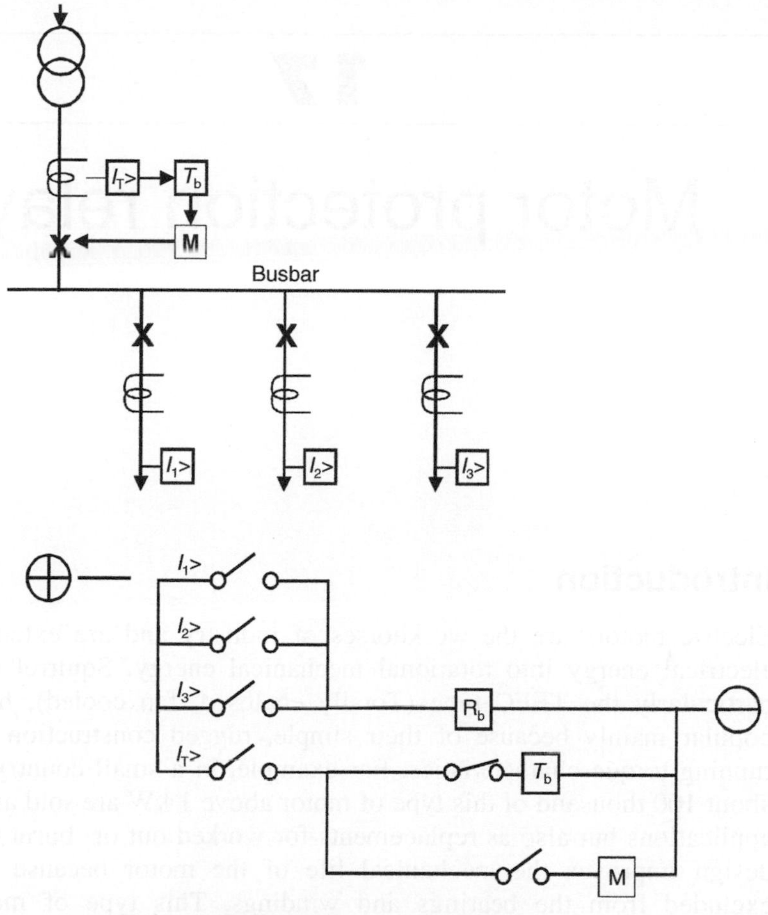

Figure 16.13
Busbar blocking scheme

17

Motor protection relays

17.1 Introduction

Electric motors are the workhorses of industry and are extensively used to convert electrical energy into rotational mechanical energy. Squirrel cage induction motors, particularly the TEFC type (Totally enclosed, fan cooled), have become extremely popular mainly because of their simple, rugged construction and good starting and running torque characteristics. For example, in a small country such as South Africa, about 100 thousand of this type of motor above 1 kW are sold annually, mainly for new applications but also as replacements for worked out or 'burnt out' motors. The TEFC design improves the mechanical life of the motor because dust and moisture are excluded from the bearings and windings. This type of motor has proved to be extremely reliable with an expected lifetime of up to 40 years when used in the correct application.

The causes of motor damage given in Figure 17.1 are taken from statistics gathered within the ABB Group. They are shown in Figure 17.2, that 81% of these failures could have been avoided by using an accurate and effective relay.

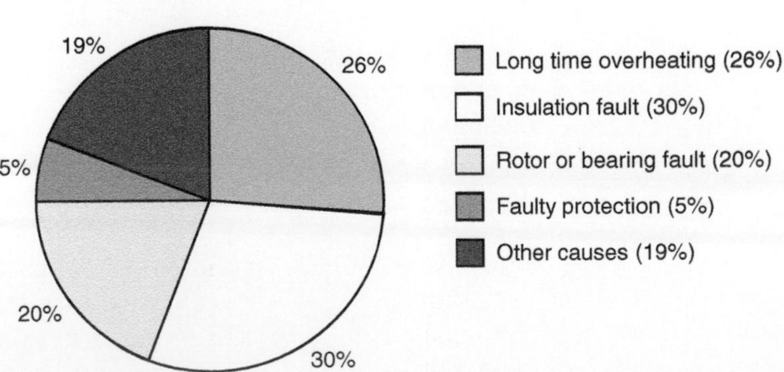

Figure 17.1
Main causes for motor damage in industrial drives

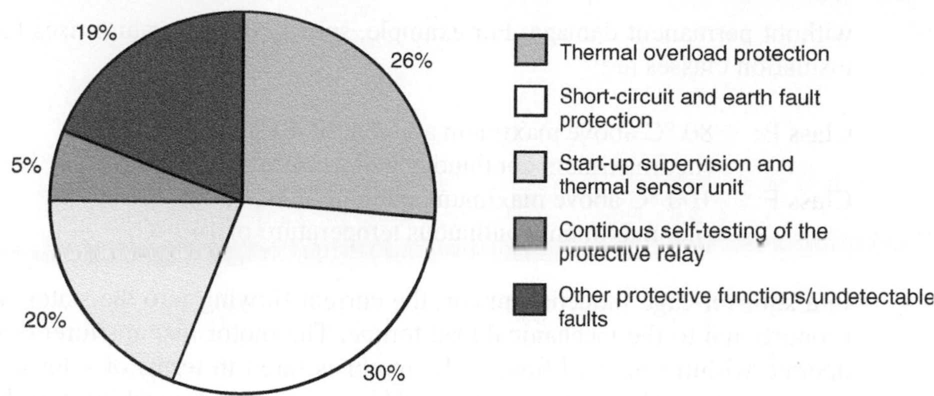

Figure 17.2
Protective functions needed to detect the motor drive faults

The life of an electric motor is determined by the shorter of the following two factors:

1. *Mechanical life*: This is the life of the mechanical parts such as bearings, shaft, fan and the frame and depends on the environment (dust, moisture, chemicals, etc.), vibration and lubrication. The mechanical life can be extended by means of regular inspection and maintenance.
2. *Electrical life*: This is the life of the electrical parts such as the stator winding and insulation, rotor winding and the cable terminations in the motor connection box. Assuming that the cable terminations are properly done and regularly checked, the electrical life may be extended by ensuring that the windings and insulation are not subjected to excessive temperatures which are usually the consequence of overloading or single phasing (loss of one-phase). The purpose of good motor protection is to continuously monitor the current flowing into the motor to detect overloading or fault conditions and to automatically disconnect the motor when an abnormal situation arises. This protection, when correctly applied, extends the useful life of the motor by preventing insulation damage through overheating.

Most people in the industry can easily understand the relatively simple mechanical aspects of an electric motor but few fully appreciate the electrical limitations and relationship of overloading to the useful life of the motor. Essentially, mechanical overloading causes excessively high currents to flow in the winding (since current in the motor is proportional to the load torque) and this results in overheating of the stator and motor windings.

These high temperatures result in the deterioration of the insulation materials through hardening and cracking, eventually leading to electrical breakdown or faults. In many cases, the motor can be repaired by rewinding the stator but this is expensive with a longer downtime. The larger the motor, the higher the cost.

There are several types of insulation materials commonly used on motors. In the IEC specifications for motors, the insulation materials are classified by the temperature rise above maximum ambient temperature, that the materials can continuously withstand without permanent damage. For example, specified temperature rises for commonly used insulation classes are:

Class B: 80 °C above maximum ambient of 40 °C
(i.e. maximum continuous temperature of 120 °C)

Class F: 100 °C above maximum ambient of 40 °C
(i.e. maximum continuous temperature of 140 °C)

In a squirrel cage induction motor, the current flowing into the stator winding is directly proportional to the mechanical load torque. The motor manufacturer designs the motor to operate within specified limits. The motor is rated in terms of kilowatts (kW) at a rated supply voltage (V) and current (I). This means that a machine can drive a mechanical load continuously up to rated torque at rated speed. Under these conditions, supply current is within the specified current and the internal heating will be within the capabilities of the specified insulation class. At full load with class B insulation, the winding temperature will stabilize at below 120 °C.

The main cause of heating in the motor windings is a function of the square of the current flowing in the stator and rotor windings. This is shown on the motor equivalent circuit of Figure 17.3 where the losses are $I^2 (R_s + R_r)$. These are often referred to as the copper losses. The stator windings have only a small mass and heat up rapidly because of the current flowing. The heat insulation and the cooling time constant is consequently quite long. Other losses also generate heat. These are referred to as the iron losses but are relatively small and are quickly dissipated into the body of the motor.

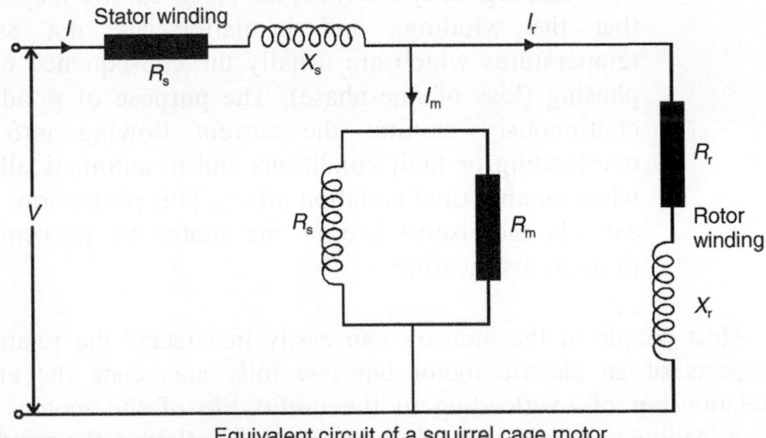

I_m = very small value, so $I = I_r$

Copper losses = $I^2(R_s + R_r)$

R_s = stator resistance

R_r = rotor resistance

X_o = magnetizing inductance

R_m = magnetizing resistance

Figure 17.3
Equivalent circuit of a squirrel cage motor

17.2 Early motor protection relays

Some of the early designs of motor protection relays have a single function whose purpose was to protect the motor against overloading by ensuring that it never draws in excess of the rated current. This was done by continuously monitoring the electrical current drawn by the motor and arranging for the motor to be disconnected when the current exceeded the rated current and remains so for a certain period of time. The higher the overload current, the shorter the permissible time before disconnection. This time delay was achieved in various ways. An example is the 'solder pot' relay, which relied on the time taken for solder in the measuring circuit to melt when the load current was passed through it. The bi-metal type relays disconnect the motor when the load current passing through a resistor heated in a bi-metallic strip sufficiently to bend it beyond a pre-set limit. This released the trip mechanism. In recent years, electronic relays utilize an analog replica circuit, comprising a combination of resistors and capacitors, to simulate the electrical characteristics of the stator and rotor. The main principle linking all these methods is the design of a replica system to simulate as closely as possible the electrical characteristics of the motor.

In the past, it has been a common practice to detect over temperature from temperature-dependent elements built into the winding of the motor.

However, this form of temperature measurement is in most cases unsatisfactory, as it is not taken directly from the current conductor. Instead, it is taken through the insulation which gives rise to considerable sluggishness. Due to insulation considerations, insertion of thermocouples in high-voltage motors can cause problems. Furthermore, after a fault (e.g. a break in the measuring lead inside the machine) high repair costs are encountered. Another problem is that no one can accurately predict, during the design, how many and where the 'hot spots' will be.

Consequently, protection is preferably based on monitoring the phase currents instead. Because the temperature is determined by the copper and iron losses, it must be possible to derive it indirectly by evaluating the currents in the motor supply leads.

The performance of a motor protection relay depends on how closely and accurately the protection simulates the motor characteristics. The ideal simulation occurs when the heating and cooling time constants of the motor windings are matched by the relay under all operating conditions. In some of the early devices, the protection could underestimate the heating time of the windings from cold and could trip before a motor/load combination with a long run-up time had reached running speed. On the other hand, during several sequential starts and stops, the device could underestimate the cooling time of the windings, allowing the motor windings to overheat. This situation can very easily arise with the bi-metallic thermal overload relays commonly used on motor starters even today. Under certain conditions, bi-metallic thermal overload relays do not provide full protection because the device does not have exactly the same thermal heating and cooling characteristics as the motor, which it is protecting. The heating and cooling time constants of a bi-metallic relay are much the same but in actual installations, it should be borne in mind that a stopped motor has a longer cooling time constant than that for a running motor. When a motor has stopped, the fan no longer provides a forced draft and cooling takes longer than when the motor is running on no load. A simple bi-metallic device is a compromise and is calibrated for normal running conditions. As soon as an abnormal situation arises, difficulties can be expected to arise.

To illustrate the point, take the case of a motor that has been running at full load for a period of time when the rotor is suddenly stalled. Figure 17.4 shows typical temperature curves of the winding temperature (solid line) compared to the heating and cooling curve of the protective device (dotted line). Starting at a normal continuous running temperature of 120 °C, the current increases for the locked rotor condition and temperature rises to 140 °C when the thermal device trips the motor after some seconds. After about 10 min, the bi-metal will have cooled to ambient, but the windings will only have reading 100 °C. With the bi-metal reset, it is then possible to attempt a restart of the motor. With the rotor still locked, high starting currents cause the temperature to quickly rise to 165 °C before the bi-metal again trips the motor.

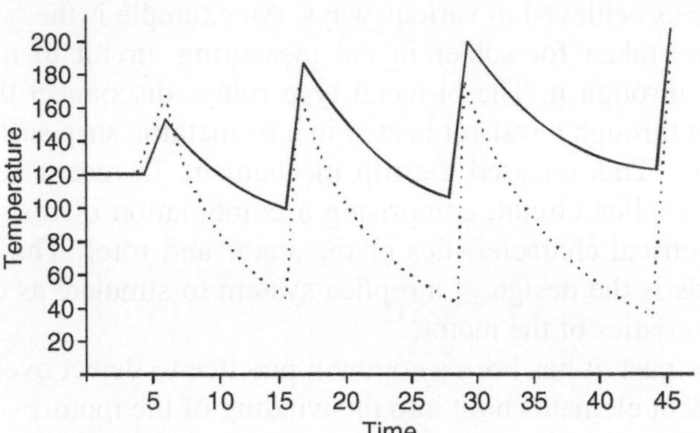

Figure 17.4
Temperature rise vs time for a motor

Consider repeating similar sequence of events as described above, where the different cooling times of the motor and bi-metal strip allow the bi-metal to reset before the windings have cooled sufficiently, and if the motor is again restarted after another 10 min, the winding temperature is likely to exceed 180 °C, the critical temperature for class B insulation materials. This illustrates the importance of an accurate simulation by the protection device in both conditions where the motor is running and when the motor is stopped.

17.3 Steady-state temperature rise

In the interest of maximum efficiency, electrical machines should be loaded as close as possible to their permitted operating temperature limit; however, excessive thermal stressing of any appreciable duration must be avoided if the life of the insulation is not to be shortened.

Under steady-state conditions, the temperature of a motor will rise exponentially, due to dissipation of the heat to the environment or cooling medium, towards its respective operative temperature. Since a motor is not a homogeneous mass, heat is dissipated in several stages. Temperature rise and fall takes place according to a series of partial time constants. Refer to Figure 17.5.

In spite of this, it is sufficient for a thermal overload relay intended for protection under steady-state conditions to be set to the mean time constant of the motor. This means that proper account is taken only of the copper losses. Measurement of the voltage would be necessary in order to include the iron losses, but is not generally possible since the voltage transformers are usually located on the busbar and not adjacent to each motor. Most modern thermal overload relays only measure current, filtering out the highest of the three-phase current. The critical cases of starting, stalling and failure of a phase are taken care by other protective functions.

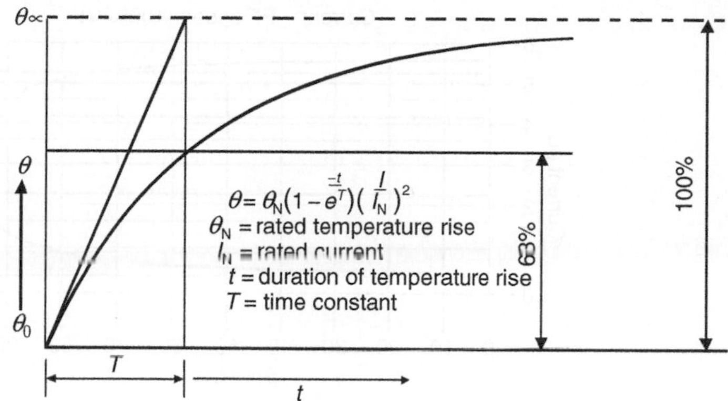

Figure 17.5
Temperature rise vs time (illustrating time constant)

17.4 Thermal time constant

The time constant T (tau) is defined (IEC 255-8) as the time in minutes required for the temperature of a body to change from an initial temperature θ_0° to 63% of the difference between θ_0° and the new steady-state temperature θ_∞.

Unfortunately, the thermal time constant T of the motor is frequently not known. Table 17.1 gives typical values in relation to motor ratings and mechanical design. The cooling time constants during operation are approximately equal to those for temperature rises, while at standstill they are 4–6 times the values given in the table.

Type \ A[mm]	355	400	450	500	560	630	710	800	900	1000	1120	1250
O	20	25	28	30	35	40	50	60	65	70		
R				45	50	55	60	70	80	90	100	110
U	30	35	40	45	50							

A = Shaft height (mm)
O = Open type (IP23)
R = Closed type with air/air heat-exchanger (IP54)
U = Fully clad with cooling finds (IP54)

Table 17.1
Mean thermal time constants of asynchronous motors from Brown Boveri in relation to motor rating and type

17.5 Motor current during start and stall conditions

As the magnitude and duration of motor starting currents and the magnitude and permissible duration of motor stalling currents are major factors to be considered in the application of overload protection, these will be discussed. It is commonly assumed that the machines started direct-on-line the magnitude of the starting current decreases linearly as the speed of the machine increases. This is not true. For normal designs, the starting current remains approximately constant at the initial value for 80–90% of the total starting time. Refer to Figure 17.6. When determining the current and time settings of the overload protection, it can be assumed that the motor starting current remains constant and equal to the standstill value for the whole of the starting time.

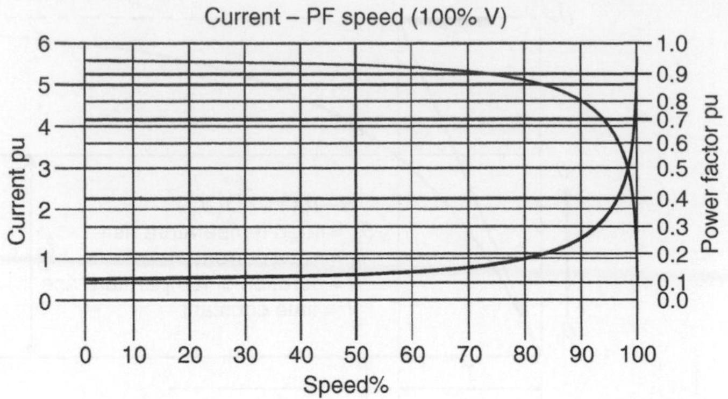

Figure 17.6
Motor current during start conditions

17.6 Stalling of motors

Refer to Figures 17.7 and 17.8. Should a motor stall when running or be unable to start (run) because of excessive load, it will draw a current from the supply equivalent to its locked rotor current. It is obviously necessary to avoid damage by disconnecting the machine as quickly as possible if this condition arises. It is not possible to distinguish this condition from a healthy starting condition on current magnitude.

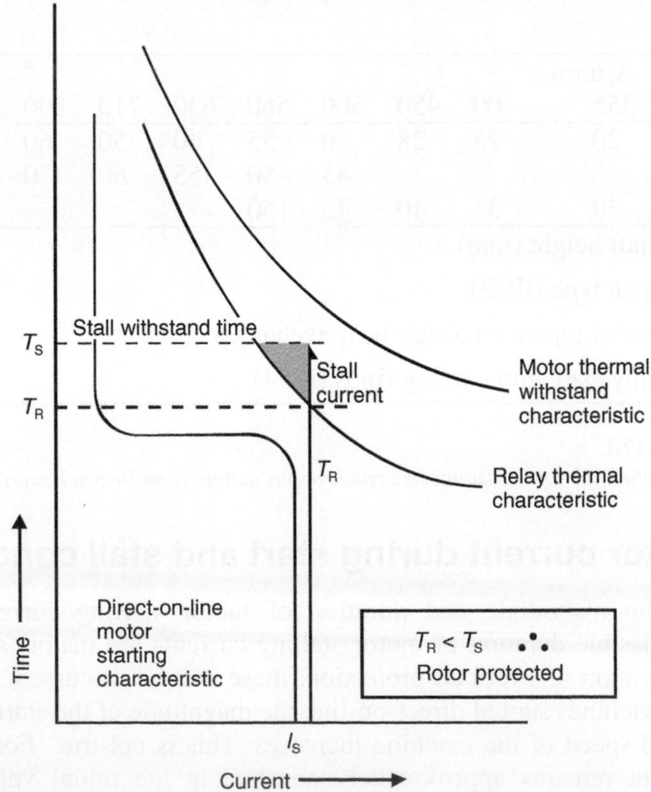

Figure 17.7
Relay operation time less than stall withstand time: relay gives stall protection

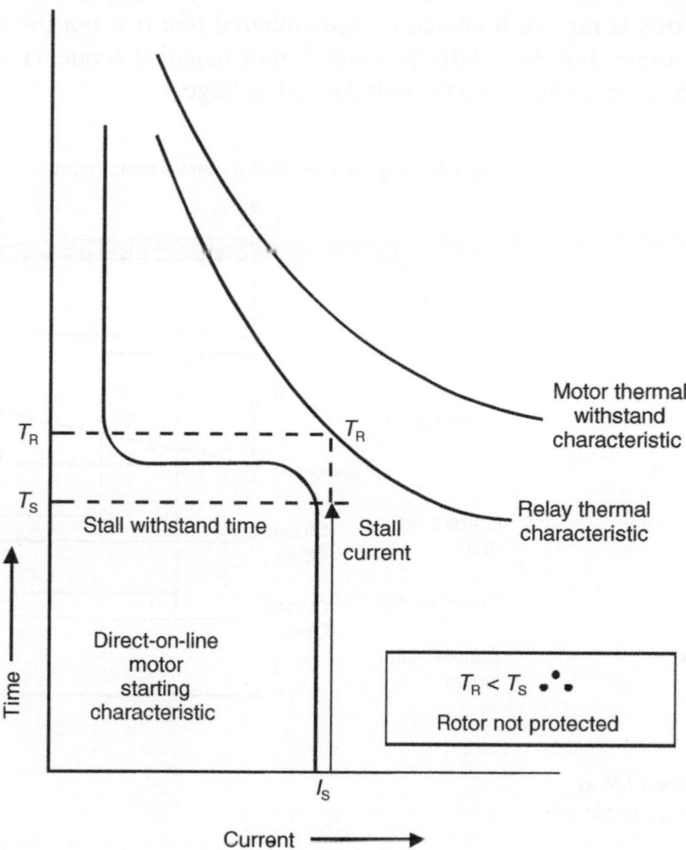

Figure 17.8
Relay operation time greater than stall with stand time: relay does not give stall protection

The majority of loads are such that the starting time of normal induction motors is about or less than 10 s, while the allowable stall time to avoid damage to the motor insulation is in excess of 15 s.

If a double cage drive is to be protected, it might be that the motor cannot be allowed to be in a stall condition even for its normal start-up time. In this case, a speed switch on the motor shaft can be used to give information about whether the motor is beginning to run-up or not. This information can be fed to suitable relays, which can accelerate their operating time. Refer to Figures 17.9(a), (b).

Whether or not additional features are required for the stalling protection, depends mainly on the ratio of the normal starting time to the allowable stall time and the accuracy with which the relay can be set to match the stalling time/current curve and still allow a normal start.

17.7 Unbalanced supply voltages

The voltage supplied to a three-phase motor can be unbalanced for a variety of reasons; single-phase loads, blown fuses in pf capacitors, etc. In addition, the accidental opening of one-phase lead in the supply to the motor can leave the motor running, supplied by two phases only.

It might seem that the degree of voltage unbalance met within a normal installation (except when one-phase is open circuited) would not affect the motor to any great extent, but this is not so. It should be remembered that it is not the unbalanced voltage which is important, but the relatively much larger negative sequence component of the unbalance current, resulting from the unbalanced voltage.

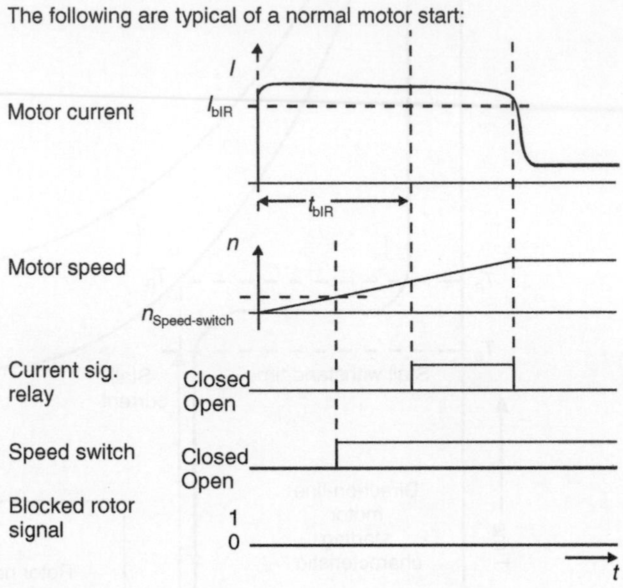

Figure 17.9(a)
Typical motor start

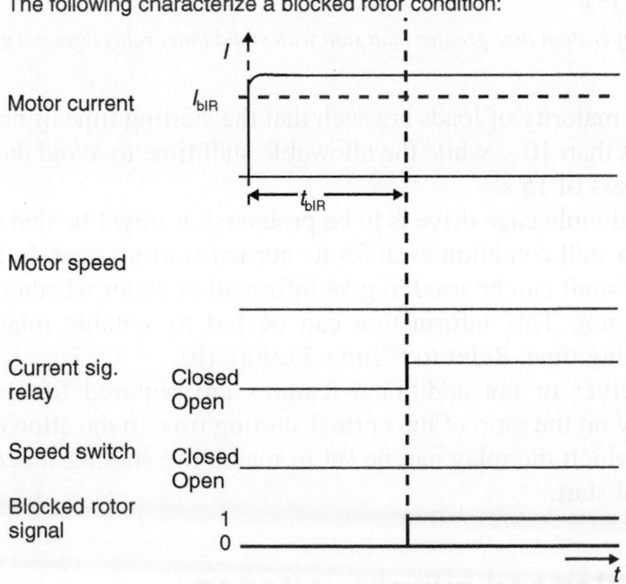

Figure 17.9(b)
Blocked rotor condition

The method of symmetrical components consists of reducing any unbalanced three-phase systems of vectors into three balanced systems: the positive, negative and zero sequence components (see Figure 17.10). The positive sequence components consist of three vectors equal in magnitude 120° out of phase, with the same phase sequence or rotation as that of the source of supply. The negative sequence components are three vectors equal in magnitude, displayed by 120° with a phase sequence opposite to the positive sequence. The zero sequence components consist of three vectors equal in magnitude and in a phase.

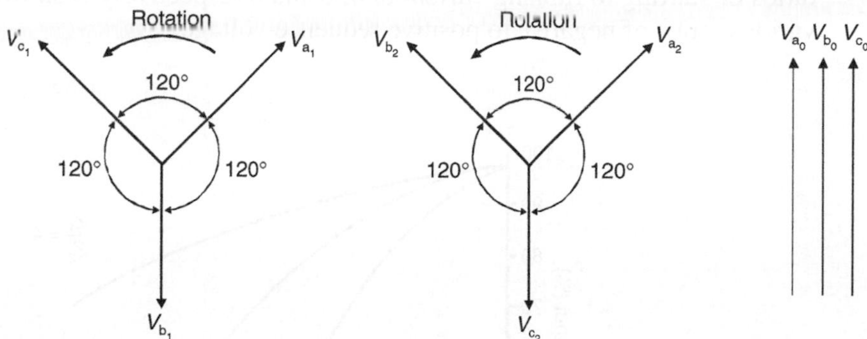

Figure 17.10
The positive, negative and zero components

Loss of one-phase represents the most dangerous case of unbalance. It is therefore essential for motors, which are protected against short circuit by fuses (limited breaking capacitor of the breaker) to be equipped with fast-operating loss of phase protection.

17.8 Determination of sequence currents

In the general case of unbalanced three-phase voltages, there is no fixed relationship between the positive and negative sequence currents; the actual value of the negative sequence current depends on the degree of unbalanced supply voltage, and on the ratio of the negative to the positive sequence impedance of the machine. The ratio can be determined from the general equivalent circuit of the induction motor.

Since in an induction motor the value of the resistance is normally small compared with the reactance, the negative sequence impedance at normal running speeds can be approximated to the positive impedance at standstill. The ratio of the positive sequence impedance to the negative sequence at normal running speeds can thus be approximated to the ratio of the starting current which will therefore be approximately equal to the product of the negative sequence voltage and the ratio of the starting current to the full-load running current.

For instance, in a motor, which has a starting current equal to 6 times rated current, a 5% negative sequence component in the supply voltage would result in approximately a 30% negative sequence component of current.

17.9 Derating due to unbalanced currents

The negative sequence component of the current does not contribute to providing the driving torque of the motor; in fact, it produces a small negative torque. The magnitude of the torque due to the negative sequence current is, however, usually less than 0.5% of the full-load-rated torque for a voltage unbalance in the order of 10% and can therefore be neglected. Hence, presence of negative sequence currents does not appreciably affect the starting characteristics.

The main effect of the negative sequence current is to increase the motor losses, mainly copper loss, thus reducing the available output of the machine if overheating of the machine windings is to be avoided. The reduction in output for the machines having ratios of starting to running current of 4, 6 and 8 respectively is shown in Figure 17.11 for various ratios of negative to positive sequence voltage.

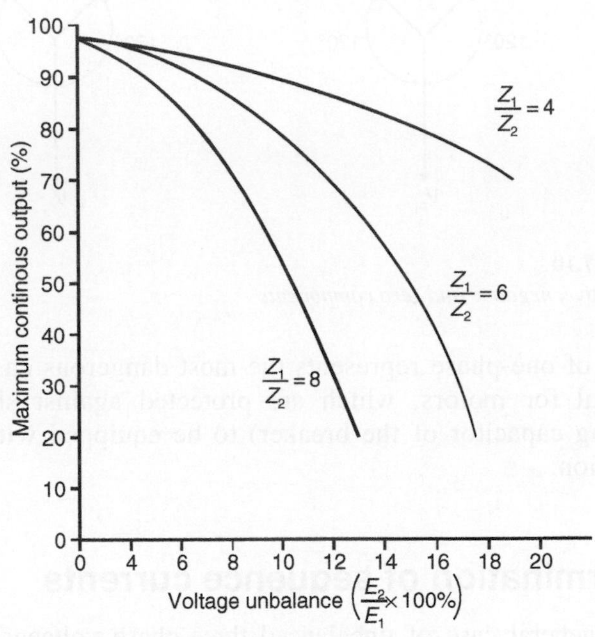

Figure 17.11
Maximum continuous output vs voltage unbalance

17.10 Electrical faults in stator windings earth faults phase–phase faults

17.10.1 Earth faults

Faults, which occur within the motor windings are mainly earth faults caused by breakdown in the winding insulation. This type of fault can be very easily detected by means of an instantaneous relay, usually with a setting of approximately 20% of the motor full-load current, connected in the residual circuit of three current transformers.

Care must be taken to ensure that the relay does not operate from spill current due to the saturation of one or more current transformers during the initial peak of the starting current; this can be as high as 2.5 times the steady-state rms value, and may cause operation, given the fast-operating speed of the normal relay. To achieve stability under these conditions, it is usual to increase the minimum operating voltage of the relay by inserting a stabilizing resistor in series with it. Refer Figure 17.12.

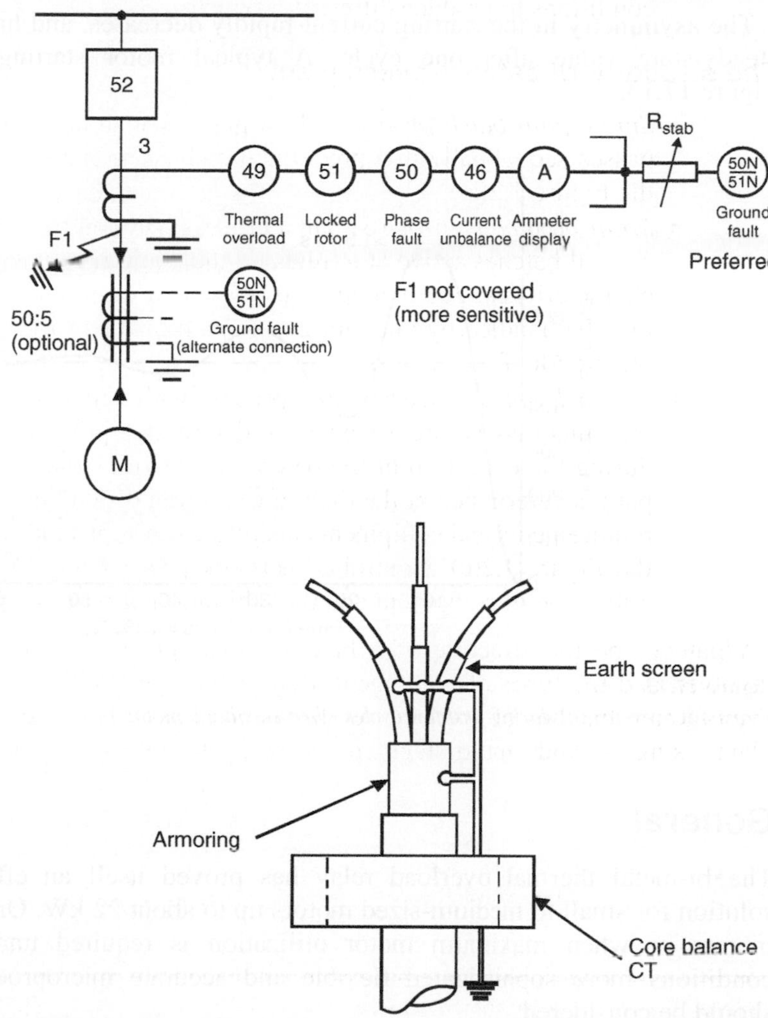

Figure 17.12
Earth fault protection

17.10.2 Phase–phase faults

Because of the relatively greater amount of insulation between phase windings, faults between phases seldom occur. As the stator windings are completely enclosed in grounded metal, the fault would very quickly involve earth, which would then operate the instantaneous earth fault protection described above.

Differential protection is sometimes provided on large (2 MW) and important motors to protect against phase–phase faults, but if the motor is connected to an earthed system there does not seem to be any great benefit to be gained if a fast-operating and sensitive earth fault is already provided.

17.10.3 Terminal faults

High-set instantaneous overcurrent relays are often provided to protect against phase faults occurring at the motor terminals, such as terminal flashovers. Care must be taken when setting these units to ensure that they do not operate on the initial peak of the motor starting current, which can be 2.5 times the steady-state rms value.

The asymmetry in the starting current rapidly decreases, and has generally fallen to its steady-state value after one cycle. A typical motor starting current is shown in Figure 17.13.

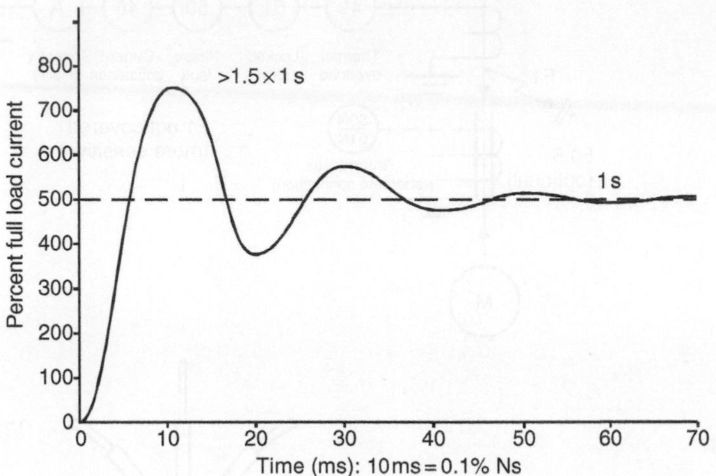

Figure 17.13
Transient overcurrent during first few cycles when starting a motor

17.11 General

The bi-metal thermal overload relay has proved itself an effective and economical solution for small to medium-sized motors up to about 22 kW. On larger, more expensive motors or when maximum motor utilization is required under varying operational conditions more sophisticated flexible and accurate microprocessor protection relays should be considered.

These relays typically include:

- Thermal overload protection, monitoring all three-phases with thermal replicas for direct and frequency convertor-controlled drives
- Short-circuit protection
- Start-up and running stall protection
- Phase unbalanced protection
- Single-phasing protection
- Earth fault protection
- Undercurrent protection
- Digital read-out of set values, actual measured values and memorized values
- Self, supervision system
- Outstanding accuracy
- Optimum philosophy.

The present day concept is use of microprocessor-based numerical relays for both HV and LV motors (say beyond 50 kW), as the relays come with lot of features which allow them to be interchangeable, ensures site settings and give valuable feedback on the load details whether a trip occurs or not.

17.12 Typical protective settings for motors

(a) Long time pick-up

- 1.15 times motor FLA times motor service factor for applications encountering 90% voltage dip on motor starting
- 1.25 times motor FLA times motor service factor for applications encountering 80% voltage dip on motor starting

(b) Long-time delay

- Greater than motor starting time at 100% voltage and the minimum system voltage
- Less than locked rotor damage time at 100% voltage and the minimum system voltage
- On high-inertia drives, it is common for the start time to be greater than the locked rotor withstand time. Under these circumstances, set the time to permit the motor to start. Supplemental protection should be added for locked rotor protection. One example of this is a speed switch set at 25% of rated speed tripping through a timer to trip if the desired speed has not been reached in a pre-determined time.

(c) Instantaneous pick-up

- Not less than 1.7 times motor LRA for medium-voltage motors
- Not less than 2.0 times motor LRA for low-voltage motors.

(d) Ground-fault protection

- Minimum pick-up and minimum time delay for static trip units
- Core-balance CT and 50 relays set at minimum for medium-voltage, low-resistance grounded systems
- Residually connected CT, and 50/51 for medium voltage, solidly grounded systems. Minimum tap and time dial equals 1 for 51 relay
- Minimum tap (not less than 5 A) for 50 relay.

18

Generator protection

18.1 Introduction

A generator is the heart of an electrical power system, as it converts mechanical energy into its electrical equivalent, which is further distributed at various voltages. It therefore requires a 'prime mover' to develop this mechanical power and this can take the form of steam, gas or water turbines or diesel engines.

Steam turbines are used virtually exclusively by the main power utilities, whereas in industry three main types of prime movers are in use:

1. *Steam turbines*: Normally found where waste steam is available and used for base load or standby.
2. *Gas turbines*: Generally used for peak-lopping or mobile applications.
3. *Diesel engines*: Most popular as standby plant.

Small- and medium-sized generators are normally connected direct to the distribution system, whilst larger units are connected to the EHV grid via a transformer (see Figures 18.1 and 18.2).

It will be appreciated that a modern large generating unit is a complex system, comprising of number of components:

- Stator winding with associated main and unit transformers
- Rotor with its field winding and exciters
- Turbine with its boiler, condenser, auxiliary fans and pumps.

Many different faults can occur on this system, for which diverse protection means are required. These can be grouped into two categories:

Electrical	Mechanical
Stator insulation failure	Failure of prime mover
Overload	Low condenser vacuum
Overvoltage	Lubrication oil failure
Unbalanced load	Loss of boiler firing
Rotor faults	Over speeding
Loss of excitation	Rotor distortion
Loss of synchronism	Excessive vibration

We will look briefly at the electrical side in this chapter.

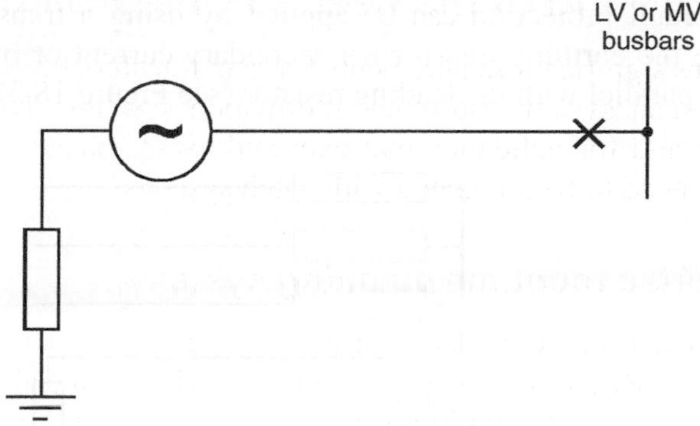

Figure 18.1
Small- and medium-sized generators

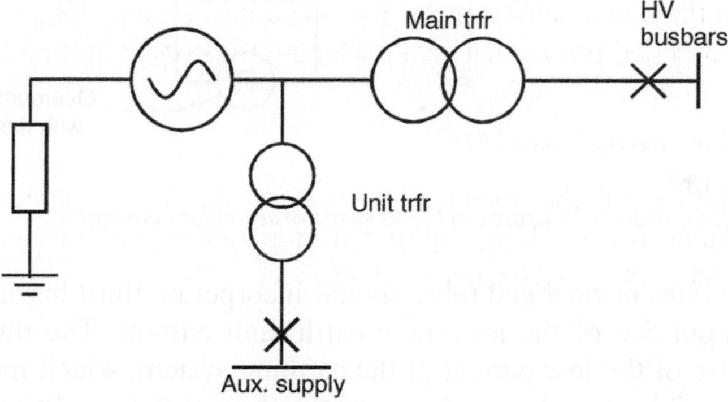

Figure 18.2
Larger generating units

18.2 Stator earthing and earth faults

The neutral point of the generator stator winding is normally earthed so that it can be protected, and impedance is generally used to limit earth fault current.

The stator insulation failure can lead to earth fault in the system. Severe arcing to the machine core could burn the iron at the point of fault and weld laminations together. In the worst case, it could be necessary to rebuild the core down to the fault necessitating a major strip-down. Practice, as to the degree of limitation of the earth fault current varies from rated load current to low values such as 5 A.

Generators connected direct to the distribution network are usually earthed through a resistor. However, the larger generator–transformer unit (which can be regarded as isolated from the EHV transmission system) is normally earthed through the primary winding of a voltage transformer, the secondary winding being loaded with a low ohmic value resistor. Its reflected resistance is very high (proportional to the turns ratio squared) and it prevents high transient overvoltages being produced as a result of an arcing earth fault.

When connected directly through impedance, overcurrent relays of both instantaneous and time-delayed type are used. A setting of 10% of the maximum earth fault current is considered the safest setting, which normally is enough to avoid spurious operations due to the transient surge currents transmitted through the system capacitance. The time delay relay is applied a value of 5%.

Earth fault protection can be applied by using a transformer and adopting a relay to measure the earthing transformer secondary current or by connecting a voltage-operated relay in parallel with the loading resistor (see Figure 18.3).

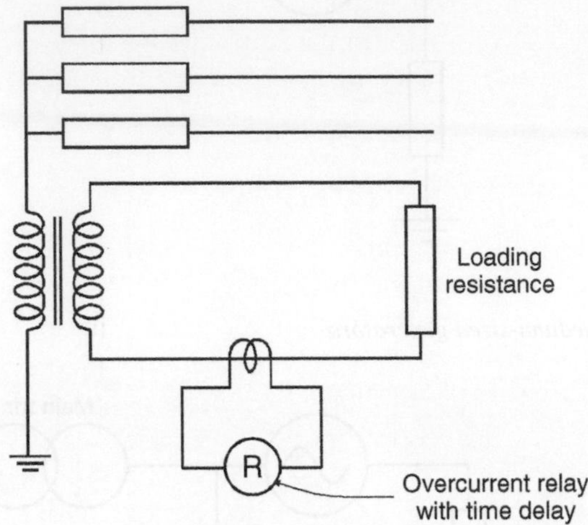

Figure 18.3
Earth fault protection using a relay to measure secondary current

The current operated relay should incorporate third harmonic filter and is normally set for about 5% of the maximum earth fault current. The third harmonic filter is required because of the low current of the earthing system, which may not be much different from the possible third harmonic current under normal conditions. The time delay is essential to avoid trips due to surges (see Figure 18.4).

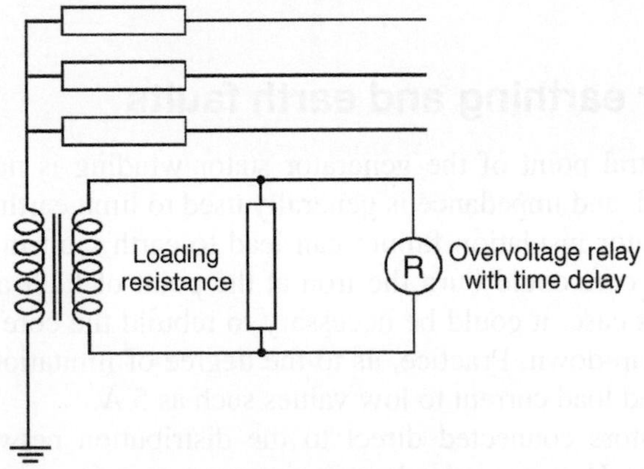

Figure 18.4
Earth fault protection using a relay in parallel with loading resistor

In the voltage-operated type, a standard induction disk type overvoltage relay is used. It is also to be noted that the relay is connected across the secondary winding of the transformer and the relay shall be suitably rated for the higher continuous operating voltage. Further, the relay is to be insensitive for third harmonic current.

Phase-to-phase faults clear of earth are less common. They may occur on the end coils or on adjacent conductors in the same slot. In the latter case, the fault would involve earth in a very short time.

18.3 Overload protection

Generators are very rarely troubled by overload, as the amount of power they can deliver is a function of the prime mover, which is being continuously monitored by its governors and regulator. Where overload protection is provided, it usually takes the form of a thermocouple or thermistor embedded in the stator winding. The rotor winding is checked by measuring the resistance of the field winding.

18.4 Overcurrent protection

It is normal practice to apply IDMTL relays for overcurrent protection, not for thermal protection of the machine but as a 'back-up' feature to operate only under fault conditions. In the case of a single machine feeding an isolated system, this relay could be connected to a single CT in the neutral end in order to cover a winding fault. With multiple generators in parallel, there is difficulty in arriving at a suitable setting so the relays are then connected to line side CTs.

18.5 Overvoltage protection

Overvoltage can occur as either a high-speed transient or a sustained condition at system frequency.

The former are normally covered by surge arrestors at strategic points on the system or alternatively at the generator terminals depending on the relative capacitance coupling of the generator/transformer, and connections, etc.

Power frequency overvoltages are normally the result of:

- Defective voltage regulator
- Manual control error (sudden variation of load)
- Sudden loss of load due to other circuit tripping.

Overvoltage protection is therefore only applied to unattended automatic machines, at say a hydroelectric station. The normal setting adopted are quite high almost equal to 150% but with instantaneous operation.

18.6 Unbalanced loading

A three-phase balanced load produces a reaction field, which is approximately constant, rotating synchronously with the rotor field system. Any unbalanced condition can be broken down into positive, negative and zero sequence components. The positive component behaves similar to the balanced load. The zero components produce no main armature reaction. However, the negative component creates a reaction field, which rotates counter to the DC field, and hence produces a flux, which cuts the rotor at twice the rotational velocity. This induces double frequency currents in the field system and rotor body.

The resulting eddy currents are very large, so severe that excessive heating occurs, quickly heating the brass rotor slot wedges to the softening point where they are susceptible to being extruded under centrifugal force until they stand above the rotor surface, in danger of striking the stator iron. It is therefore very important that negative phase sequence protection be installed, to protect against unbalanced loading and its consequences.

18.7 Rotor faults

The rotor has a DC supply fed onto its winding which sets up a standing flux. When this flux is rotated by the prime mover, it cuts the stator winding to induce current and voltage therein. This DC supply from the exciter need not be earthed. If an earth fault occurs, no fault current will flow and the machine can continue to run indefinitely, however, one would be unaware of this condition. Danger then arises if a second earth fault occurs at another point in the winding, thereby shorting out portion of the winding. This causes the field current to increase and be diverted, burning out conductors.

In addition, the fluxes become distorted resulting in unbalanced mechanical forces on the rotor causing violent vibrations, which may damage the bearings and even displace the rotor by an amount, which would cause it to foul the stator. It is therefore important that rotor earth fault protection be installed. This can be done in a variety of ways.

18.7.1 Potentiometer method

The field winding is connected with a resistance having center tap. The tap point is connected to the earth through a sensitive relay R. An earth fault in the field winding produces a voltage across the relay.

The maximum voltage occurs for faults at end of the windings. However, there are chances that the faults at the center of the winding may get undetected. Hence, one lower tap is provided in the resistance. Though normally, the center tap is connected, a pushbutton or a bypass switch is used to check for the faults at the center of winding. A proper operating procedure shall be established to ensure that this changeover is done at least once in a day (see Figure 18.5(a)).

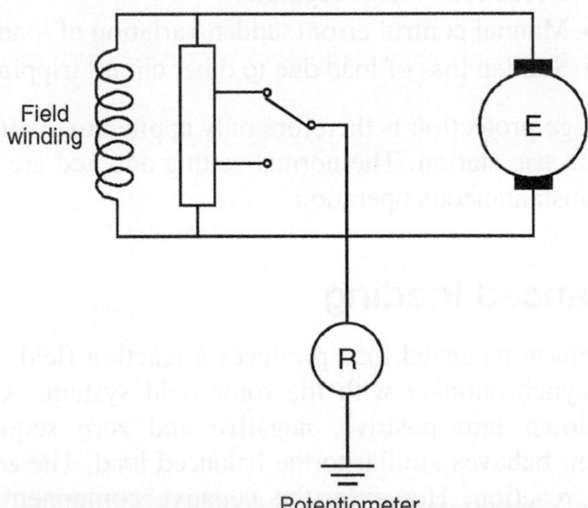

Figure 18.5(a)
Potentiometer

18.7.2 AC injection method

This method requires an auxiliary supply, which is injected to the field circuit through a coupling capacitance. The capacitor prevents the chances of higher DC current passing through the transformer. An earth fault at any part of the winding gives rise to the field current, which is detected by the sensitive relay. Care should be taken to ensure that the bearings are insulated, since there is a constant current flowing to the earth through the capacitance (see Figure 18.5(b)).

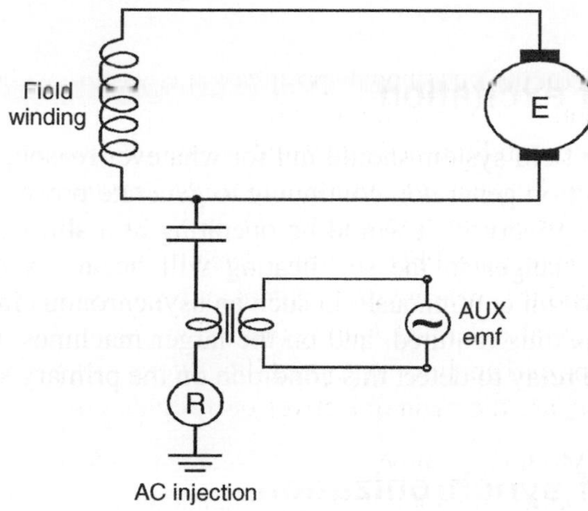

Figure 18.5(b)
AC injection

18.7.3 DC injection method

This method avoids the capacitance currents by rectifying the injection voltage adopted in the previous method. The auxiliary voltage is used to bias the field voltage to be negative with respect to the earth. An earth fault causes the fault current to flow through the DC power unit causing the sensitive relay to operate under fault conditions (see Figure 18.5(c)).

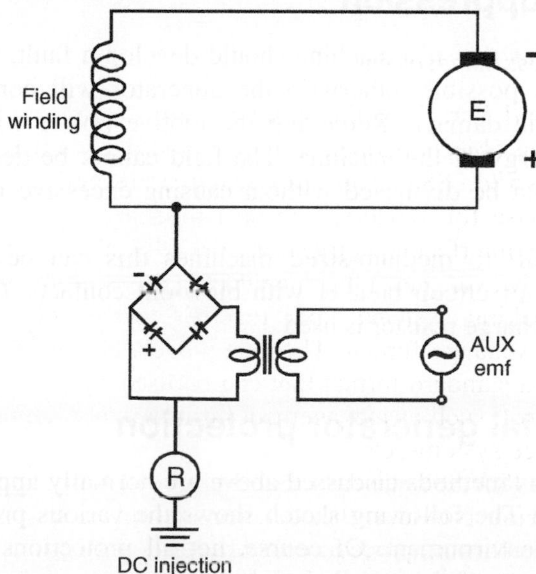

Figure 18.5(c)
DC injection

18.8 Reverse power

Reverse power protection is applicable when generators run in parallel, and to protect against the failure of the prime mover. Should this fail then, the generator would motor by taking power from the system and could aggravate the failure of the mechanical drive.

18.9 Loss of excitation

If the rotor field system should fail for whatever reason, the generator would then operate as an induction generator, continuing to generate power determined by the load setting of the turbine governor. It would be operating at a slip frequency and although there is no immediate danger to the set, heating will occur, as the machine will not have been designed to run continuously in such an asynchronous fashion. Some form of field failure detection is thus required, and on the larger machines, this is augmented by a mho-type impedance relay to detect this condition on the primary side.

18.10 Loss of synchronization

A generator could lose synchronism with the power system because of a severe system fault disturbance, or operation at a high load with a leading power factor. This shock may cause the rotor to oscillate, with consequent variations of current, voltage and power factor. If the angular displacement of the rotor exceeds the stable limit, the rotor will slip a pole pitch. If the disturbance has passed, by the time this pole slip occurs, then the machine may regain synchronism otherwise it must be isolated from the system. Alternatively, trip the field switch to run the machine as an asynchronous generator, reduce the field excitation and load, then reclose the field switch to resynchronize smoothly.

18.11 Field suppression

It is obvious that if a machine should develop a fault, the field should be suppressed as quickly as possible, otherwise the generator will continue to feed its own fault and increase the damage. Removing the motive power will not help in view of the large kinetic energy of the machine. The field cannot be destroyed immediately and the flux energy must be dissipated without causing excessive inductive voltage rise in the field circuit.

For small- to medium-sized machines this can be satisfactorily achieved using an automatic air circuit breaker with blow-out contacts. On larger sets above say 5 MVA a field discharge resistor is used.

18.12 Industrial generator protection

The various methods discussed above are normally applicable for an industrial generator protection. The following sketch shows the various protection schemes employed in an industrial environment. Of course, not all protections are adopted for every generator since the cost of the installation decides the economics of protection required. Note that the differential relay (though not discussed separately in this chapter) is normally necessary for generators in the range of megawatts (see Figure 18.6).

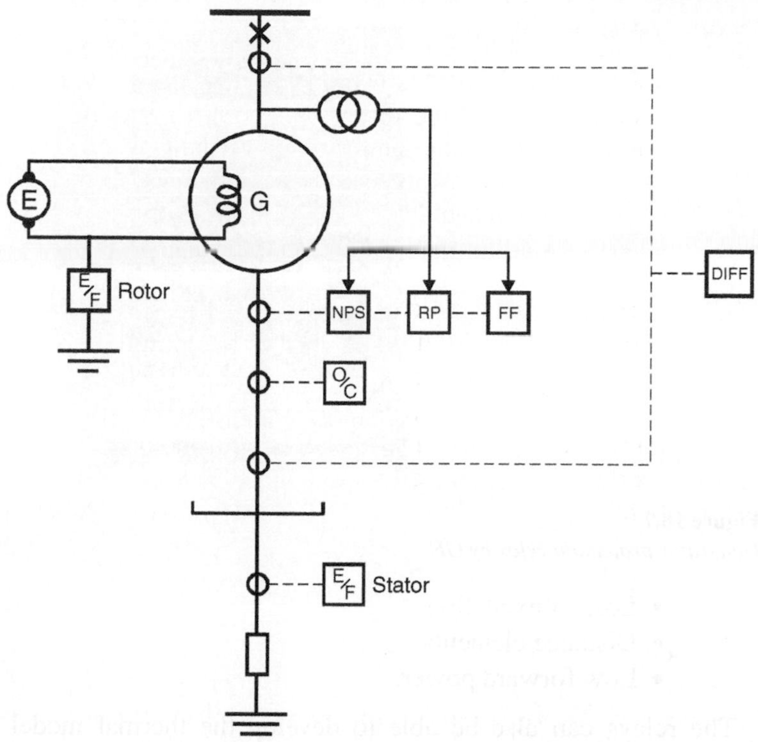

Figure 18.6
Typical protection scheme for industrial generator

18.13 Numerical relays

The above paragraphs described use of individual relays for different fault conditions. However, the modern numerical relays combine most of the above functions in a single relay with programing features that make them useful for any capacity generator.

The numerical relays are manufactured by all the leading relay manufacturers. The various protections functions that are available in a typical numerical relay are as below (see Figure 18.7).

- Inverse time overcurrent
- Voltage restrained phase overcurrent
- Negative sequence overcurrent
- Ground overcurrent
- Phase differential
- Ground directional
- High-set phase overcurrent
- Undervoltage
- Overvoltage
- Volts/hertz
- Phase reversal
- Under frequency
- Over frequency
- Neutral overvoltage (fundamental)
- Neutral undervoltage (3rd harmonic)

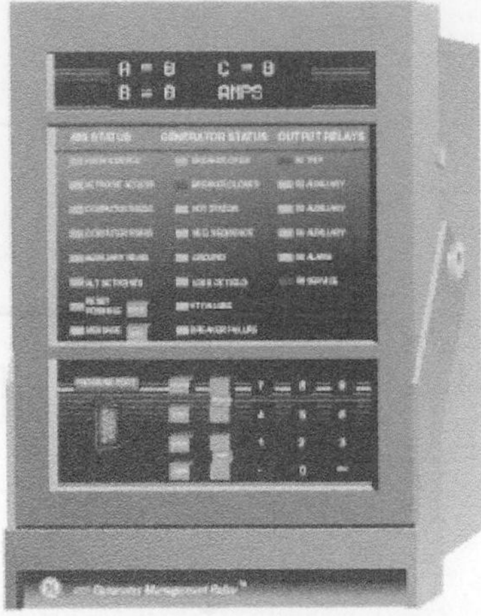

Figure 18.7
Generator protection relay by GE

- Loss of excitation
- Distance elements
- Low forward power.

The relays can also be able to develop the thermal model for the generators being protected, based on the safe stall time, previous start performances, etc., which is used to prevent the restart attempt of the generator under abnormal conditions or after a few unsuccessful starts/trips.

In addition to the various protection functions, these numerical relays also record the generator output figures like voltage, current, active power, reactive power, power factor, temperature of stator/rotor windings, etc. on a continuous basis. Hence, the numerical relays are finding increasing applications in modern industries.

18.14 Parallel operation with grid

In modern industries and continuous process plants, it is customary to have the plant generators (gas/steam turbine or diesel engine driven) to be operated in parallel with the grid to ensure uninterrupted power to essential loads. The basic protection employed in such systems are use of reverse power relays, which are used basically, to protect the grid from the faulty generators operating as motors.

It is also quite common to see that these systems are provided with 'islanding' feature, which enables the unstable grid to be isolated from the stable generating sets due to transmission disturbances. The protection employed in such cases are under frequency and dv/df, which are basically the effects of grid disturbances.

It is also common that power is exported to the grid from the industrial generators, when the power is generated in excess of the demand. The protective systems employed in all such cases shall be discussed with supply authorities to ensure that all protective functions as required per local regulations are met.

19

Management of protection

19.1 Management of protection

A protective system is considered 100% perfect if the number of circuit breakers opened under a fault are as per the design configuration. However, there are occasions when a few protective relays incorrectly operate or fail to operate. There could be many reasons but the principal reasons could be:

- The internal faults in the relays
- Defects in the wiring to the relays
- Wrong and poorly coordinated settings
- Unforeseen faults at the design stage
- Mechanical failures.

Protection systems must be kept 100% operational at all times as one never knows when or where faults are likely to occur. The systems must therefore be maintained and managed properly to ensure safe and efficient operation of the power network.

Although the relays are tested prior to commissioning a system, it is most likely that the relays may not be operating due to the soundness of the system. However, it cannot be assumed that the relay did not operate because of the system healthiness. Hence, it is very vital that the relays should be periodically checked and tested at fixed intervals. It is also important that the records must be kept about the tests being conducted and the details of results for future reference and records.

The functions required for good maintenance are listed on the following schedule A and it is important that good records are kept of the system parameters, wiring schematics, relay settings and calculations, CT magnetization curves and so on. Some suggested formats of test sheets are attached to give some idea of the sort of information that should be kept on file.

19.2 Schedule A

Schedule A is nothing, but the basic functions that are considered essential to ensure that the relays are kept in good form during their life. The tests will give the idea about any internal parts that are to be corrected or replaced. The records will also give an idea on the frequency of failures expected in typical relays and the replacements that are needed at regular intervals. Such frequent replacement parts can be kept as spares so that the relays can be put back in perfect conditions immediately on noticing the defects. Table 19.1 generally outlines this schedule.

Functions of Maintenance
Routine inspection and testing
Annual trip testing (random) Full scheme test every 4th year
Investigations Defects Incorrect operations
Spares and repairs
Performance assessment
Modifications
Refurbishment
Replacements/up-grading

Table 19.1
Schedule A

Protection management also involves addressing some of the following issues listed in schedule B (see Table 19.2).

Issues
Technology
Organization
Privatization
Skilled technical staff
Environment
Access for work

Table 19.2
Schedule B

19.3 Schedule B

The technology has been changing at a rapid rate in recent times and it is important that staffs are trained to be skilled in the area and are kept up to date. Good forward planning is essential to get access to plant for maintenance.

If the budget cannot carry permanent staff, then bring in specialist private companies to do the annual checks. Above all, make sure that the relays are:

- Applied correctly for job
- Commissioned properly
- Set correctly
- Maintained in good condition and working order.

19.4 Test sheets

A typical test sheet standard format can be as seen below and the format can be redesigned based on the relay type and the tests needed.

Test Certificate	
Station:	Circuit:
Customer: Circuit: Relay:	
Test Details: Injection Current/Voltage: Fault Simulated: Results Obtained:	
Date:	Tested By:
Engineer:	Date:

A typical test format for a motor protection relay with various protective functions could be as below.

P & B Golds Motor Protection						
Station:			Circuit:			
Date:			Panel No.:			
CT Ratio:			Class:			
Relay Type:			Serial No.:			
Rated: A Min		Tap Set: %		Load to Trip: %		
Aux Volts: V	R Stab:		Inst O/CL A		E/F: A	

Thermal Checks: (100% Tap) Three Current Elements in Series						
Check center pointer read '0'	Yes	Yes	No	No	Adj	Adj
Check outer elements are central	Yes		No		Adj	

Inject Ir	A	Load Reads	%	Adjusted		

1.	Inject		XIr =		A	Elements move together	
						Adjusted	
2.	Time (from cold):	Rated	S	Actual	S	Error Allowed	

Check Running Load Values:			
80%	%		
Check 90% Tap @ 90% Ir	Load Reads:	%	
Check 80% Tap @ 80% Ir	Load Reads	%	

Instantaneous Checks							
Overcurrent:		Phase	Set	A	Operate	A	
		Phase	Set	A	Operate	A	
Earth Fault:			Set	A	Operate	A	
			Set	A	Operate	V	

Remarks:	
Date:	Tested By:
Engineer:	Date:

When a system is put into service, it is necessary that proper records should be available for the various tests to be conducted. Above all, a checklist is mandatory to ensure that all basic tests are carried out before putting the system into use. Following table is a typical commissioning checklist, which should be planned, well in advance before taking up the commissioning of any electrical system, whether simple or complex.

Commissioning Check List			
Station:		Circuit:	
1. Current transformer tests	☐	5. Functional tests (DC)	☐
Magnetization curve	☐	Tripping circuits	☐
Polarity	☐	Closing circuits	☐
Ratio	☐	Supervisory circuits	☐
Megger	☐	Fuse ratings	☐
SEC resistance	☐	6. Phasing tests	
2. Voltage transformer tests	☐	Primary circuits	☐
Polarity	☐	VT secondaries	☐
Ratio	☐	Auxiliary supplies	☐
3. Primary injection tests		7. On load checks	☐
Protection CTs	☐	8. TRFR Buch alarm/trip	☐
Metering CTs	☐	TRFR Wdg temp alarm/trip	☐
Bus Zone CTs	☐	TRFR pressure alarm/trip	☐
4. Secondary injection tests		TRFR oil temp alarm/trip	☐
Relays	☐	9. NEC Buch alarm/trip	☐
Metering	☐	NEC temp alarm/trip	☐
Comments:			
Engineer:		Date:	

Primary and secondary injection tests are the most common tests applicable for voltage and current sensing relays, whose functions depend on the correct sensing characteristics. Following table shows typical test sheet for such a purpose.

A secondary injection test serves the purpose when there is no possibility to apply the primary voltage or pass the primary current to the voltage and current transformers, connected to a relay. For e.g. a 110 V supply and a 5 A current source would be able to complete most of the functional tests of typical relays.

Primary Injection Test						
Station:				Circuit:		
Ratio:				Function:		
Phases Injected	Primary Current	Secondary Red	Secondary Yellow	Secondary Blue	Secondary Neutral	
R/R						
R/Y						
R/B						
Ratio:				Function:		
Phases Injected	Primary Current	Secondary Red	Secondary Yellow	Secondary Blue	Secondary Neutral	
R/R						
R/Y						
R/B						
Ratio:				Function:		
Phases Injected	Primary Current	Secondary Red	Secondary Yellow	Secondary Blue	Secondary Neutral	
R/R						
R/Y						
R/B						
Notes/Remarks:						
Engineer:					Date:	

C.T. Secondary Injection Test Sheet

Station:			Circuit:		
Details:					
Type:					
Serial No.:	R: 1	W: 1		B: 1	
Serial No.:	R: 2	W: 2		B: 2	
Serial No.:	R: 3	W: 3		B: 3	

Core:		1	2	3
Ratio:				
Class:				
Res:	R–W			
	W–B			
	B–R			
	R–N			
	W–N			
	B–N			

	mA	V_R	V_W	V_B
1				
2				
3				

Notes/Remarks:

Engineer: Date:

Index

THIS BOOK WAS DEVELOPED BY IDC TECHNOLOGIES

WHO ARE WE?

IDC Technologies is internationally acknowledged as the premier provider of practical, technical training for engineers and technicians.

We specialise in the fields of electrical systems, industrial data communications, telecommunications, automation & control, mechanical engineering, chemical and civil engineering, and are continually adding to our portfolio of over 60 different workshops. Our instructors are highly respected in their fields of expertise and in the last ten years have trained over 50,000 engineers, scientists and technicians.

With offices conveniently located worldwide, IDC Technologies has an enthusiastic team of professional engineers, technicians and support staff who are committed to providing the highest quality of training and consultancy.

TECHNICAL WORKSHOPS

TRAINING THAT WORKS

We deliver engineering and technology training that will maximise your business goals. In today's competitive environment, you require training that will help you and your organisation to achieve its goals and produce a large return on investment. With our "Training that Works" objective you and your organisation will:

- Get job-related skills that you need to achieve your business goals
- Improve the operation and design of your equipment and plant
- Improve your troubleshooting abilities
- Sharpen your competitive edge
- Boost morale and retain valuable staff
- Save time and money

EXPERT INSTRUCTORS

We search the world for good quality instructors who have three outstanding attributes:

1. Expert knowledge and experience – of the course topic
2. Superb training abilities – to ensure the know-how is transferred effectively and quickly to you in a practical hands-on way
3. Listening skills – they listen carefully to the needs of the participants and want to ensure that you benefit from the experience

Each and every instructor is evaluated by the delegates and we assess the presentation after each class to ensure that the instructor stays on track in presenting outstanding courses.

HANDS-ON APPROACH TO TRAINING

All IDC Technologies workshops include practical, hands-on sessions where the delegates are given the opportunity to apply in practice the theory they have learnt.

REFERENCE MATERIALS

A fully illustrated workshop book with hundreds of pages of tables, charts, figures and handy hints, plus considerable reference material is provided FREE of charge to each delegate.

ACCREDITATION AND CONTINUING EDUCATION

Satisfactory completion of all IDC workshops satisfies the requirements of the International Association for Continuing Education and Training for the award of 1.4 Continuing Education Units.

IDC workshops also satisfy criteria for Continuing Professional Development according to the requirements of the Institution of Electrical Engineers and Institution of Measurement and Control in the UK, Institution of Engineers in Australia, Institution of Engineers New Zealand, and others.

CERTIFICATE OF ATTENDANCE

Each delegate receives a Certificate of Attendance documenting their experience.

100% MONEY BACK GUARANTEE

IDC Technologies' engineers have put considerable time and experience into ensuring that you gain maximum value from each workshop. If by lunch time of the first day you decide that the workshop is not appropriate for your requirements, please let us know so that we can arrange a 100% refund of your fee.

ONSITE WORKSHOPS

All IDC Technologies Training Workshops are available on an on-site basis, presented at the venue of your choice, saving delegates travel time and expenses, thus providing your company with even greater savings.

OFFICE LOCATIONS

AUSTRALIA • CANADA • IRELAND • NEW ZEALAND • SINGAPORE • SOUTH AFRICA • UNITED KINGDOM • UNITED STATES

idc@idc-online.com • www.idc-online.com

Visit our Website for FREE Pocket Guides

IDC Technologies produce a set of 4 Pocket Guides used by thousands of engineers and technicians worldwide.

Vol. 1 - **ELECTRONICS**

Vol. 2 - **ELECTRICAL**

Vol. 3 - **COMMUNICATIONS**

Vol. 4 - **INSTRUMENTATION**

To download a **FREE copy** of these internationally best selling pocket guides go to:

www.idc-online.com/freedownload/

Printed and bound by CPI Group (UK) Ltd, Croydon, CR0 4YY
03/10/2024
01040338-0018